Metal Hydrides

NATO ADVANCED STUDY INSTITUTES SERIES

A series of edited volumes comprising multifaceted studies of contemporary scientific issues by some of the best scientific minds in the world, assembled in cooperation with NATO Scientific Affairs Division.

Series B: Physics

Recent Volumes in this Series

This series is published by an international board of publishers in conjunction with NATO Scientific Affairs Division

A Life Sciences	Plenum Publishing Corporation
B Physics	London and New York
C Mathematical and Physical Sciences	D. Reidel Publishing Company Dordrecht, Boston, and London
D Behavioral and Social Sciences	Sijthoff & Noordhoff International Publishers
E Applied Sciences	Alphen aan den Rijn, The Netherlands, and Germantown, U.S.A.

Metal Hydrides

Edited by
Gust Bambakidis

Wright State University
Dayton, Ohio

SPRINGER SCIENCE+BUSINESS MEDIA, LLC

Library of Congress Cataloging in Publication Data

Main entry under title:

Metal hydrides.

(NATO advanced study institutes series. Series B, Physics ; v. 76)
"Proceedings of a NATO Advanced Study Institute on Metal Hydrides, held June 17-
27, 1980, in Rhodes, Greece" — Verso t.p.
"Published in cooperation with NATO Scientific Affairs Division."
Includes bibliographies and index.
1. Hydrides — Congresses. 2. Transition metal hydrides — Congresses. I. Bambakidis,
Gust. II. NATO Advanced Study Institute on Metal Hydrides (1980 : Rhodes, Greece)
III. North Atlantic Treaty Organization. Division of Scientific Affairs. IV. Series.
QD181.H1M48   546'.6   81-17761
ISBN 978-1-4757-5816-0   ISBN 978-1-4757-5814-6 (eBook)   AACR2
DOI 10.1007/978-1-4757-5814-6

Proceedings of a NATO Advanced Study Institute on Metal Hydrides
held June 17–27, 1980, in Rhodes, Greece

© 1981 Springer Science+Business Media New York
Originally published by Plenum Press, New York in 1981

PREFACE

In the last five years, the study of metal hydrides has expanded enormously due to the potential technological importance of this class of materials in hydrogen based energy conversion schemes. The scope of this activity has been worldwide among the industrially advanced nations.

There has been a consensus among researchers in both fundamental and applied areas that a more basic understanding of the properties of metal/hydrogen systems is required in order to provide a rational basis for the selection of materials for specific applications.

The current worldwide need for and interest in research in metal hydrides indicated the timeliness of an Advanced Study Institute to provide an in-depth view of the field for those active in its various aspects. The inclusion of speakers from non-NATO countries provided the opportunity for cross-fertilization of ideas for future research. While the emphasis of the Institute was on basic properties, there was a conscious effort to stimulate interest in the application of metal hydrides to solar/hydrogen energy conversion schemes in land areas where solar energy has promise as a primary energy source.

In addition to the lectures, several seminars were given which treated topics of special interest in greater detail.

The course began with an overview of electronic states in transition metal hydrides with emphasis on superconductivity in palladium based hydrides. In addition to electrical conductivity in both the normal and superconducting state, the electronic structure enters in a direct way into several experimentally accessible properties such as magnetic susceptibility and specific heat. Current methods were presented for calculating the electronic states in the presence of hydrogen atoms distributed randomly in the metal lattice. The magnetic properties of metallic compounds containing hydrogen were also reviewed. These include transition metal-rare earth intermetallic hydrides, which are presently a very promising class of metal hydrides for various applications.

A series of talks addressed the questions of how hydrogen atoms
are arranged in the metal lattice, and on their affect on the arrangement of the metal atoms themselves. This is intimately related to the
thermodynamic phase diagram of the hydride. The techniques of X-ray
diffraction, neutron diffraction, electrical resistivity and calorimetry and their theoretical interpretation for determining the phase
diagram were presented.

The long-range motion of hydrogen through the hydride is important in most applications since cyclic absorption and desorption to
or from the metal is usually involved. Various techniques (nuclear
magnetic resonance, quasielastic neutron scattering and internal
friction) for determining the diffusion parameters of hydrogen and
its isotopes were treated. Likewise, the effect of the metal surface
on absorption/desorption kinetics, which has previously received
little attention, was discussed. The formation of hydrides was also
covered, emphasizing absorption under high hydrogen gas pressure and
properties of the hydrides thus formed.

The behavior of hydrogen in alloys has only recently come under
investigation. In particular, the effect of alloying on the hydrogen
solvus and of hydrogen on the alloy phase diagram have been the object
of recent studies and were presented.

The Director wishes to express his deep appreciation to the
various lecturers, seminar speakers and students, all of whom contributed to the success of the Institute. Special thanks go also to
Mrs. Elli Bambakidis and Mrs. Donna Aurand for help with preparation
of the manuscript.

G. Bambakidis
Dayton, Ohio
January 31, 1981

CONTENTS

THE LECTURES

THE SEMINARS

SUPERCONDUCTIVITY IN PALLADIUM BASED HYDRIDES

E. N. Economou

Department of Physics, University of Virginia
Charlottesville, VA 22901, USA
and University of Athens, Athens, Greece

ABSTRACT

A simple version of the theory of superconductivity is presented. The final result for T_c is very similar to McMillan's formula, which expresses T_c in terms of electronic quantities and the distribution of the eigenmodes of the lattice vibrations. Taking this distribution from neutron scattering experiments and the electronic properties from first principle band structure calculations, T_c values for Palladium based hydrides and deuterides are obtained in good agreement with the experimental data.

INTRODUCTION

In 1970 Satterthwaite and Toepke[1] found that the critical temperature, T_c, below which superconductivity appears, is approximately $9^{\circ}K$ for Th_4H_{15} and Th_4D_{15}; the pure Th metal becomes superconducting only for temperatures lower than $T_c = 1.37^{\circ}K$.[2] In 1972 Skoskiewicz[3] observed superconductivity in the PdH_x system for $x \geq 0.8$; this was an impressive finding in view of the fact that Pd is not superconducting. Subsequently, Stritzker and Buckel[4], by ion implantation techniques, increased the H concentration and obtained $T_c \approx 8.8^{\circ}$ K in PdH and $T_c = 10.7^{\circ}$ in PdD. An even higher T_c was achieved[5] in systems of the type $Pd_{1-y}M_yH_x$, where M stands for Au, Ag, or Cu; e.g. $T_c \approx 16^{\circ}K$[6] for M = Ag and $y \approx 0.20-0.25$. More recently an increased T_c upon hydrogenation was found in $A\ell$[7,8] and TaS_2[9]. Many other elements and compounds have been hydrogenated in search for a high T_c system[2]; the results were negative.[2]

1

The above experimental observations raise the following questions: (a) How H enhances T_c in systems like $Pd_{1-y}M_yH_x$, Th_4H_{15} or, to a lesser degree, $A\ell H_x$? (b) Why H has little effect on T_c for most other systems? In this article we attempt to provide an answer to the first question as far as the Pd based systems are concerned. We attribute the H-induced high T_c in Pd and Pd based hydrides to a rather unusual combination of (1) a "soft" H sublattice and (2) the Fermi energy (E_F) electrons spending considerable time around the H sites. This combination allows the electrons at E_F to experience a H-sublattice mediated attraction somehow analogous to the attraction between two persons walking on a soft mattress. The higher the attraction the higher T_c is. In another article in this volume, M. Gupta shows that this favorable combination is not present in various other systems and consequently H is not expected to raise T_c in these systems.

This article is organized as follows: In the next section we present a new simplified version of the BCS theory of superconductivity; this version may have some pedagogic value. Then, in the following section, the theory is applied to the Pd based hydrides and quantitative results are presented and compared with experimental data.

A SIMPLIFIED VERSION OF BCS THEORY OF SUPERCONDUCTIVITY

Introductory Remarks

The basic effects associated with the superconducting state of a metal - absence of resistance to the flow of DC electrical current, expulsion of magnetic fields from the interior of a superconductor, magnetic flux quantization, Josephson effect and quantum interference in a macroscopic scale - can be explained if one assumes that the electrons (or at least a finite fraction of them) participate in an organized strongly correlated motion which can be described by a single rigid wavefunction $\psi(\vec{r}) = a(\vec{r})\exp[i\phi(\vec{r})]$, where the amplitude $a(\vec{r})$ equals to the square root of the superconducting electron density n. Having $\psi(\vec{r})$ one can easily calculate the current $\vec{j}$ which is given by $\vec{j} = q\psi^*\vec{v}\psi$. The momentum $\vec{p} = -i\hbar\nabla$ consists of the particle momentum $m\vec{v}$ and the electromagnetic field momentum $(q/c)\vec{A}(\vec{r})$, where q is the charge of the particle. Combining the above relations we obtain for the current

$$\vec{j}(\vec{r}) = -\frac{q^2 n}{mc}\vec{A}(\vec{r}) + \frac{\hbar n q}{m}\nabla\cdot\phi(\vec{r}). \tag{1}$$

In the bulk of a superconductor the phase $\phi(\vec{r})$ is constant and hence $\vec{j} \sim \vec{A}$, which accounts for the zero resistance and the Meissner effect. If Eq. (1) is applied to a closed superconducting

loop one easily derives that the magnetic flux through the loop
is quantized in units of hc/q. Comparing with the experimental
value for the quantum of the magnetic flux one obtains that q=2e,
which shows clearly that the electrons pair up and thus they can
undergo the quantum condensation-implied by the single $\psi(\vec{r})$ -
without violating Pauli's principle.

Phonon Mediated Attraction

    The attraction required to overcome the Coulomb repulsion
and bind the partners of each pair together can only be provided
by the polarizable medium in which the electrons are embedded.
The analogy of two persons on a mattress (which plays the role
of the polarizable medium) suggests that the attractive interaction V_a between two electrons is proportional to the interaction
of each electron with the polarizable medium, V_{e-m}, and proportional to how easily the medium is polarizable which can be
characterized by the inverse of a typical eigenfrequency, ω_m,
of the medium. Thus V_a is given by

$$V_a \sim \frac{V_{e-m}^2}{\hbar\omega_m} \qquad (2)$$

In all well understood cases the polarizable medium is the ionic
lattice; then $V_{e-m} = V_{e-p}$ is the electron-phonon interaction
and ω_m is a typical phonon frequency which can be taken equal to
the Debye frequency ω_D. Many efforts have been made to find a
V_a mediated by degrees of freedom other than the ionic ones but
no positive results have been established. Thus, from now on,
we will take $V_{e-m} = V_{e-p}$ and $\omega_m = \omega_D$. In general V_{e-p} depends on
the momenta $\vec{p}_1$, $\vec{p}_2$ of each electron and on the wave number $\vec{q}$ of
the eigenoscillation of the lattice. In all cases we shall
examine, the momenta $\vec{p}_1$, $\vec{p}_2$ are on the Fermi surface. Thus V_{e-m}^2
in Eq. (2) must be replaced by an appropriate Fermi surface
average, $<V_{e-p}^2>$. Furthermore, there is not simply one eigenfrequency as was assumed in Eq. (2) but a continuum of eigen-
frequencies characterized by a distribution $F(\omega)$. Thus Eq. (2)
must be replaced by

$$V_a = \int d\omega F(\omega) \frac{<V_{e-p}^2>}{\hbar\omega} \qquad (3)$$

We shall see later than quantities of physical interest depend
on the product $\rho_F \cdot V_a$ which shall be denoted by λ. The quantity
ρ_F is the single electron density of states (DOS) per spin at
the Fermi level with band structure effects included but with no
many body effects (electron-electron or electron-phonon interactions) incorporated. Usually one introduces the quantity
$\alpha^2(\omega) \equiv \rho_F \cdot <V_{e-p}^2>/2\hbar$ and expresses λ as follows

$$\lambda = 2 \int d\omega F(\omega) \frac{\alpha^2(\omega)}{\omega} \tag{4}$$

The phonon mediated electron-electron attraction V_a depends on the square of the electron-phonon interaction, V^2_{e-p}. But the same quantity, V^2_{e-p}, determines the scattering probability of an electron by the lattice vibrations and consequently the phonon contribution to the electrical resistivity r. Thus one expects the materials with high lattice resistivity to have a high λ and hence to be high T_c superconductors. Such a correlation does indeed exist. As a matter of fact one can show that at high temperatures the derivative of the resistivity is given by

$$\frac{dr}{dt} = \frac{8\pi^2 k}{\hbar} \frac{1}{\omega_p^2} \lambda_{tr}$$

where k is Boltzmann's constant and ω_p is the plasma frequency

Table 1. Comparison of λ_{tr} with λ as given in Ref. 9

Metal	λ_{tr}	λ
Li	.40	.41±.15
Na	.16[a]	.16±.04
K	.14	.13±.03
Rb	.19	.16±.04
Cs	.26[a]	.16±.06
Mg	.32	.35±.04
Zn	.67	.42±.05
Cd	.51	.40±.05
Al	.41	.43±.05
Pb	1.79	1.55
In	.85	.805
Hg	2.3[b]	1.6
Cu	.13	.14±.03
Ag	.13	.10±.04
Au	.08[a]	.14±.05
Nb	1.11	.9±.2

[a] Experimental value for ω_p was used.
[b] Free electron value for ω_p was used.

which can be expressed (in eV) as

$$\omega_p = 19.59 \left[<v_F^2> \rho_F' \right]^{\frac{1}{2}}. \tag{6}$$

In Eq. (6) v_F is the velocity at the Fermi level in units of 10^8
cm/sec, the average is over the Fermi surface, and ρ_F' is the DOS
per Ryd per spin per volume. If band structure effects are
negligible ω_p reduces to the free electron value $(4\pi e^2 n/m)^{\frac{1}{2}}$.
The quantity λ_{tr} is obtained from λ by substituting the quantity
$<v_{e-p}^2>$ by $<v_{e-p}^2 (1-\cos\theta)>$, where θ is the angle between $\vec{p}_1$ and
$\vec{p}_2$. In Table 1 we compare the value λ_{tr} with the best estimates[9]
of λ for various materials. The quantity dr/dT was taken as
$[r(295°K) - r(250°K)]/45°K$ and explicit values were obtained from
ref. 10; ω_p was obtained from APW band structure calculations[11],
unless otherwise stated. The overall agreement between λ_{tr} and λ
is impressive.

From Eq. (3) one can see that a large V_a can be obtained
if there are many low frequency phonons. This means that a
soft, not so stable, lattice may imply a high T_c. Indeed most
high T_c materials are not so stable structurally.

To obtain the total interaction potential V between two
electrons one must subtract from the phonon mediated attraction
V_a a quantity V_c^* which is proportional to the screened Coulomb
repulsion V_c. The actual calculation of V_c^* is rather complicated.[12,13] The total interaction V is

$$V = V_a - V_c^* \tag{7}$$

Binding the pair

The problem of two particles interacting via a potential,
which is a function of their relative coordinates only, can be
separated to the center of mass motion (which is characterized
by the total momentum $\vec{P}$ and the total energy $E = \vec{P}^2/2m$) and to
the relative motion. The latter is equivalent to the motion of
a single particle in the presence of an external potential.
Here we are interested in the bound states associated with this
potential. The most general way to treat this problem is by
employing Green's function techniques.[14] A crude outline of
such an approach will be given here. A more careful presentation
can be found in ref. 14.

Consider the Hamiltonian $H = H_o + V$, where H_o is its
unperturbed part and V is the external potential. Let us define
the operator $G(E) \equiv (E-H)^{-1}$; when $E = E_o$, where E_o is a bound discrete level, G blows up. Thus the bound levels, if any, will
appear as poles of $G(E)$. The operator $G(E)$ can be expressed as
follows: $G(E) = (E-H_o-V)^{-1} = \{(E-H_o)[1-(E-H_o)^{-1}V]\}^{-1} =$
$(1-G_oV)^{-1}G_o$, where $G_o(E) \equiv (E-H_o)^{-1}$. Hence the bound levels

will be solutions of the equation

$$G_o(E)V = 1 \tag{8}$$

By introducing the eigenstates of H_o, $\{|n>\}$, we can reexpress G_o as $G_o(E) \equiv (E-H_o)^{-1} = \sum_n |n><n| (E-E_n)^{-1}$. The summation of the states $\sum_n$ can be transformed to an integration over the levels $\int dE' D_o(E')$, where $D_o(E')$ is the DOS. Thus we obtain

$$G_o(E) = \int \frac{dE' D_o(E')}{E-E'} \tag{9}$$

Eqs. (8) and (9) solve the problem of finding the bound levels in an external potential V. For shallow potentials the bound level, if any, would be very close to the band edge and consequently its existence and behavior would be controlled by the behavior of $G_o(E)$ near the band edge. The behavior of $G_o(E)$, according to Eq. (9), depends on the behavior of the unperturbed DOS, $D_o(E)$, which thus becomes the decisive factor in this problem. If, for E near the band edge $E_B = 0$, the DOS is approaching zero continuously, i.e. $D_o(E) \sim E^s$ ($s > 0$), then from Eq. (9), it follows that

$$G_o(E) \sim |E|^s + \phi(E) \tag{10}$$

where $\phi(E)$ is analytic around $E = 0$. Hence for a sufficiently weak potential V, there is no solution of Eq. (8); in other words V must exceed a critical value (which depends on $\phi(0)$) in order to have a bound level. An elementary example of this behavior is the case of a 3-d potential well in an otherwise free space. Indeed, in this case the unperturbed DOS is proportional to the volume in momentum space over dE, i.e. $D_o(E) \sim p^2 \, dp/dE \sim p^2 \, dp/pdp = p \sim E^{\frac{1}{2}}$.

On the other hand if $D_o(E) \sim E^{-s}$ ($0 < s < 1$), it follows from Eq. (9) that

$$G_o(E) \sim |E|^{-s} \tag{11}$$

Then Eq. (8) tells us that there is always a bound level which for weak V has a binding energy $E_b \sim |V|^{1/s}$. An elementary realization of this behavior is the 1-d potential well. In this case $D_o(E) \sim dp/dE \sim 1/p \sim E^{-\frac{1}{2}}$.

The borderline case s=0, where the DOS jumps discontinuously at zero by an amount equal to δD_o is very interesting. Eq. (9) in this case gives

$$G_o(E) \sim \delta D_o \ln|E| \tag{12}$$

Substituting in Eq. (8) we find that there is always a bound level the binding energy of which (for weak V) is given by

$$E_b \sim \exp\left[-\frac{1}{|V|\delta D_o}\right] \tag{13}$$

Thus a discontinuity δD_o in the unperturbed DOS leads to a binding energy given by Eq. (13). An elementary application of this result is a 2-d potential well, where $D_o(E) \sim pdp/dE = \text{const.} = \delta D_o$.[14]

We return now to the problem of binding two electrons together in a solid. The attractive potential V has already been obtained. Thus one needs to calculate the unperturbed DOS for the relative motion; this DOS is proportional to the volume of the momentum space $\vec{p}_1$, $\vec{p}_2$ subject to the restriction $\vec{P}=\vec{p}_1+\vec{p}_2$ (conservation of momentum) and $E = \varepsilon(\vec{p}_1) + \varepsilon(\vec{p}_2)$ (conservation of energy), i.e.

$$D_o(E;\vec{P}) \sim \int d^3p_1 d^3p_2 \delta(E-\varepsilon(\vec{p}_1)-\varepsilon(\vec{p}_2))\delta(\vec{P}-\vec{p}_1-\vec{p}_2)\theta(\varepsilon(\vec{p}_1)-E_F)$$

$$\theta(\varepsilon(\vec{p}_2)-E_F) \tag{14}$$

The two step functions $\theta(\varepsilon(\vec{p}_1)-E_F)$ and $\theta(\varepsilon(\vec{p}_2)-E_F)$ were included because the Pauli principle does not allow either electron of the pair to occupy states below the Fermi energy E_F associated with all the other electrons. For finite temperatures the product $\theta \cdot \theta$ has to be replaced by $[1-f(\varepsilon(\vec{p}_1))][1-f(\varepsilon(\vec{p}_2))]$ where f is the Fermi distribution. Thus we see that the presence of the other electrons in the solid is very important because, due to Pauli principle, drastically modifies the DOS and consequently the binding energy. One modification in the DOS (for T=0) is that the band edge (i.e. the point below which the DOS is zero) moves from E=0 to E=2E_F. Furthermore the DOS depends now on the total momentum $\vec{P}$. In particular for E close to the band edge $2E_F$ (this is the region of importance for the binding problem) the DOS approaches zero continuously as $E \to 2E_F^+$ for $\vec{P} \neq 0$; but for $\vec{P} = 0$ it is easy to see from Eq. (14) that

$$D_o(E;0) = \tfrac{1}{2}\rho(E/2)\cdot\theta(E/2 - E_F) \tag{15}$$

which drops discontinuously to zero at $E = 2E_F$. The discontinuity is $\rho_F/2$. Hence according to Eq. (13) the binding energy is

$$E_b \sim e^{-\frac{2}{\lambda-\mu^*}} \tag{16}$$

where $\lambda = \rho_F V_\alpha$, $\mu^* = \rho_F V_c^*$ and $\lambda-\mu^* = \rho_F V$. The conclusion is that the $\vec{P} = 0$ pair is always bound no matter how small the attractive potential V is. The binding energy is given by

Eq. (16). On the other hand for $\vec{P} \neq 0$ the DOS drops continuously
to zero and consequently a critical value of V must be exceeded
before a bound pair appears. The binding energy for $\vec{P} \neq 0$ (if it
exists at all) will be smaller than the one for $\vec{P} = 0$. Now if the
pairs are condensed to a single quantum state the lowest energy
will be achieved if each pair has a total momentum $\vec{P} = 0$. We have
thus succeeded in deriving from a simple a general argument the
basic features of the superconducting state.

Overcoming some difficulties

One problem with the simple minded theory outlined above is
that the factor of 2 in the exponent of the RHS of Eq. (16) is
spurious. Another more serious problem is that the above theory,
if applied to a system of non interacting fermions moving in
an external static potential V, would conclude incorrectly that
bound states exist just below the Fermi energy. This conclusion
is reached because the available DOS is $\rho(E)[1-f(E)]$, which at
T=0 exhibits a discontinuity, ρ_F, and, hence, should give a binding
energy proportional to $\exp[-1/|\vec{V}|\rho_F]$.

The origin of these difficulties is our omission of some
indirect processes. The DOS for the single fermion problem was
found to be $\rho(E)[1-f(E)]$ because it was implicitly assumed that
the considered fermion undergoes successive scattering processes
(direct processes), while all the other fermions of the Fermi
sea play no role other than the passive one dictated by the Pauli
principle. However, the fermions of the Fermi sea make possible
additional indirect processes: A Fermi sea fermion can jump
to the final state and the fermion under consideration fills
up the created hole. Obviously the DOS for this indirect process
involves the occupied levels and equals to $\rho(E)f(E)$. Adding
this DOS to the direct processes DOS $\rho(E)[1-f(E)]$ we obtain
that the total DOS is $\rho(E)$, which is independent of $f(E)$ as it
must be.

It is worthwhile to mention parenthetically the problem of
independent electrons undergoing spin flip scattering by external
local moments. In this case the direct process DOS is proportional
to $S_{\mp}S_{\pm}\rho$ (1-f), while the indirect DOS is proportional to
$S_{\pm}S_{\mp}\rho f$, where $S_{\pm}$ is the raising and lowering spin operators of
the local moment. Note the different order of the S_+, S_-
between the direct and indirect processes due to reversal of
the time sequence of the two spin flips. Now the total DOS
is proportional to $S_{\mp}S_{\pm}\rho + (S_{\pm}S_{\mp} - S_{\mp}S_{\pm})\rho f$. The quantity in
parenthesis is not zero and consequently the discontinuity at
E_F (for T=0) survives. This should lead to a bound electronic
state around the local moment. Such a bound level actually
exists up to a critical temperature T_c below which the fermi
factor f does vary sharply enough to produce a $G_o(E)$ which will

satisfy Eq. (8). Above T_c an electron is no longer bound to a
local moment but still it can spend a long time around it before
it propagates further on. This so called resonance scattering
implies an enhanced resistivity, which will increase with decreasing temperature until one reaches T_c below which the resistivity
is not expected to vary significantly with T. This behavior is
actually observed in the so called Kondo systems.[15]

Returning now to our problem of the $\vec{P} = 0$ electronic pair
we see that the omitted indirect processes (which involve two
Fermi sea electrons jumping to the final states while the original
electrons fill up the resulting holes) have a DOS equal to
$\frac{1}{2}\rho(E/2)f(E/2)f(E/2)$. This DOS must be <u>subtracted</u> from the direct
process DOS. The subtraction rather than addition of the two
DOS has to do with the antisymmetry of the pair function under
particle exchange. However, I was not able to find a simple
physical explanation for this feature. Taking into account both
the direct and the indirect DOS we obtain for the total effective
DOS

$$D_0(E;0) = \tfrac{1}{2}\rho(E/2)[1-2f(E/2)] \tag{17}$$

Eq. (17) shows that the discontinuity at $E = 2E_F$ for $T = 0$ is
ρ_F i.e. twice as big as that of Eq. (15). Hence Eq. (16) will
become now

$$E_b \sim e^{-\frac{1}{\lambda-\mu^*}} \tag{18}$$

which is the standard BCS result. Using Eqs. (17) and (9) one
can find the critical temperature T_c which is the highest temperature for which Eq. (8) still has a solution. The integral in
Eq. (9) extends from $2E_F - 2\hbar\omega_D$ to $2E_F + 2\hbar\omega_D$ because according
to the BCS assumption this is the range where the phonon mediated
attraction exists. Performing the integration and substituting
in Eq. (8) we find

$$T_c = \frac{2e^C}{\pi}\frac{\hbar\omega_D}{k}\exp\left[-\frac{1}{\lambda-\mu^*}\right]$$

$$= 1.13\,\Theta_D\exp\left[-\frac{1}{\lambda-\mu^*}\right] \tag{19}$$

where C is Euler's constant. This is exactly the BCS result for
T_c, since $\lambda-\mu^* = \rho_F V$.

Strong coupling modifications

Up to now we have implicitly assumed that the electron-
phonon interaction, V_{e-p}, and the Coulomb repulsion, V_c, are so
weak that they have no effect on the propagation of the individual

electrons from which the pairs are made up. Actually both V_{e-p} and V_C modify the properties of each electron; it is these modified electrons (which are called dressed or quasi electrons) which combine to make up the pairs. One modification of importance is that the discontinuity at the Fermi level is reduced by a factor $w_F \equiv (1 - \partial\Sigma/\hbar\partial\omega)^{-1}$ for each electron of the pair (see, e.g. ref. 14 pp. 180, 212), where $\Sigma(\omega,\vec{p})$ is the so called self energy. One way to understand this reduction is by taking into account that only a fraction w_F of each electron propagates as a quasi-electron, while the rest, $1-w_F$, has not a well defined energy-momentum relation and thus does not produce any discontinuity at E_F. The net result is to multiply the DOS given by Eq. (17) by a factor w_F^2.

Another important effect of V_{e-p} and V_C is that they change the electron velocity at the Fermi level from its unperturbed value v_F to $\vec{v}'_F = (\vec{v}_F + \partial\Sigma/\partial\vec{p})w_F$. Since the single particle DOS is inversely proportional to the magnitude of the velocity, it follows that the DOS ρ_F must be replaced by $(v_F/v'_F)\rho_F$.

The result of the above two effects together is to multiply the quantity $\lambda-\mu^*$ in Eq. (19) by a factor equal to $w_F^2(v_F/v'_F)$. To calculate this factor explicitly one needs to obtain the self energy $\Sigma(\omega,\vec{p})$ which depends both on V_{e-p} and V_C. It is usually assumed that the Coulomb interaction, V_C, has no significant effects on $\partial\Sigma/\partial\omega$; furthermore, calculations employing the Hubbard dielectric function give that the effect of V_C on $\partial\Sigma/\partial\vec{p}$ is negligible for the usual electronic densities $(r_s \approx 2.5)$.[13] Thus, in calculating Σ we usually keep only V_{e-p} and we employ second order perturbation theory to obtain[12,13,16]

$$\Sigma(\omega,\vec{p}) = -\lambda\hbar\omega, \tag{20}$$

from which we got $w_F = (1+\lambda)^{-1}$, $v_F/v'_F = (1+\lambda)$ and $w_F^2(v_F/v'_F) = (1+\lambda)^{-1}$. Hence the expression for T_c becomes

$$T_c = p \exp\left[-\frac{1+\lambda}{\lambda-\mu^*}\right], \tag{21}$$

where the prefactor p is not equal to that of Eq. (19) because of contributions due to the non-quasiparticle smooth background of each electron propagator.

A more rigorous analysis based on the Eliashberg gap equations[13,16] gives the following expression for T_c[16,17]

$$T_c = p \cdot \exp\left[-\frac{1.04(1+\lambda)}{\lambda-\mu^*(1+0.62\lambda)}\right] \tag{22}$$

which is remarkably close to our simplified result. According to McMillan[16] the prefactor in Eq. (17) is given by $p=\Theta_D/1.45$ or more accurately by

$$p = \bar{\omega}/1.2 \qquad (23)$$

where

$$\bar{\omega} = 2 \int d\omega \alpha^2(\omega) F(\omega)/\lambda \qquad (24)$$

Allen and Dynes[17] have obtained the following expression for the prefactor in Eq. (22)

$$p = f_1 f_2 \omega_{log}/1.2 \qquad (25)$$

where

$$\omega_{log} = \exp \left[\frac{2}{\lambda} \int \frac{d\omega}{\omega} \alpha^2(\omega) F(\omega) \ln\omega \right] \qquad (26)$$

$$f_1 = \{1+[\lambda/(2.46 + 9.35\mu^*)]^{3/2}\}^{1/3} \qquad (27)$$

$$f_2 = 1 + \frac{(\bar{\omega}_2/\omega_{log} - 1)\lambda^2}{\lambda^2 + (1.82+11.5\mu^*)^2 \bar{\omega}_2^2/\omega_{log}^2} \qquad (28)$$

$$\bar{\omega}_2 = \left[2 \int d\omega\omega\alpha^2(\omega) F(\omega)/\lambda \right]^{\frac{1}{2}} \qquad (29)$$

It is worthwhile to mention that it is usually convenient to rewrite λ as

$$\lambda = \frac{\eta}{M\bar{\omega}^2} \qquad (30)$$

$$\bar{\omega}^2 = \langle\omega\rangle/\langle\omega^{-1}\rangle; \quad \langle\omega^m\rangle = \int d\omega\omega^m F(\omega) \qquad (31)$$

where M is the ionic mass and $M\bar{\omega}^2$ is an effective spring constant measuring the softness of the lattice. The quantity η was found to be almost independent of the phonon spectrum.[16] Furthermore, as we will see in the next section, η can be expressed rather accurately in terms of quantities which are obtained from a band structure calculation. Thus Eq. (30) is very convenient because it separates λ into a phonon factor $(M\omega^{-2})^{-1}$ and an electronic factor η which can be calculated from first principles.

The quantity μ^*, which was not discussed in any detail here, is of the order of 0.1 for most materials.[16,12] Bennemann and Garland[18] obtained the following expression for μ^*

$$\mu^* = \frac{.52 \, n_F}{1+2. \, n_F} \qquad (32)$$

where n_F is the DOS per eV per unit cell per spin.

We conclude this subsection by mentioning that in the case
of more than one atom per unit cell λ is given by summing the
contributions of each atom[19], i.e.

$$\lambda = \sum_i \lambda_i \qquad (33)$$

where

$$\lambda_i = 2 \int_0^\infty d\omega \, \frac{a_i^2(\omega)}{\omega} \, F_i(\omega) = \frac{\eta_i}{M_i \omega_i^{-2}} \qquad (34)$$

Calculating η_i
___

The quantity η_i can be calculated if one knows how the
crystalline potential is modified upon displacing the ith atom.
The rigid muffin tin approximation (RMTA) obtains this modification by assuming that the crystalline muffin tin potential U_{MT}
moves rigidly with the displaced atom. By employing, in
addition, the so called local and spherical approximations[20]
one finds[20,21] that

$$\eta_i = \frac{E_F}{\pi^2 \rho_F} \sum_{\ell=0}^\infty \frac{2(\ell+1)\rho_{\ell+1,i}\rho_{\ell,i}}{\rho_{\ell+1,i}^{(1)} \rho_{\ell,i}^{(1)}} \sin^2(\delta_{\ell+1,i}-\delta_{\ell,i}). \quad (35)$$

In obtaining Eq. (35) the wave function inside the muffin-tin
has been analyzed in spherical harmonics; $\rho_{\ell,i}$ is the ℓ-partial
DOS at E_F for the ith atom; $\rho_{\ell,i}^{(1)}$ is the corresponding quantity
for a single isolated muffin-tin potential; $\delta_{\ell,i}$ is the ℓ-phase
shift for the ith atom. In Eq. (35) we have the product of two
consecutive ℓ's because η_i depends on the matrix element of the
vector operator $\partial U_{MT}(r)/\partial \vec{r}$; the sin terms appear because
$\int_0^\infty R_{\ell,i} dU_{MT}/dr \, R_{\ell+1,i} r^2 dr = \sin(\delta_{\ell,i}-\delta_{\ell+1,i})$ where R_ℓ is the
radial part of the wave function. Eq. (35) is very convenient
because all quantities in the RHS are obtained from a standard
APW band structure calculation (see, e.g. the article by D. A.
Papaconstantopoulos[22] in this volume). It must be pointed out
that the various approximations used in deriving Eq. (35) are well
justified for d-band metals.[20,21]

RESULTS FOR Pd BASED HYDRIDES
___

The results to be summarized in this section were obtained
as follows:

The quantity μ^* was usually calculated from Eq. (32). The
phonon distribution function $F_i(\omega)$ was obtained from the eigen-
frequencies and eigenmodes of the lattice vibrations, which were
expressed in terms of the spring constants $k_{ij}^{\alpha\beta}$ between a given

atom and its neighbors. The quantities $k_{ij}^{\alpha\beta}$ were determined by fitting the calculated eigenfrequency spectrum to that determined by inelastic neutron scattering experiments. The quantities η_i, ρ_F, etc. were obtained from ab initio APW band structure calculations.[22]

Pd

The band structure calculations[22,23] show that the Fermi energy for Pd is at the center of a sharp d-like peak of the DOS. The quantity n_F equals[22] to 1.1 states per eV per unit cell per spin which yields for μ^* through Eq. (32), $\mu^* = 0.18$. It is generally believed that such a high d-character n_F implies strong attractive electron-hole correlations which will introduce an additional repulsive contribution[24] λ_p so that λ will be replaced by $\lambda - \lambda_p$. The value of η determined from Eq. (35) is $\eta = 3.40$ eV/$\overset{\circ}{A}^2$; the quantity $M\omega^{-2}$ was obtained as explained above by fitting the experimental data of Miller and Brockhouse[25]: $M\overline{\omega}^{-2} = 7.546$ eV/$\overset{\circ}{A}^2$. The resulting value of λ is $\lambda = .45$. This value is in good agreement with $\lambda_{tr} \approx .50$ which was obtained by using the band structure[11] value of $\omega_p = 7.3$ eV. Taking into account that $\lambda \approx .45$ and that Pd is not superconducting we find that $\lambda_p > .1$.

PdH and PdD

The band structure of PdH, which was assumed identical to that of PdD, shows[26] that the Fermi level is now above the sharp Pd d-like peak and that the quantity n_F has dropped to $n_F \approx .24$ states per eV per unit cell per spin. Eq. (32) is then giving $\mu^* = .085$. The calculated values for the η's are: $\eta_{Pd} = .865$ eV/$\overset{\circ}{A}^2$ and $\eta_{H(D)} = .392$ eV/$\overset{\circ}{A}^2$. To obtain $F_{Pd}(\omega)$, $F_H(\omega)$, and $F_D(\omega)$ one first observes that, because of the large mass difference between Pd and H or D, F_{Pd} is essentially associated with the acoustic part of the spectrum and F_H or F_D with the optical part. $F_{Pd}(\omega)$ was found by fitting the linear extrapolation[26] of the experimentally determined acoustic spectrum of Pd[25] and PdD$_{.63}$[27,28]. $F_D(\omega)$ was obtained by fitting the optical spectrum of PdD$_{.63}$[27,28]. If the harmonic approximation were exact F_H could be easily obtained by rescaling F_D: $F_H(\omega) = (M_H/M_D)^{\frac{1}{2}}F_D[\omega(M_H/M_D)^{\frac{1}{2}}]$. This rescaling means that the force constants $M_H\langle\omega^2\rangle_H$ for H is the same as the force constant for D, $M_D\langle\omega^2\rangle_D$. However, the experimental data[28] reveal that there is considerable anharmonicity so that $M_H\langle\omega^2\rangle_H \approx 1.2\ M_D\langle\omega^2\rangle_D$. Thus the rescaling was done as to satisfy this experimentally observed relation

$$F_H(\omega) \approx .913\ (M_H/M_D)^{\frac{1}{2}}F_D[.913(M_H/M_D)^{\frac{1}{2}}\omega] \tag{36}$$

Having $F_{Pd}(\omega)$, $F_D(\omega)$, and $F_H(\omega)$ we obtain that $\overline{M\omega^2}$ is 4.95, .87, and 1.06 eV/$\mathring{A}^2$ for Pd, D, and H respectively. The corresponding λ's are .175, .450 and .370. Thus the total λ for PdH is λ=.545 and for PdD is λ=.625. The resulting values for T_c, using Eqs. (22,25), are: T_c = 10.4°K for PdD and 9°K for PdH in very good agreement with the experimentally observed values[29,30] of 9.5-11°K for PdD and 7.5-9 for PdH. This agreement is very impressive if one takes into account that no adjustable parameters were used in the calculation.

The high T_c in PdH and PdD is due to the large H or D contribution to λ. This large contribution comes mostly from the low value of $\overline{M\omega^2}$, i.e. from the softness of the H (or D) sublattice. As a matter of fact, because of anharmonic effects, the D sublattice is softer than the H sublattice and this accounts for the observed inverse isotope effect.

PdH_x and PdD_x

In the calculation[26] of T_c for the substoichiometric Pd hydride and deuteride the very weak x-dependence of the lattice factors $\overline{M\omega^2}$ has been omitted. To obtain η_{Pd} and $\eta_{H(D)}$ one has to find the electronic structure of a random system, since each H(D) site can be either occupied with probability x or can be empty with probability 1-x. Since an empty H(D) site does not contribute to $\eta_{H(D)}$, we can write $\eta_{H(D)} = x \cdot \bar{\eta}_{H(D)}$ where $\bar{\eta}_{H(D)}$ is a configurational average under the condition that the H(D) site under consideration is occupied. Both $\bar{\eta}_{H(D)}$ and $\bar{\eta}_{Pd}$ are functions of x both explicitly and implicitly through $E_F(x)$. The rigid band approximation (RBA), which assumes that the bands do not change with varying x, allows an easy determination of the x-dependence of $\bar{\eta}_{H(D)}$ and $\bar{\eta}_{Pd}$, since within the RBA only $E_F(x)$ changes in order to accomodate the varying electronic density. In general the RBA is not accurate at all for PdH(D)$_x$. However, for the present purposes and for the limited range $.7 \lesssim x \leq 1$ the RBA gives reasonable results.[26] This has been confirmed by calculating $\bar{\eta}_{H(D)}$ and $\bar{\eta}_{Pd}$ within the coherent potential approximation (CPA)[20,31], which constitutes the state of the art for obtaining the DOS of disordered systems. The reasons for the unexpected success of the RBA in the present case are: (a) the total number of states below the center of the highest Pd d-like peak does not vary significantly with x and (b) the DOS above this peak varies smoothly with x.[26] It was found that $\lambda_{Pd} = \bar{\eta}_{Pd}/M_{Pd}\overline{\omega^2}_{Pd}$ = .175 for $.75 \leq x \leq 1$. On the other hand the quantity $\lambda_{H(D)} = x\bar{\eta}_{H(D)}/M_{H(D)}\overline{\omega^2}_{H(D)}$ changes drastically with x as shown in Table 2. This change is due mostly to a drop of $\bar{\eta}_{H(D)}$ with decreasing x, which was traced to the ℓ=0 term in Eq. (35). The physical meaning of the last observation is that, as x decreases, the Fermi level moves towards lower energies,

Table 2: The dependence of μ^*, λ's, and T_c's on x for PdH(D)$_x$. Experimental values for T_c are also shown.

x	μ^{*c}	λ_D	λ_{tot}	PdD$_x$ T_c(calc)^c	T_c(expt)^a	T_c(expt)^b	λ_H	λ_{tot}	PdH$_x$ T_c(calc)^c	T_c(expt)^a	T_c(expt)^b
1.00	0.085	0.45	0.62	10.4	9.8	10.3	0.37	0.54	9	8.0	9.1
0.96	0.087	0.39	0.56	7.6	7.8	9.1	0.32	0.49	6.3	6.3	6.6
0.92	0.091	0.35	0.52	5.7	6.4	7.1	0.29	0.46	4.6	4.8	4.8
0.89	0.093	0.32	0.49	4.4	5.0	5.1	0.26	0.43	3.3	3.5	3.1
0.85	0.093	0.27	0.44	2.8	3.7	3.3	0.22	0.39	2.0	2.2	2.1
0.81	0.097	0.24	0.41	1.8	2.5	...	0.20	0.37	1.3	1.3	...
0.77	0.101	0.19	0.36	.7	...	...	0.16	0.33	.5	...	...

[a]Reference 27
[b]Reference 28
[c]Using Eq. (22)

where the H s-like character of the states is smaller; as a
result the Fermi level electrons do not couple anymore so effectively with the soft H (or D) sublattice and the H(D) mediated
attraction tends to disappear. As shown in Table 2 the agreement
between theory and experiment is very good. Note that the
calculated values of T_c in Table 2 are slightly different than
those in ref. 26 because there[26] the more accurate Eliashberg
equations were used to obtain T_c.

$Pd_{1-y}Ag_yH(D)_x$ and $Pd_{1-y}Rh_yH(D)_x$

By alloying Pd with Ag the Fermi level moves towards higher
energies corresponding to more H s-like character eigenstates.
On the basis of the argument given above one expects that T_c
would increase with increasing Ag concentration. On the other
hand alloying with Rh would lower the Fermi level and consequently
would decrease T_c. This is exactly the behavior observed experimentally.[6,2]

The results of ref. 32 which will be briefly summarized here
were obtained as follows: The quantities f_1 and f_2 (Eqs. (27-28))
were taken equal to one. The quantity ω_{log} was approximated[17]
by $\omega_{log} = 2<\omega>-<\omega^2>^{\frac{1}{2}}$. The average $<\omega^n>_{H(D)}$ for n=-1, 1, 2
was taken as in PdH(D). The quantity $<\omega^n>_{Met}$ was obtained from
the approximate relation $<\omega^n>_{Met} \approx <\omega^n>_{Pd}[y+(1-y)\Theta_M^n/\Theta_{Pd}^n]$
where M stands for Ag or Rh and Θ_M is the corresponding Debye
temperature; μ^* was taken[32] as .13. The virtual crystal
approximation (VCA) was employed in order to handle the disorder
associated with the random substitution of Pd by Ag or Rh. The
H(D) associated randomness was treated both within the RBA and,
in order to check the validity of the RBA, the CPA. The band
structure calculations were not self-consistent.[22]

It must be stressed that the various assumptions and approximations restrict the range of validity of the results to a portion
of the x-y plane. The assumption of a NaCl structure is valid
up to a maximum value of y beyond which lattice instabilities seem
to take place.[33] The quantity $<\omega^n>_{H(D)}$ is approximately independent of y up to a maximum value $y_0 \approx .3$. The RBA is valid for
x>.7. On the other hand x in $Pd_{1-y}A_yH(D)_x$ cannot exceed an upper
bound $x_0(y) \approx 1-y/2$ beyond which the hydrogen is totally lost.[2]

In Table 3 we show the dependence of calculated λ's and T_c's
on y for $Pd_{1-y}Ag_yH(D)_x$. The H(D) concentration x was kept equal
to 1-y/2, which seems to be close to the maximum possible value
of x at least for small y. For larger y the maximum possible
value of x may be considerably lower than 1-y/2, which would
account for the rather sharp drop of T_c with increasing y for
$y \gtrsim .35$. In the range $0 < y \lesssim .3$, where the assumptions of the

Table 3. The dependence of λ's and T_c's on y for $Pd_{1-y}Ag_yH(D)_x$ and for $x = 1-y/2$

y	x=1-y/2	λ_H	λ_D	λ_{Met}	T_{cH}	$T_{cH}(exp)$[a]	T_{cD}	$T_{cD}(exp)$[a]
0	1	.37	.45	.175	9	8.5	10.4	10.1
.15	.925	.52	.63	.18	14	14	14.6	14.5
.30	.85	.53	.65	.18	15.2	15	15.6	13.5
.50	.75	.55	.67	.17	15.8	--	16.2	--

[a]From ref. 6

theory are valid, the agreement with the experimental values is very good. Similarly good agreement with the experiment exists for $Pd_{.9}Rh_{.1}H(D)$, where we obtain $T_{cH} = 2.3°K$ and $T_{cD} = 3.4°K$ vs $4°K$ and $5°K$ for the experimental values.

In all cases the variation of T_c with x and y was mainly due to changes in $\eta_{H(D)}$, which in turn were determined almost entirely by the ratio ρ_{sH}/ρ_F. This strongly supports the previously proposed physical explanation: The variations of T_c in the Pd based hydrides and deuterides are determined by the probability that the electrons at the Fermi energy will find themselves in the vicinity of the H(D) sites; this allows them to take advantage of the strong attraction which the soft H(D) sublattice can mediate.

REFERENCES

1. C. B. Satterthwaite and I. L. Toepke, Superconductivity of Hydrides and Deuterides of Thorium, Phys. Rev. Lett. 25, 741 (1970).
2. See e.g. B. Stritzker and H. Wühl, Superconductivity in Metal-Hydrogen Systems, in "Topics in Applied Physics", Vol. 29, G. Alefeld and J. Völkl, ed., Springer, Berlin (1978).
3. T. Skoskiewicz, Superconductivity in the Palladium-Hydrogen and Palladium-Nickel-Hydrogen Systems, Phys. Status Solidi (a) 11, K123 (1972).
4. B. Stritzker and W. Buckel, Superconductivity in the Palladium-Hydrogen and the Palladium-Deuterium Systems, Z. Physik 257, 1 (1972).

5. W. Buckel and B. Stritzker, Superconductivity in the Palladium-Silver-Deuterium System, Phys. Lett. 43A, 403 (1973).

6. B. Strizker, High Superconducting Transition Temperatures in the Palladium-Noble Metal-Hydrogen System, Z. Phys. 268, 261 (1974).

7. G. Deutscher and M. Pasternak, Effect of argon and hydrogen coating on the superconducting transition temperature of granular aluminum, Phys. Rev. B10, 4042 (1974).

8. A. M. Lamoise, J. Chaumont, F. Meunier, and H. Bernas, Superconducting Properties of Aluminum Thin Films after Ion Implantation at Liquid Helium Temperatures, J. Physique Lett. 36, L-271 (1975); Resistivity Annealing Properties of Aluminum Thin Films after Ion Implantation at Liquid Helium Temperatures, J. Physique Lett. 36, L-305 (1975).

9. G. Grimvall, The Electron-Phonon Interaction in Normal Metals, Phys. Scripta 14, 63 (1976).

10. G. T. Meaden, "Electrical Resistance of Metals", Plenum Press, New York (1965).

11. D. A. Papaconstantopoulos, private communications.

12. J. R. Schrieffer, "Theory of Superconductivity", Benjamin, Reading (1964).

13. D. J. Scalapino, The Electron-Phonon Interaction and Strong-Coupling Superconductors in "Superconductivity", R. D. Parks, ed., Dekker, New York (1969).

14. E. N. Economou, "Green's Functions in Quantum Physics", Vol. 7 in Springer Series in Solid-State Sciences, Springer-Verlag, Berlin (1979).

15. J. Kondo, Resistance Minimum in Dilute Magnetic Alloys, Progr. Theor. Phys. (Japan) 32, 37 (1964).

16. W. L. McMillan, Transition Temperature of Strong-Coupled Superconductors, Phys. Rev. 167, 331 (1968).

17. P. B. Allen and R. C. Dynes, Transition temperature of strong-coupled superconductors reanalyzed, Phys. Rev. B12, 905 (1975).

18. K. H. Bennemann and J. W. Garland, Theory for Superconductivity in d-Band Metals, AIP Conf. Proc. 4, 103 (1972).

19. B. M. Klein and D. A. Papaconstantopoulos, On calculating the electron-phonon mass enhancement λ for compounds, J. Phys. F6, 1135 (1976).

20. H. Rietschel, A Non-Local Extension of the Gaspari-Gyorffy Theory for Superconductors, Z. Phys. B30, 271 (1978).

21. G. D. Gaspari and B. L. Gyorffy, Electron-Phonon Interactions, d Resonances, and Superconductivity in Transition Metals, Phys. Rev. Lett. 28, 801 (1972).

22. D. A. Papaconstantopoulos, Electronic Structure of Metal Hydrides present volume.

23. D. A. Papaconstantopoulos and B. M. Klein, Superconductivity in the Palladium-Hydrogen System, Phys. Rev. Lett. 35, 110 (1975).

24. N. F. Berk and J. R. Schrieffer, Effect of Ferromagnetic Spin Correlations of Superconductivity, Phys. Rev. Lett. 17, 433 (1966).

25. A. P. Miiller and B. N. Brockhouse, Crystal Dynamics and Electronic Specific Heats of Palladium and Copper, Can. J. Phys. 49, 704 (1971).

26. D. A. Papaconstantopoulos, B. M. Klein, E. N. Economou, and L. L. Boyer, Band Structure and Superconductivity of PdD_x and PdH_x, Phys, Rev. B17, 141 (1978).

27. J. M. Rowe, J. J. Rush, H. G. Smith, M. Mostoller, and H. E. Flotow, Lattice Dynamics of a Single Crystal of $PdD_{0.63}$, Phys. Rev. Lett. 33, 1297 (1974).

28. A. Rahman, K. Sköld, C. Pelizzari, S. K. Sinha, and H. Flotow, Phonon Spectra of nonstoichiometric palladium hydrides, Phys. Rev. B14, 3630 (1976).

29. J. E. Schirber and C. J. M. Northrup, Jr., Concentration dependence of the superconducting transition temperature in PdH_x and PdD_x, Phys. Rev. B10, 3818 (1974).

30. R. J. Miller and C. B. Satterthwaite, Electronic model for the Reverse Isotope Effect in Superconducting Pd-H(D), Phys. Rev. Lett. 34, 144 (1975).

31. D. A. Papaconstantopoulos, B. M. Klein, J. S. Faulkner, and L. L. Boyer, Coherent-potential-approximation calculations for PdH_x, Phys. Rev. B18, 2784 (1978).

32. D. A. Papaconstantopoulos, E. N. Economou, B. M. Klein, and L. L. Boyer, Electronic structure and superconductivity in Pd-Ag-H and Pd-Rh-H alloys, Phys. Rev. B20, 177 (1979).

33. B. N. Ganguly, Superconductivity in Palladium-Noble Metal-Hydrogen Systems, Z. Phys. B22, 127 (1975).

MAGNETISM OF LANTHANIDE AND ACTINIDE INTERMETALLIC HYDRIDES[*]

W. E. Wallace

Department of Chemistry, University of Pittsburgh

Pittsburgh, PA 15260

ABSTRACT

The search for new hydrogen storage materials (needed for the development of hydrogen as a fuel) is impeded by a paucity of information about metal-metal and metal-hydrogen bonding. Information bearing upon these types of bonds is provided by appropriate magnetic studies.

When the elemental lanthanides are hydrogenated to saturation, metallic electrical conduction is lost and magnetic ordering is eliminated or greatly suppressed. Since electrical conduction and the exchange interaction which leads to magnetic ordering both require the presence of conduction electrons, it appears that hydrogenation results in a complete depopulation of the host metal conduction band. Thus hydrogen in these materials is essentially anionic and hence the bonding in the lanthanide hydrides is essentially ionic.

Intermetallic compounds involving Fe, Co or Ni combined with Th or the lanthanides hydrogenate to form very hydrogen-rich systems. They have proton densities exceeding that of liquid hydrogen by up to a factor of 4. Hydrogenation to this extent profoundly modifies the electronic nature of the host metal in many cases. Among the effects observed: (1) Strengthening of exchange to convert a paramagnetic system into a ferromagnet (Th_6Mn_{23}); (2) Suppression of ferromagnetism (Y_6Mn_{23}); (3) Transforming a

[*]The present work was supported by a series of grants from the Army Research Office.

superconductor into a ferromagnet (Th_7Fe_3); (4) Generation of a spin-glass system (Er_6Mn_{23}). Interpretation of these experimental results requires adoption of the concept that hydrogen is acting as an electron acceptor. Hydrogen is probably largely anionic in these systems and the ionic contribution appears to be an important part, perhaps the dominant part, of the bonding.

I. INTRODUCTION

Hydrides of intermetallic compounds attract attention because of two noteworthy features: (1) striking electronic changes are often encountered during the hydrogenation process (vide infra) and (2) the formation of hydrides is related to the development of hydrogen as a fuel. The developed world is heavily dependent upon petroleum as an energy source. Within a 50-year period it is clear that petroleum will either be exhausted or prohibitively expensive. If wise policies prevail, petroleum will be reserved for better purposes - for example, the petrochemical industry - in the much nearer future.

Metal hydrides connect with the development of hydrogen as a fuel through the issue of hydrogen storage. The present prime candidates for hydrogen storage are TiFe, Mg_2Ni and $LaNi_5$. Other systems which are rich in hydrides are listed in Table 1. The present prime candidates are good but not excellent for hydrogen storage purposes. They are either uneconomical or too heavy or fail to release hydrogen at normal temperatures, or some combination of these.

From an economic point of view a material such as Ca_5Fe or $CaFe_5$ would be better, but such materials do not exist for reasons which are yet to be elucidated. $CaAl_2$ exists and would seem to offer a promise for hydrogen storage; however, it does not absorb hydrogen (1,2).

In seeking new hydrogen storage materials very much more information is needed about metallic bonding and also an improved understanding of the metal-hydrogen bond in metal hydride systems is an urgent necessity. These are very challenging problems, as challenging as any confronting Materials Scientists at the present time. Magnetic studies can help with the elucidation of metal-metal and metal-hydrogen bonding. In the program at Pittsburgh studies have been made of the magnetic properties of metals and metal hydrides for a number of years, in part to elucidate these important features of the metallic state. In the present chapter, work pertaining to the magnetic features of lanthanide hydrides and hydrides of lanthanide and actinide intermetallics is summarized and related to electronic structure and bonding in these materials. Section II summarizes the information pertaining to

Table 1

Hydrogen Capacity of Some Representative Metallic Hosts

Host	x in (Host)·H_x	r^a	Ref.
TiFe[b]	1.9	1.4	3
LaNi$_5$[b]	6	1.4	4
Mg$_2$Ni[b]	1.3	1.3	5
RFe$_2$[c]	∿4	∿1.9	6
RFe$_3$	2.7 – 4.2	0.9 – 1.3	7
Th$_7$Fe$_3$	28	3.9	8
RCo$_3$	4.2 – 4.6	1.4 – 1.5	9
RCo$_5$	2.5 – 4.2	.65 – .99	10
Th$_7$Co$_3$	30	4.2	11
La$_3$Ni	8.8	3.1	12
Y$_6$Mn$_{23}$	25	1.2	12,13,14
Th$_6$Mn$_{23}$	25	1.2	12,14

^{a}r = the number of hydrogen atoms/volume relative to
the number density for liquid hydrogen (4.2 x 10^{22}
cm^{-3}).

^{b}These are the present prime candidates for hydrogen
storage.

^{c}R = lanthanide element.

the electronic makeup of lanthanide hydrides which has been derived; this has come largely from magnetic studies. In Section
III the types of intermetallic compounds formed between the 3d
transition elements and the lanthanide elements are exemplified.
In the final sections information is provided concerning the magnetic properties of hydrides of thorium intermetallic compounds
with iron, cobalt and nickel and of various hydrides which occur
in systems involving the lanthanides combined with iron, cobalt
and nickel. This information is utilized to develop information
about the electronic status of these complex hydrides.

II. HYDRIDES OF THE ELEMENTAL LANTHANIDES

Magnetic information has provided important insight into the
electronic makeup of the lanthanide hydrides. In the elemental
lanthanides, as well as in their hydrides, magnetic effects originate with their incomplete f-shells. The f-shells have a rather
limited spatial extension. Accordingly, f-shells centered on adjacent atoms do not overlap to any significant degree. Hence exchange, which is necessary for the magnetic ordering of the elemental lanthanides, does not occur through overlap of the f-or-
bitals centered on adjacent atoms. Instead, the interaction is

transmitted through the conduction of electrons. Delocalized electrons are accordingly essential for the magnetic interaction.

     Several investigators observed (15-17) a number of years ago
that during hydrogenation the elemental lanthanide system loses
its metallic conduction. The resistivity increases by several
orders of magnitude (see Table 2) and there is a change in the
sign of the temperature coefficient of resistivity from positive,
which is characteristic of the metal, to negative, which is characteristic of an insulator or semiconductor. Wallace (18) suggested
a number of years ago that these features implied that the conduction electron was being depopulated during the hydrogenation process. He proposed that in the fully hydrogenated material the conduction band was fully depopulated. If so, magnetic ordering
should either be eliminated or greatly suppressed during the hydrogenation process. Experiments confirmed this (18) for the fully
hydrogenated material (see Table 3). The dihydrides, which are
metallic conductors, do exhibit magnetic ordering. Thus the magnetic behavior of lanthanide hydrides is entirely consistent with
the concept of band depopulation during hydrogenation.
     This simple perspective on lanthanide hydrides proved to be
inconsistent with certain of their features, specifically the Hall
coefficients observed (16) for the extensively hydrogenated systems. To account for the observed Hall coefficient's behavior it
was necessary to recognize that during hydrogenation a new band
was formed, lying at a lower energy than the conduction band of
the host metal (see Fig. 1). During the hydrogenation process
electrons are progressively transferred from the conduction band
of the host metal into this new band which is mainly hydrogenic in
nature. Detailed consideration of the band structure of these

Table 2

Resistivity of RH_x (R = La,Ce) Powders[a]

	ρ (ohm cm)	
x	LaH_x	CeH_x
2.0	0.08	0.03
2.3	0.09	0.04
2.5	0.2	0.06
2.7	0.8	0.3
2.8	3.0	0.9
2.85	9.5	9.0
2.92	900	900

[a]From Stalinski, ref. 15.

Table 3

Ordering Temperatures (K) of Rare Earth Metals
and Hydrides

	R	RH_2	RH_3
Pr	n^a	n	n
Nd	$19(N)^b$	$9.5(C)^b$	n
Sm	106(N)	n	n
Eu	88(N)	24(C)	$---^c$
Gd	291(C)	21(N)	n
Tb	235(N)	16(N)	n
Dy	184(N)	8(N)	n
Ho	135(N)	8(N)	n
Er	88(N)	n(N)	n
Tm	56(N)	n(N)	n

[a] n indicates that the sample does not magnetically order at 4 K.
[b] N and C indicate antiferromagnetic and ferromagnetic ordering, respectively.
[c] EuH_3 does not form.

systems suggested that hydrogen in the lanthanide hydrides is tending toward the anionic model (19)

An opposing point of view was advanced by Schreiber and Cotts (20) based upon interpretation of NMR results obtained for the hydrides of lanthanum. They proposed that band-filling effects rather than band-depletion was occurring during the hydrogenation process. This implied an approach toward the protonic model for

Conduction Band Depleted ; New Band Formed

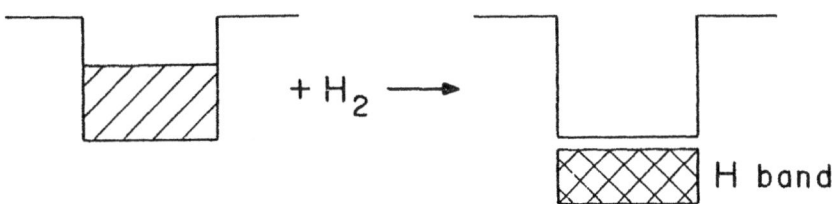

H band comes from antibonding H ls orbitals

Fig. 1

metallic hydrides rather than the anionic model as suggested by the magnetic measurements alluded to above. To provide information pertaining to the validity of these two opposing points of view detailed studies of PrH_2 were made some years back by Wallace and Mader (21). This material crystallizes in the fluorite structure. It is well known that the lanthanide ions are affected by the crystal field interaction and this produces an effect in the bulk magnetism of systems of lanthanide ions (22). In the specific case of PrH_2 the ground state multiplet is split, as shown in Fig. 2. Calculations show that if hydrogen is bearing a positive charge, the ground state is a Γ_1 singlet. If, on the other hand, hydrogen is bearing a negative charge the ground state will be a Γ_5 triplet. In the first case, it is to be expected that at low temperatures the PrH_2 will exhibit the Van Vleck temperature-independent paramagnetism. Experiments of Wallace and Mader clearly indicated the Van Vleck paramagnetism does not develop at reduced temperatures (see Fig. 3). In fact, the susceptibility is continuously increased at the lowest measured temperatures, firmly

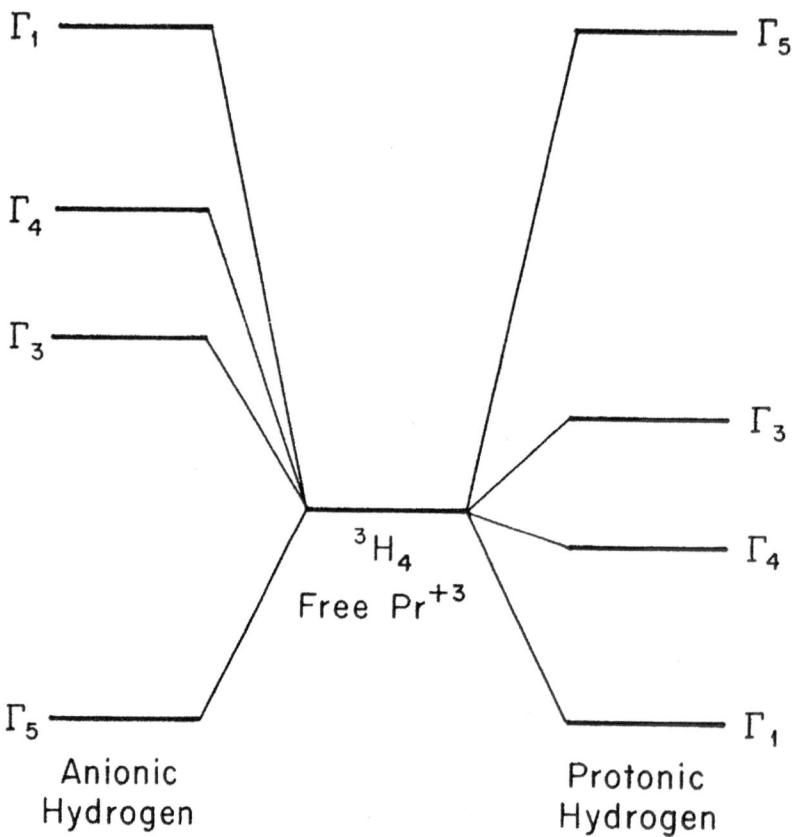

Fig. 2. Crystal field splitting for the 9-fold degenerate ground state of Pr^{3+} in PrH_2. The patterns to the left and right are those expected if hydrogen is anionic (H^-) or protonic (H^+), respectively.

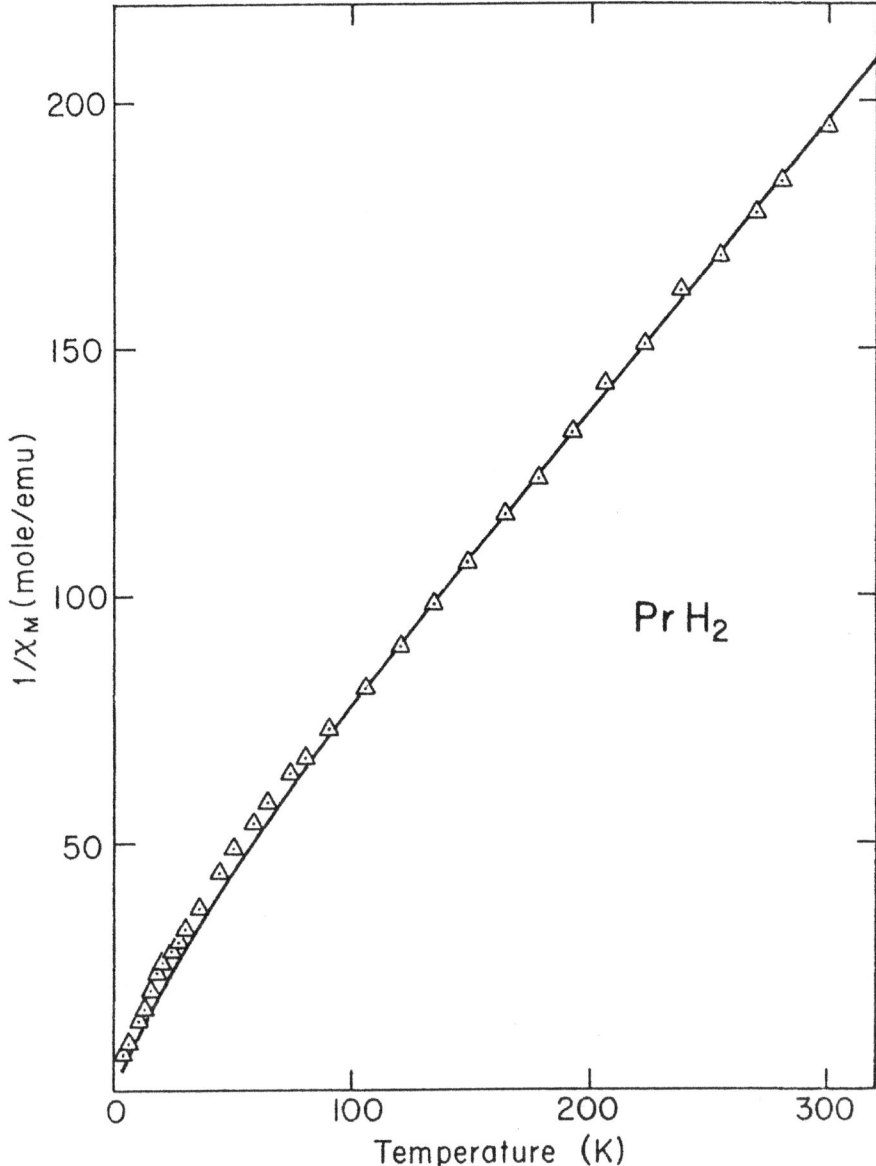

Fig. 3. The points give the experimental values for reciprocal
susceptibility of PrH$_2$ (ref. 21). The curve is the
calculated χ^{-1} versus temperature if hydrogen is anionic
and the crystal field splitting is as shown to the left
in Fig. 2.

establishing that the Γ_5 state is the ground state and also indicating that hydrogen is bearing a negative charge.

Information in regard to the specifics of the crystal field
splitting can also be obtained by inelastic neutron scatterin
measurements. Work of this nature has been recently carried out
by Knorr and Fender (23). They observe, in agreement with the

susceptibility studies, that the ground state of the Pr ion in PrH_2 is the Γ_5 state and these results "can be explained by a negative charge on the hydrogen."

To a first approximation then the lanthanide hydrides can be described as saline hydrides with electron transfer from the lanthanide element to hydrogen to render the latter essentially anionic; bond, therefore, is largely ionic. This concept of these materials is consistent with their magnetic properties, their electrical properties (variation of electrical conductivity upon hydrogenation), their stoichiometries and with neutron inelastic scattering measurements. Specific information bearing upon the latter point is confined to the special case of the hydrides of Pr. However, there is no reason to suppose this result lacks generality.

III. INTERMETALLIC COMPOUNDS INVOLVING THE LANTHANIDE ELEMENTS

The lanthanide elements are very reactive chemically. They form compounds with most of the elements in the Periodic Table. They do not react, of course, with the inert gases nor with elements in the chromium and vanadium group. They form, however, many intermetallic compounds. Over 1000 of these have been described in the monograph by Wallace (24).

The lanthanide intermetallics occur in a number of structural types. These are illustrated in Table 4. A very large number of systems form in the $MgCu_2$ structure (C-15) and in the $CaCu_5$ structure ($D2_d$).

A number of the systems absorb hydrogen in very large quantities (25) and very rapidly. The capacity of a few selected systems is indicated in Table 1. Certain of these absorb hydrogen with great rapidity, reaching saturation in less than five minutes at room temperature in favorable cases (26). The features which are responsible for the rapid uptake of hydrogen have been described in considerable detail for $LaNi_5$ in a recent publication

Table 4

Rare Earth Intermetallic Compounds

Structural Types	Examples
CaCl	GdAg
$MgCu_2$	$PrAl_2$
CrB	CeNi
Fe_3C	Er_3Ni
$CaCu_5$	$LaNi_5$
Th_2Ni_{17}	Ho_2Co_{17}

by Wallace, Karlicek and Imamura (27). The essence of the matter
is that a d-transition metal must exist in the surface of the materials. In the particular case of $LaNi_5$ the surface is oxidized
into La_2O_3 and elemental nickel. Molecular hydrogen chemisorbs on
the surface of the nickel nodules and then is broken down into
atomic hydrogen, after which it diffuses rapidly into the interior
of the material. The factors responsible for large capacity for a
host metal are very much less clear. This, as is noted in the
Introduction, is related to the details of the rather imperfectly
understood metal-hydrogen bonding.

IV. HYDRIDES OF COMPOUNDS OF Th WITH Fe, Co AND Ni

Th_7Fe_3 and its hydride have recently been studied in considerable detail (28,29). The parent material is a superconductor with
a transition temperature $\sim$2 K (30), but upon hydrogenation it is
transformed into a ferromagnet with a Curie temperature slightly
above room temperature. This is unusual in that superconductivity
and ferromagnetism are normally antithetical. The behavior of
this system during hydrogenation can be rationalized in terms of a
concept as follows: When the compound Th_7Fe_3 is formed from the
elements, electrons are transferred from Th to Fe to fill the 3d-
band of the latter element, thus rendering Fe nonmagnetic. This
results in a material which is a Pauli paramagnet at room temperature and enables superconductivity to develop when the temperature
is reduced below $\sim$2 K. When hydrogen enters the lattice it engages
in competition with Fe for electrons. The tendency for electron
capture by hydrogen, which is evident in the lanthanide hydrides,
also appears in this system. Hydrogen absorbs some of the electrons and this results in the formation of vacancies in the 3d-
band of Fe. Hence in the hydride iron regains its localized magnetic moment and these moments interact in such a way as to produce ferromagnetism at normal temperatures and below. Thus the
behavior of the hydride of Th_7Fe_3 is consistent with the concept
in which hydrogen behaves as an electron acceptor in the metallic
system. It is to be surmised that the principal bonding between
thorium and hydrogen in this system is ionic in nature.

More recently, studies have been made of the hydrides of the
corresponding nickel and cobalt systems (29). It was of interest
to ascertain whether these materials also develop ferromagnetism
when hydrogenated. The three Th_7T_3 systems where T = Fe, Co or Ni
are similar in that all are Pauli paramagnets at room temperature
and they all develop superconductivity at low temperatures, the
superconducting transition temperature being nearly the same in
each case (30). However, it has been found that only the iron
compound develops ferromagnetism when hydrogenated. Th_7Co_3 hydride
shows only an enhanced susceptibility compared to the parent metal;
it does not exhibit ferromagnetism. Also, hydrogenation of Th_7Ni_3

does not lead to ferromagnetism. In fact, the hydride has a lower susceptibility than Th_7Ni_3.

The varying behavior of hydrides of Th_7Fe_3, Th_7Co_3 and Th_7Ni_3 is a consequence of the varying number of electrons supplied by the 3d element. This number is larger for the Co and Ni systems. Hence electron capture by hydrogen is insufficient to begin de-population of the d-band. Or if vacancies are produced during hydrogenation they are too few to lead to magnetic ordering.

The behavior of the Th_7T_3-H systems is consistent with the concept that hydrogen acts as an electron acceptor in these sys-tems and the coulombic interaction is a significant contribution to the binding energy of system.

V.   HYDRIDES OF RFe_3 and RFe_2

The hydride systems containing Fe combined with the lanthanide elements have received the largest amount of attention to date. RFe_3-H and RFe_2-H systems have been studied. Both have been studied by conventional bulk magnetic methods. The RFe_2 hydrides have been studied by neutron scattering techniques as well.

The RFe_3-H systems with R = Gd, Dy and Ho were studied by Malik, Takeshita and Wallace (31); YFe_3-H was studied by Buschow (32). The results obtained (see Table 5) indicate a hydrogen-induced rise in Fe moment from 1.67 to 1.90 μ_B/atom for YFe_3 and 1.71 to 1.87 for $GdFe_3$. (The latter data emerge when antiparallel Gd-Fe coupling is assumed and the Gd moment is taken to be 7 μ_B.) A rise in Fe moment is suggested for the $DyFe_3$-H and $HoFe_3$-H sys-tems; however, results for these systems are not conclusive be-cause of difficulties in achieving saturation and uncertainty about the Ho and Dy moments. (See the behavior of $TmFe_2$ and $ErFe_2$ hydrides in regard to saturation described in the following sec-tion.) They, unlike Gd, may sustain a substantial reduction in moment from gJ because of quenching by the crystal field; thus there is uncertainty about the magnitude of the Ho and Dy moments.

In YFe_3 and $GdFe_3$ and their hydrides the Fe moment is well below the value for elemental Fe, 2.2 μ_B. This is a consequence of electron transfer from Y or Gd to Fe. These hydrides behave in a fashion similar to the Th_7Fe_3 hydride described above, in that hydrogenation increases the Fe moment. Thus H works in opposition to Y or Gd or Th. These elements give electrons to Fe, increasing the d-band population, whereas the observed magnetization data in-dicate that hydrogen takes electrons from Fe. Thus hydrogenation leads to depopulating of the d-band.

MAGNETISM OF LANTHANIDE BASED HYDRIDES

Table 5

Influence of Hydrogenation on the Magnetism of RFe_3

R	RFe₃			RFe₃H₃*		
	T_c (K)	T_{comp} (K)	Moment (μ_B/formula unit)	T_c (K)	T_{comp} (K)	Moment (μ_B/formula unit)
Y[32][a]	549	---	5.01	545	---	5.70
Gd[7]	725	615	1.87	---[b]	170	1.39[c]
Dy[7]	605	545	4.25	---[b]	175	2.2[d]
Ho[7]	575	395	4.42	---[b]	112	2.53[c]

*Nominal compositions. Actual compositions were YFe₃H₅, GdFe₃H₃.₁, DyFe₃H₃.₀ and HoFe₃H₃.₆.
[a]Numbers in brackets indicate the reference.
[b]T_c could not be observed due to loss of hydrogen.
[c]Measurements made for field strengths up to 21 kOe and extrapolated to H = ∞.
[d]Moment at 21 kOe. This compound, unlike GdFe₂H₃ and HoFe₂H₃, was far from saturation at 21 kOe.

The data in Table 5 show that the compensation temperature is sharply depressed by hydrogenation. This implies a weakening of the R-R exchange, a point which will be considered further in discussion of results obtained for RFe_2-H systems (see below).

The bulk magnetic characteristics of several RFe_2 systems and their hydrides are summarized in Table 6. Fe moments for the nine systems studied to date and their hydrides are listed; of these only the Fe moment for YFe_2H_4 is considered reliable. This follows since saturation could not be achieved for the $ErFe_2$ and $TmFe_2$ systems at fields of 120 kOe (see Fig. 4) and presumably this is the case for all of the other systems except YFe_3. When saturation is not achieved, the Fe moment is overestimated by assuming antiferromagnetic R-Fe coupling. The overestimate can be very substantial. The Fe moments listed in column 9 of Table 6 are computed assuming (1) Y and Ce are non-magnetic, (2) free ion moments for the other rare earths and (3) antiparallel coupling between rare earth and Fe moments. It is likely that neither of these latter two assumptions is strictly correct. Except for Gd^{3+} the moment of the rare earth ion could be sharply reduced by the crystal field interaction. Additionally, the expected antiparallel coupling may break down at strong applied fields due to weakening of R-Fe exchange by hydrogenation. This weakening is clearly revealed by the results of neutron scattering experiments described below.

From the comments in the preceding paragraph it is clear that reliable information in regard to the influence of hydrogenation on the Fe moment in Fe intermetallics cannot be acquired from bulk magnetic measurements except in special cases, e.g., YFe_2. It becomes necessary to examine the magnetism of the Fe sublattice by other means - neutron scattering (diffraction) or Mössbauer spectroscopy.

The most reliable evaluation of the effect of hydrogenation on the Fe moment in RFe_2 systems has come from the neutron diffraction studies of Rhyne et al. (33-35). Results are available for $HoFe_2$ and $ErFe_2$ and their hydrides RFe_2H_4. Data obtained are shown in Figs. 5 and 6. The reduction of T_c of the RFe_2 systems by hydrogenation (see Table 6) indicates an accompanying weakening of exchange. This is further indicated by the data shown in Figs. 5 and 6. The rare earth moment diminishes rapidly with increasing temperature and in the case of $ErFe_2H_4$ it vanishes about 150 C below T_c. Thus there is substantial weakening of the R-Fe exchange by hydrogenation and some decrease in the Fe-Fe exchange. A decrease in R-R exchange upon hydrogenation is further indicated by results obtained on hydrogenated $GdNi_2$, described in the next section.

The data for $HoFe_2$ indicate a rise in Fe moment (from 1.5 to 1.9 μ_B) by hydrogenation, whereas in the case of $ErFe_2$ the insertion of hydrogen in the lattice leaves the moment essentially

Table 6

Magnetism of RFe_2 Systems and Hydrogenated RFe_2-H Systems

R	T_c(K)	T_{comp}(K)	RFe_2 Moment (μ_B/formula unit)	Fe Moment (μ_B)	T_c(K)	T_{comp}(K)	RFe_2H_4 Moment (μ_B/formula unit)	Fe Moment (μ_B)
Y	545	---[a]	2.90	1.45	308	---	3.66[b]	1.83
Ce	230	---	2.60	1.30	358	---	4.14[b]	2.07(?)[d]
Sm	700	---	2.5		333	---	3.2	1.6
Gd	785	---	3.80	1.65	388	---	4.10[c]	1.45(?)[d]
Tb	705		4.72		303		4.6	
Dy	638		5.50		385		4.9	
Ho	614	---	6.1	1.95	387	60	2.35[e]*	
Er	596	480	4.75	2.13	280	42	5.60[e]*	1.7(?)[d]
Tm	610	225	2.52	2.24	270	18	6.45	0.28(?)[d]

*Nominal compositions. The actual compositions used by Gualtieri et al. were $HoFe_2H_{4.47}$, $ErFe_2H_{3.9}$ and $TmFe_2H_{4.3}$.

[a]No entry indicates that there is no compensation point (T_{comp}).

[b]Moment at saturation at 4.2 K.

[c]Moment at 4.2 K and 18 kOe.

[d]This quantity is uncertain.

[e]Moment at 4.2 K and 120 kOe.

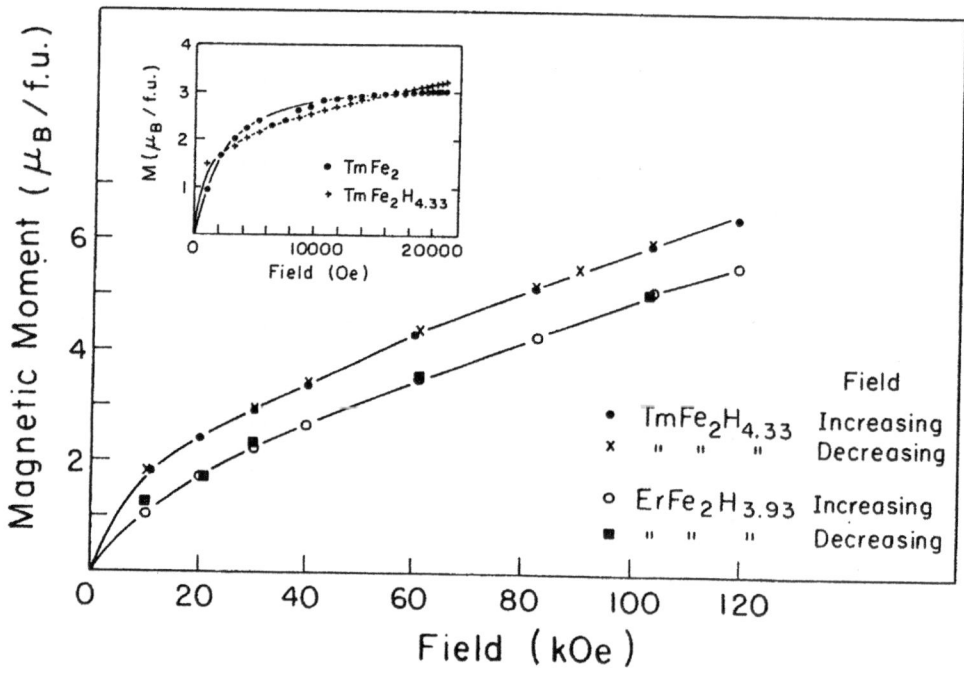

Fig. 4. The variation of magnetization with applied field for hydrogenated $ErFe_2$ and $TmFe_2$ at 4.2 K. The insert compares the
magnetization of hydrided $TmFe_2$ and $TmFe_2$.

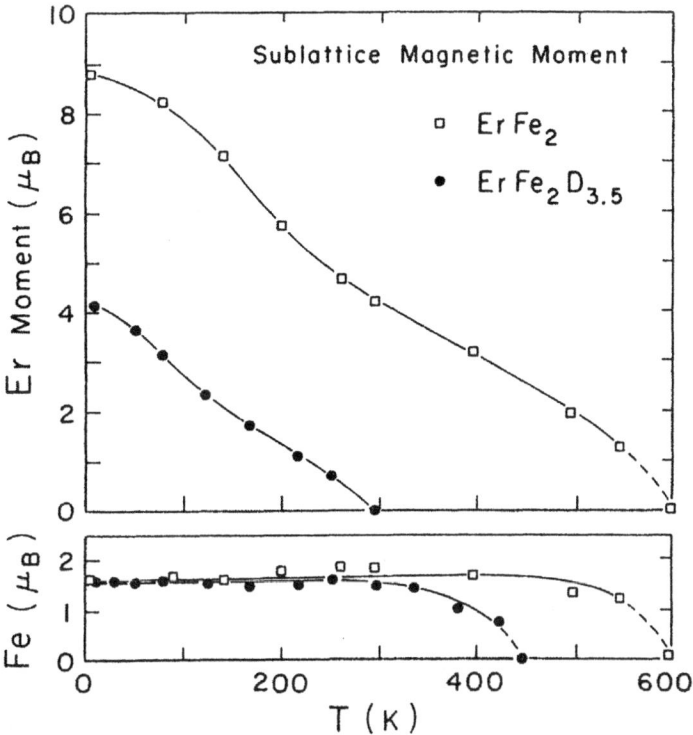

Fig. 5. Temperature dependence of the sublattice magnetizations
of $ErFe_2$ and $ErFe_2D_{3.5}$.

unaffected. The moment of Fe in the RFe_2 systems and in the RFe_3
system is, because of electron transfer, invariably less than the
2.2 μ_B for elemental Fe. The rise in moment observed for Fe in
$HoFe_2H_4$ is undoubtedly due to electron transfer from the Fe d-band
to a lower-lying hydrogenic band which is analogous to that which
forms in the RH_3 systems.

The difficulties experienced in saturating the RFe_2H_4 systems
(see Fig. 4) find a ready interpretation in terms of the neutron-
scattering results. The low value of the rare earth moment and
the diminished R-Fe exchange in the hydride makes it appear likely
that "fanning" takes place as depicted in Fig. 7. The fanning is
a consequence of greatly weakened R-Fe exchange in these systems
and is brought about by spatial variation in the crystal field interaction. The latter originates with random site occupancy by

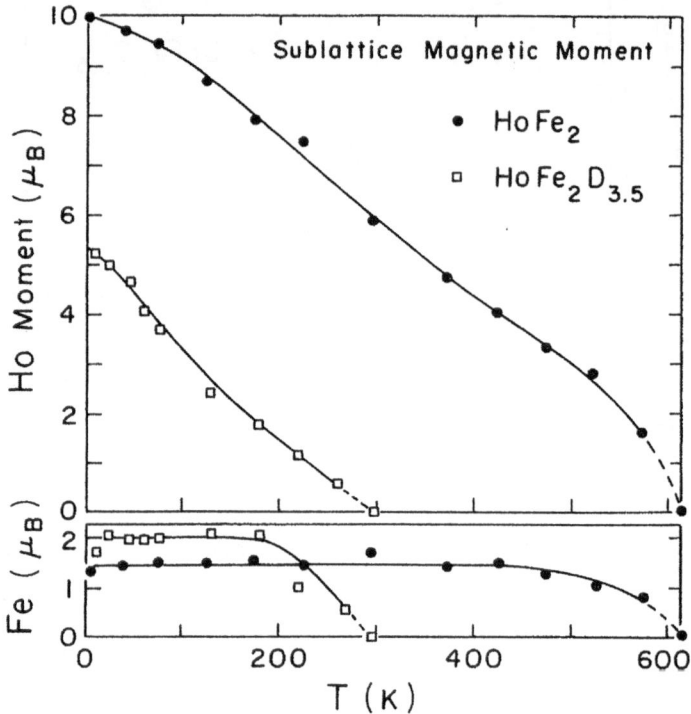

Fig. 6. Temperature dependence of the sublattice magnetizations
of $HoFe_2$ and $HoFe_2D_{3.5}$.

hydrogen of the many-fold more abundant lattice sites. Application of a field biases the fanning, as shown in Fig. 7c; under
these circumstances the material behaves as a metamagnet. The
antiparallel R-Fe coupling is beginning to break down under the
influence of an applied field because of the weakened R-Fe exchange. The modification of coupling toward parallel coupling
increases in proportion to the strength of the applied field and
so magnetization continually rises as the applied field strength
increases. Accordingly, inferences about the Fe moment in RFe_2
systems drawn from bulk measurements of magnetism can be without
validity.

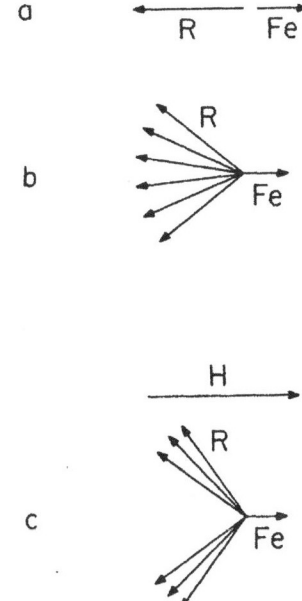

Fig. 7.   (a) In RFe_2 systems when R-Fe exchange is strong, the
               moments couple antiparallel.
          (b) As exchange is weakened, fanning occurs due to
               spatial variation of anisotropy.
          (c) Fanning is biased by an applied field.

VI.   LANTHANIDE-COBALT HYDRIDES AND $GdNi_2$ HYDRIDE

A.   Cobalt-Containing Systems

(1) Effect of Hydriding on the Co Moment. Kuijpers (36)
noted that progressive hydrogenation of rare earth-cobalt systems
led to a monotonous decline in cobalt moment. The data in Table 7
taken from his work illustrate this point. Buschow (37) has discussed these results in terms of electron transfer from H to Co,
regarding the fall in moment as an indication of a filling of the
Co d-band. In view of the small difference in electronegativity
between Fe and Co it seems unlikely that the direction of electron
transfer would differ between Fe and Co intermetallics. It seems
more likely that the reduced Co moment is a consequence of the reduction of exchange which accompanies hydrogenation. The behavior

Table 7

Cobalt Moments in Several Rare Earth-Cobalt Systems[a]

	Co Moment(μ_B/atom)
RCo_5	1.50
$NdCo_5H_{0.3}$	1.45
$SmCo_5H_{2.5}$	1.2
$LaCo_5H_{3.35}$	1.15
$LaCo_5H_{4.3}$	0.3

[a]Data from ref. 36.

of the host rare earth-cobalt systems indicates that the Co moment is sensitive to the exchange field acting upon it. Cobalt is often non-magnetic when united with a non-magnetic partner, e.g., YCo_2 or $LuCo_2$ (38). It is magnetic when united with a magnetic partner, e.g., $GdCo_2$ (39) or when it is in a Co-rich system even if its partner is non-magnetic, e.g., $LaCo_5$ (38). Thus, the cobalt moment is a strong function of exchange in the system.

The dependence of the Co moment on the strength of exchange has been evident for many years. Bleaney (40) appears to have been the first to draw attention to the fact that an increase in exchange produced a corresponding increase in the magnitude of the Co moment. This effect was experimentally demonstrated in the work of Leon and Wallace (41) in their study of the $Pr_{1-x}Dy_xCo_2$ and $Pr_{1-x}Ho_xCo_2$ ternaries. The Co moment in these systems increased with x because of the increasing strength of exchange. Lemaire (38) had earlier observed a similar effect in the $Y_{1-x}Gd_xCo_2$ system. Accordingly, it seems likely that the effect of hydrogenation on the magnetism of Co in Co-containing intermetallics is a matter of reduced exchange rather than a transfer of electrons from hydrogen to cobalt.

Evidence for suppression of exchange by hydrogenation rests on three observations: (1) T_c values for the hydride of the lanthanide cobalt intermetallics are sharply reduced from the value for the parent intermetallic, e.g., for $GdCo_2H_4$ T_c is $\sim$90 K compared to $\sim$400 K for $GdCo_2$; (2) T_c for $GdNi_2$ is reduced from 80 to 8 K by hydrogenation (42) and (3) suppression of exchange is clearly indicated in RFe_2-H systems suggests that this behavior in Fe systems should find its counterpart in the Co systems.

(2) $GdCo_3$-H, $DyCo_3$-H and $HoCo_3$-H Systems. The pressure-composition isotherms indicate (43) the existence of three phases in the RCo_3-H systems - an α or primary solid solution phase, and β and γ phases corresponding to the approximate compositions RCo_3H_2 and $RCo_3H_{4.5}$, respectively. Bulk magnetic studies on the

three ternary systems with R = Gd, Dy and Ho by Malik et al. showed (44) that exchange in the β hydride is comparable with that in the parent intermetallic. However, exchange in the γ hydride is sharply reduced. The results obtained are summarized in Table 8. The sharp reduction in exchange in the fully hydrogenated material is analogous to the behavior of the elemental rare earths discussed above in Section II. An indicated there, the behavior of the RH_3 systems has been ascribed to depopulation of the rare earth conduction band by hydrogenation. A similar effect is probably operative in the RCo_3-H systems. If so, this again implies a transfer of electrons from the metallic d-band to the lower-lying hydrogenic band, constituting electron capture by hydrogen.

B. Hydride of $GdNi_2$

The only Ni-containing rare earth intermetallic which has been well studied in regard to the effect of hydrogenation on magnetic properties is the $GdNi_2H_4$ system. Malik and Wallace studied this system and observed (42) the magnetization-temperature behavior shown in Fig. 8. The Curie temperature is reduced from 81 to 8 K by hydrogenation, indicating an appreciable weakening of exchange. In this respect $GdNi_2$ and its hydride resemble Gd and GdH_3. The magnetic ordering temperature is reduced by ∼300 K to < 4 K when Gd is hydrogenated. This has been ascribed to electron transfer from Gd to H. It seems likely that this also occurs in

Table 8

Magnetic Properties of RCo_3 Hydrides[a]

	σ(emu/g)	μ_{Co}(μ_B)	T_c(K)
$GdCo_3$	47.7	1.38	612
$GdCo_3H_{2.2}$	56.1	1.2[b]	---[c]
$GdCo_3H_{4.6}$	64.7	1.0[b]	28
$DyCo_3$	86.5		450
$DyCo_3H_{4.3}$	62.0		18
$HoCo_3$	94.6		418
$HoCo_3H_{4.2}$	51.0		15

[a]Data from ref. 44.
[b]Computed assuming antiferromagnetic coupling and a Gd moment of 7.0 μ_B.
[c]T_c > 300 K. It could not be determined because of loss of hydrogen.

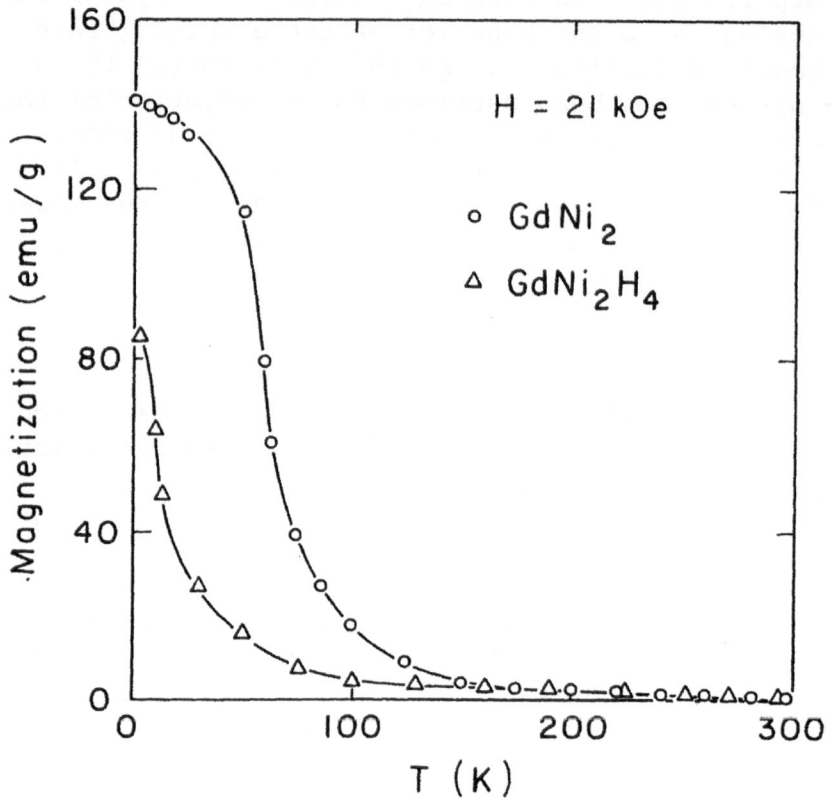

Fig. 8. Temperature dependence of GdNi$_2$ and GdNi$_2$H$_4$ in an applied
         field of 21 kOe.

GdNi$_2$. Support for the concept of hydrogen as an electron acceptor
is also provided by the [155]Gd Mössbauer spectroscopy studies of
Bauminger et al. (45) on hydrogenated Gd$_{0.1}$La$_{0.9}$Ni$_5$. They found
from the isomer shift that the s electron density on the Gd nucleus is reduced when the material is hydrogenated. This suggested
electron transfer from R to H.

VII. LANTHANIDE-MANGANESE HYDRIDES

A. RMn$_2$ Hydrides

Hydrides of YMn$_2$ have been studied by Malik, Takeshita and
Wallace (14), by Buschow and Sherwood (46) and by Oesterreicher

and Bittner (47). The latter authors have also studied $GdMn_2$ hydride. Malik et al. found that YMn_2 remains (48) a Pauli paramagnet upon hydrogenation to $YMn_2H_{2.9}$; the only change is that there is a slight decline in susceptibility. Buschow and Sherwood find that with increasing hydrogen content the material is first transformed into an unusual seemingly ordered magnetic state and with further hydrogen uptake it regains the Pauli paramagnetism of the host metal. Oesterreicher and Bittner also observe magnetic behavior which depends upon hydrogen concentration with the fully hydrogenated material exhibiting Pauli paramagnetism. Buschow (49) reports that the low temperature magnetism of partially hydrogenated YMn_2 is field-dependent; it is small when cooled in the absence of a field but large when cooled in a field. He surmises that this is due to a situation "in which ferromagnetic clusters are imbedded in a magnetically disordered or spin-glass like matrix."

Oesterreicher and Bittner report (47) that $GdMn_2H_{4.6}$ is not magnetically ordered at 4 K, whereas $GdMn_2$ orders antiferromagnetically at $\sim$85 K. Thus hydrogenation has in this case reduced exchange, as was observed for $GdNi_2$ and $GdCo_3$. It is tempting to conclude that hydrogen absorbs electrons from the d-band in all of these Gd intermetallics and in this respect they bear a strong resemblance to elemental Gd.

B. R_6Mn_{23} and Th_6Mn_{23} Hydrides

The R_6Mn_{23} compounds are cubic, belonging to space group Fm3m. They have 4 formula units (116 atoms) per unit cell. These materials, including Th_6Mn_{23}, very readily hydrogenate to form rather hydrogen-rich systems (see Table 1), with about 25 H per R_6Mn_{23} formula unit.

The 6:23 systems exhibit a number of unusual features: (1) In the isostructural R_6Fe_{23} systems a given R_6Fe_{23} compound has a Curie temperature (T_c) essentially identical with that of the corresponding R_6Mn_{23}. Yet when Fe replaces Mn in R_6Mn_{23} or when Mn replaces Fe in R_6Fe_{23} there is a dramatic decline in T_c (49). In the system in which R = Y, there is a large composition region for which there is no evidence of magnetic ordering even in the liquid helium range, this despite the fact that T_c for Y_6Mn_{23} or Y_6Fe_{23} is quite high, nearly 500 K (50,51). (2) Although Y_6Mn_{23} is a ferrimagnetic material (vide infra) with T_c = 486 K, its hydride is a Pauli paramagnet (52). In contrast, isostructural Th_6Mn_{23} is a Pauli paramagnet but becomes ferro or possibly ferrimagnetic upon hydrogenation (52). Mn is obviously non-magnetic in $Y_6Mn_{23}H_{25}$ since the compound exhibits (52,53) only Pauli paramagnetism.

To establish the behavior of the corresponding systems in
which Y is replaced by a magnetic species, results were recently
acquired (54,55) for hydrides of R_6Mn_{23}, where R = Gd, Tb, Dy, Ho
or Er. Only the hydrides of Gd_6Mn_{23} and Tb_6Mn_{23} gave indication
of magnetic ordering. Presumably because of the smaller deGennes
factor and weakened exchange, hydrides of R_6Mn_{23} with R = Dy, Ho
or Er failed to order at temperatures extending into the helium
regime. Data for Gd_6Mn_{23}, Er_6Mn_{23} and their hydrides are shown in
Figs. 9-12 . These data are interpreted to indicate that
$Gd_6Mn_{23}H_{22}$ is magnetically ordered whereas $Er_6Mn_{23}H_{23}$ is not.

Data for the various R_6Mn_{23} systems and their hydrides are
collected in Table 9. From the Pauli paramagnetism of $Y_6Mn_{23}H_{25}$
it is inferred that Mn in these hydrides is non-magnetic, or at
least is not magnetically ordered. Thus, the magnetic ordering
observed involves only the R sublattice.

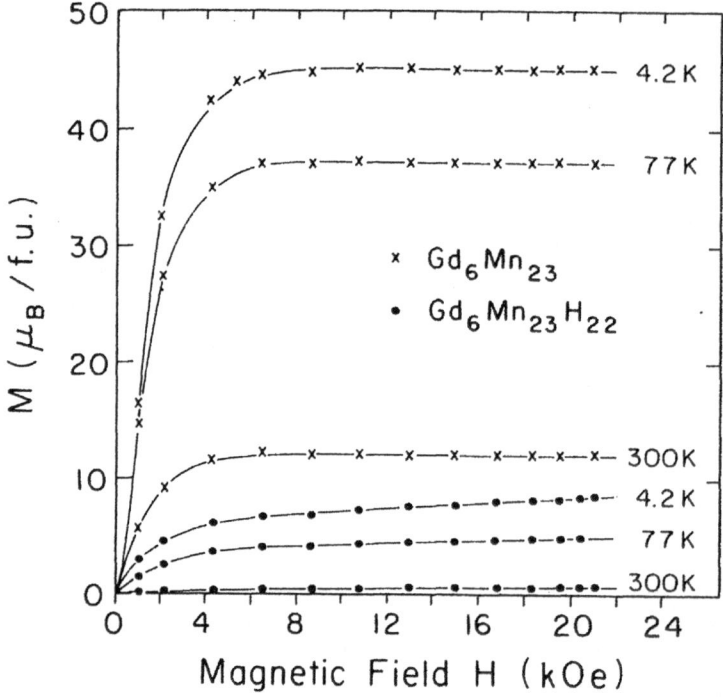

Fig. 9. Magnetization vs. field for Gd_6Mn_{23} and its hydride
         measured at 4.2, 77 and 300 K.

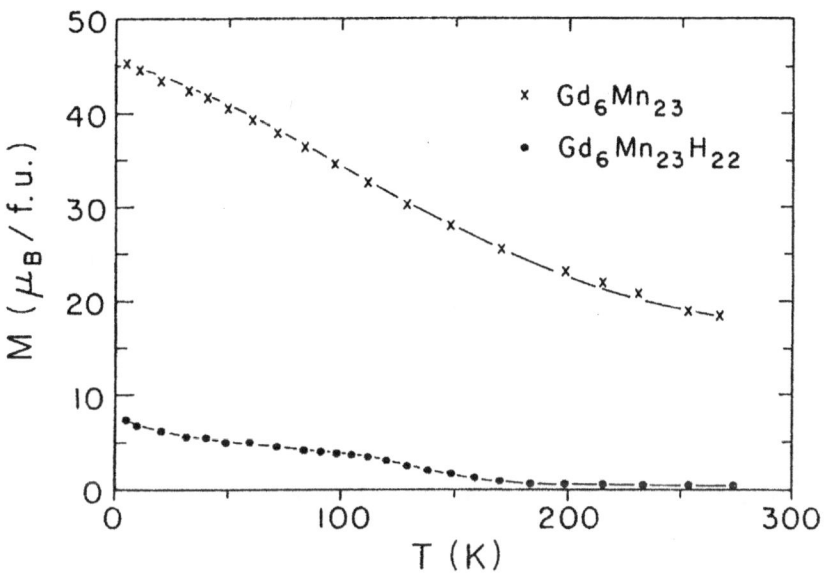

Fig. 10. Magnetization vs. temperature for Gd_6Mn_{23} and its hydride measured in an applied field of 12 kOe.

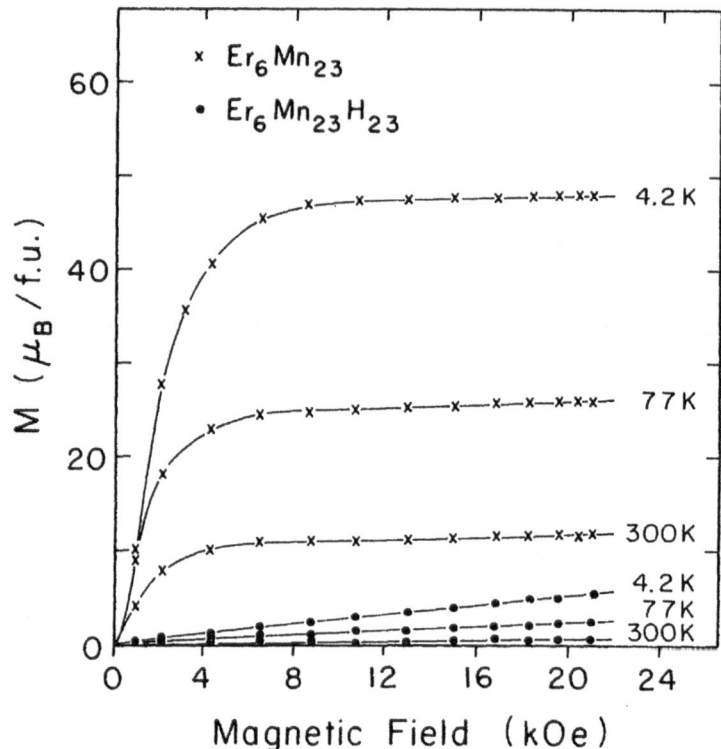

Fig. 11.   Magnetization vs. field data for Er_6Mn_{23} and its
           hydride measured at 4.2, 77 and 300 K.

The most remarkable feature of the results in Table 9 is the
low moment measured for magnetically ordered hydrides $Gd_6Mn_{23}H_{22}$
and $Tb_6Mn_{23}H_{23}$. The small moment of $Tb_6Mn_{23}H_{23}$ was initially
ascribed to the quenching of the Tb moment by the crystal field
interaction. This interaction, however, cannot be responsible for
the magnetization observed for the Gd_6Mn_{23} hydride. Ferrimagnetism of $Gd_6Mn_{23}H_{22}$ originating from the antiferromagnetic coupling
of the inequivalent Gd and Mn sublattices is one possible explanation for the low observed moment. This, however, seems rather unlikely since the results obtained for hydrided Y_6Mn_{23} indicate
(52,53) that Mn does not carry a moment in this system.

Table 9

Lattice Parameters and Magnetic Characteristics
of R_6Mn_{23} Systems and Their Hydrides

Compound	a(Å)	$\sim \frac{\Delta V}{V}$ %	T_c (K)	$\mu(4.2)$ [b] (μ_B/f.u.)	Ref.
Gd_6Mn_{23}	12.592	10	468[a]	55[b]	56
$Gd_6Mn_{23}H_{22}$	12.970		∿180	8.4[c]	
Tb_6Mn_{23}	12.484	12	455[a]	49[c]	54
$Tb_6Mn_{23}H_{23}$	13.017		∿220	17.7[c]	54
Dy_6Mn_{23}	12.439	11	443[a]	49.6[c]	54,56
$Dy_6Mn_{23}H_{23}$	12.898		----	----	54
Ho_6Mn_{23}	12.383	12	434[a]	59.8[c]	
$Ho_6Mn_{23}H_{23}$	12.852		----	----	
Er_6Mn_{23}	12.321	9.7	415[a]	36[b]	56
$Er_6Mn_{23}H_{23}$	12.796		----	----	
Y_6Mn_{23}	12.451	9.7	486[a]	13.8	50,51
$Y_6Mn_{23}H_{25}$	12.842		----	----	52,53

[a]Taken from ref. 49.
[b]Determined with an applied field of 50 kOe.
[c]Determined with an applied field of 21 kOe.

It seems likely that the hydrides of Gd_6Mn_{23} and Tb_6Mn_{23} involves either ferrimagnetism on the R sublattice or some type of non-collinear magnetic structure, with the latter regarded as somewhat more probable. The low magnetic moment probably originates in one of these two ways. It is to be noted (Fig. 9) that the Gd_6Mn_{23} hydride gives no indication of saturation at the highest field employed. Most likely the magnetic structure in this system and in the Tb_6Mn_{23} hydride, which behaves in an entirely analogous manner, is being progressively modified as the applied field is increased. Presumably it is this modification which gives rise to the steadily increasing magnetization.

There is already a suggestion of these features postulated for the hydride occurring in the host metals. Delapalma et al., using polarized neutron-scattering equipment, have found (56) a ferrimagnetic Mn sublattice structure for Y_6Mn_{23}. This material contains four crystallographically distinguishable types of Mn in the 4b, 24d, $32f_1$ and $32f_2$ sites. Mn moments on the 4b and 24d sites are aligned in a positive direction and antiparallel to the moments of Mn in the two 32f sites. It is observed (57), moreover, that

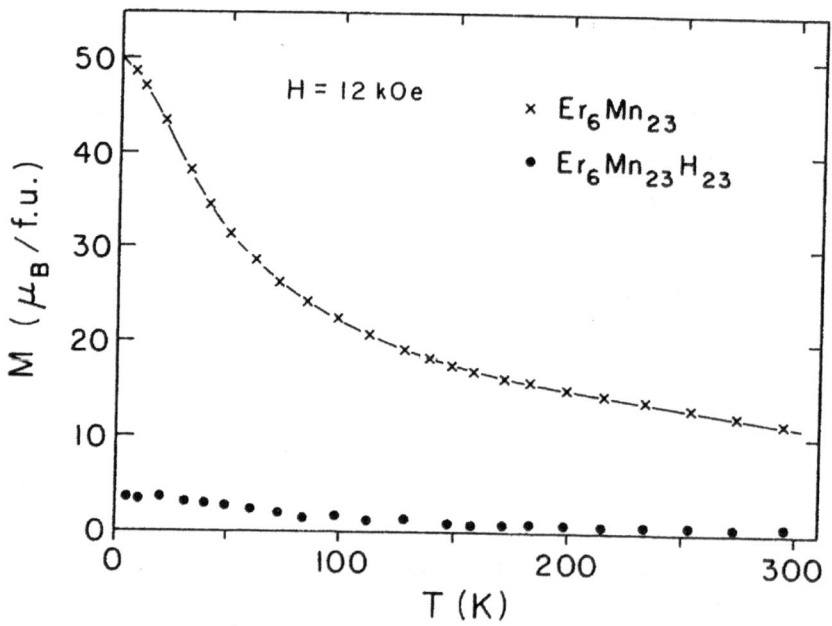

Fig. 12.   Temperature dependence of Er_6Mn_{23} and its hydride measured
in an applied field of 12 kOe.

Y_6Mn_{23} saturates at an applied field of about 2 kOe. This indicates that the Mn sublattice is nearly isotropic magnetically.

Measurements on Gd_6Mn_{23} single crystals show a saturation moment of ~ 55 μ_B/f.u. (57). This is approximately the sum of the Y_6Mn_{23} moment, ~ 14 μ_B, and that of the 6 Gd^{3+} ions, which indicates that _under_ _the_ _influence_ _of_ _a_ _field_ Gd couples ferromagnetically with resultant of the Mn sublattice moment. This implies a weak R-Mn interaction compared to the R-Co and R-Fe interactions which in themselves are fairly weak.

Studies of single crystals Er_6Mn_{23} give (57) a magnetization (with field along <100>) which is considerably less than that of $6Er^{3+}$ + Y_6Mn_{23}, viz. 68 μ_B/f.u. The measured value at 50 kOe is ~ 38 μ_B/f.u. The reduction has been interpreted on a crystal field model. It was postulated that 1/3 of the Er moments are directed along each of the mutually perpendicular crystallographic axes. Hence a field along <100> would produce a moment of (18+14) μ_B/f.u. There is in addition a small contribution from the polarization of the 6 Er^{3+} moments directed along <010> and <001>. In general terms there is, then, a reduction in moment due to a non-collinear arrangement of rare earth moments, which is a manifestation of the crystal field interaction. A generally similar situation obtains (56) for Dy_6Mn_{23}.

The low moment observed for the hydrides in Table 10 implies non-collinearity of the R moments. However, this cannot be entirely due to the crystal field interaction since the effect is observed in $Gd_6Mn_{23}H_{22}$. The non-collinearity must therefore be a consequence of exchange, perhaps competition between exchange involving nearest and next-nearest neighbors. The true situation may turn out to be describable as a spin glass.

The surprising alterations in behavior of Y_6Mn_{23} and Th_6Mn_{23} when hydrogenated has recently been interpreted by Wallace. He has made use of the molecular field analysis of the $Er_6Mn_{23-x}Fe_x$ system recently carried out by Hilscher and Rais (58). The essence of this interpretation is that the molecular field, H_{Mn}, acting upon Mn is electron concentration dependent. It is maximal at the electron concentration existing in R_6Mn_{23} (see Fig. 13). In Th_6Mn_{23} the electron concentration is higher and H_{Mn} is significantly diminished. It is reduced to the point that magnetic ordering does not occur. When hydrogenated, electron concentration in the Mn d-bands is reduced, H_{Mn} is increased and ordering develops. When Y_6Mn_{23} is hydrogenated, H_{Mn} is also reduced and ordering is suppressed.

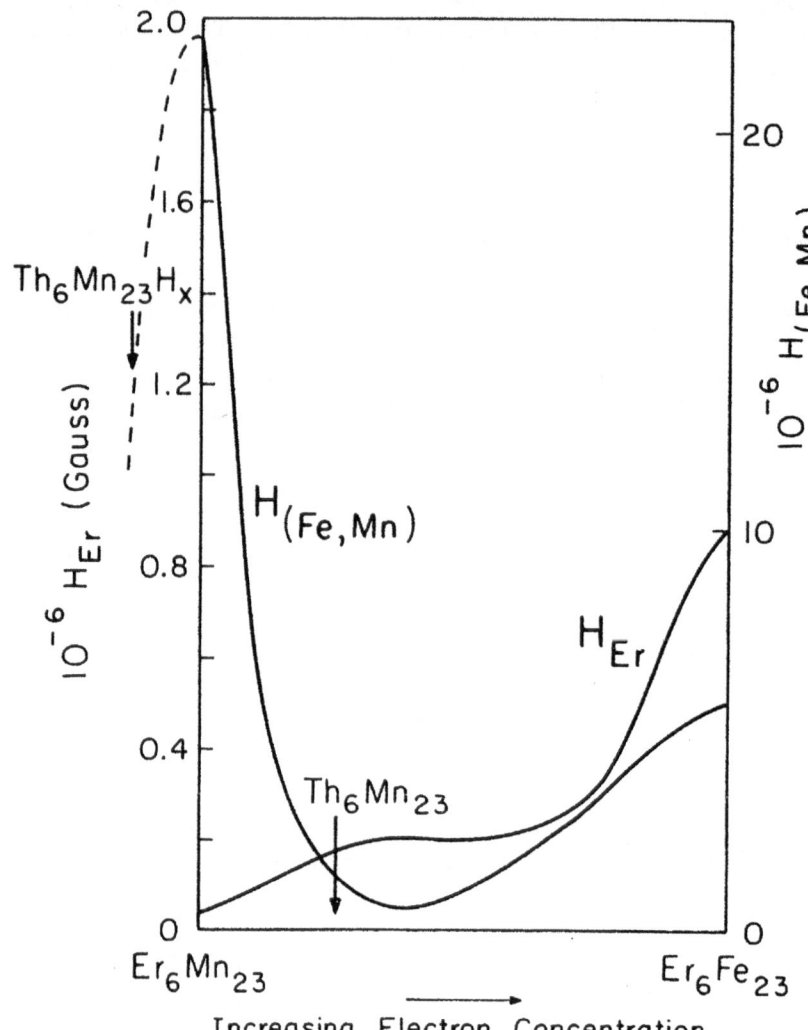

Fig. 13. Molecular fields in $Er_6(Mn_{1-x}Fe_x)_{23}$ ternaries. The dashed line shows the presumed trend for electron concentrations lower than for Er_6Mn_{23}. Arrows show the estimated molecular field for Mn in Th_6Mn_{23} and its hydrides.

VIII.   CONCLUDING STATEMENT

The results described above are all described in a model which involves hydrogen acting as an electron acceptor, withdrawing electrons from the d-band of the host metal. This was the concept originally held for the hydrides of the elemental lanthanides (18). The band structure calculations of Switendick revealed that this concept is too simple (19). Undoubtedly future work will reveal that the situations for hydrided intermetallics are more complex than those considered in the present treatment. A proper analysis will require band structure information for the intermetallic hydrides analogous to that provided by Switendick for the simple hydrides. At present there is no such information. However, a small beginning has been made by Malik, Arlinghaus and Wallace. They have made self-consistent APW calculations on $SmCo_5$, YCo_5 and $GdCo_5$ (59) and on $LaNi_5$ and $GdNi_5$ (60). Their work will have to be continued and expanded to acquire the information needed to interpret properly magnetism in intermetallic hydrides.

REFERENCES

1.   Richard L. Beck, Summary Report to the U.S. Atomic Energy
     Commission from the Metallurgy Department, University of
     Denver, Oct. 15, 1962.
2.   F. Pourarian and W. E. Wallace, unpublished measurements.
3.   J. J. Reilley and R. H. Wiswall, Jr., Inorg. Chem. 13, 218
     (1974).
4.   J. H. N. Van Vucht, F. A. Kuijpers and H. C. A. M. Bruning,
     Philips Res. Repts. 25, 133 (1970).
5.   J. J. Reilley and R. H. Wiswall, Jr., Inorg. Chem. 7, 2254,
     (1968).
6.   D. M. Gualtieri, K. S. V. L. Narasimhan and W. E. Wallace,
     AIP Conf. Proc. 34, 219 (1976).
7.   S. K. Malik, T. Takeshita and W. E. Wallace, Mag. Lett. 1,
     33 (1976).
8.   K. H. J. Buschow, H. H. Van Mal and A. R. Miedema, J. Less-
     Common Met. 42, 163 (1975).
9.   C. A. Bechman, A. Goudy, T. Takeshita, W. E. Wallace and
     R. S. Craig, Inorg. Chem. 15, 2184 (1976).
10. F. A. Kuijpers, Ph.D. Thesis, Technische Hogeschool, Delft
     (1973).
11. See ref. 8.
12. H. V. Van Mal, K. H. J. Buschow and A. R. Miedema, J. Less-
     Common Met. 49, 473 (1976).
13. H. Oesterreicher and H. Bittner, phys. stat. sol.(a) 41, K101
     (1977).
14. S. K. Malik, T. Takeshita and W. E. Wallace, Solid State
     Commun. 23, 599 (1977).

15. B. Stalinski, Bull. Acad. Sci.(Poland) 5, 100 (1957) and 7, 269 (1959).

16. R. Heckman, J. Chem. Phys. 40, 2958 (1964) and 46, 2158 (1967).

17. G. G. Libowitz and J. G. Pack, J. Chem. Phys. 50, 3557 (1969).

18. W. E. Wallace et al., J. Appl. Phys. Suppl. 33, 1348 (1962); J. Appl. Phys. 34, 1348 (1963); J. Chem. Phys. 39, 1285 (1963); Advances in Chemistry Series 39, 122 (1963).

19. A. C. Switendick, Solid State Commun. 8, 1463 (1970) and Int. J. of Quantum Chem. 5, 459 (1971).

20. D. S. Schreiber and R. M. Cotts, Phys. Rev. 131, 1118 (1969).

21. W. E. Wallace and K. H. Mader, J. Chem. Phys. 48, 84 (1968).

22. For a comprehensive discussion of the crystal field interaction and associated phenomena see W. E. Wallace, S. G. Sankar and V. U. S. Rao, in Structure and Bonding, 33, 1. Springer-Verlag (1977).

23. K. Knorr and B. E. F. Fender in Crystal Field Effects in Metals and Alloys, A. Furrer, ed. Plenum Press, New York (1977), p. 42.

24. W. E. Wallace, Rare Earth Intermetallics. Academic Press Inc., New York (1973).

25. W. E. Wallace, R. S. Craig and V. U. S. Rao, Advances in Chemistry, to appear.

26. D. M. Gualtieri, K. S. V. L. Narasimhan and T. Takeshita, J. Appl. Phys. 47, 3432 (1976).

27. W. E. Wallace, R. F. Karlicek, Jr. and H. Imamura, J. Phys. Chem. 83, 1708 (1979).

28. S. K. Malik, W. E. Wallace and T. Takeshita, Solid State Commun. 28, 359 (1978).

29. E. B. Boltich, S. K. Malik and W. E. Wallace, J. Less-Common Metals, in press.

30. B. T. Matthias, V. B. Compton and E. Corenzwit, J. Phys. Chem. Solids 17, 130 (1961).

31. See ref. 7.

32. K. H. J. Buschow, Solid State Commun. 19, 421 (1976).

33. J. J. Rhyne, S. G. Sankar and W. E. Wallace, in The Rare Earths in Science and Technology, eds. G. J. McCarthy and J. J. Rhyne. Plenum Press, N.Y. (1978), p. 63.

34. J. J. Rhyne, G. E. Fish, S. G. Sankar and W. E. Wallace, J. de Physique, coll C5, Suppl. No. 5, 40, C5-209 (1979).

35. G. E. Fish, J. J. Rhyne, S. G. Sankar and W. E. Wallace, J. Appl. Phys. 30(3), 2003 (1979).

36. See ref. 8.

37. K. H. J. Buschow, in Hydrides for Energy Storage, A. F. Andresen and A. J. Maeland, eds. Pergamon Press, Inc. (1978), p. 273.

38. R. Lemaire, Cobalt 32, 132 (1966).

39. See ref. 24, p. 147.

40. B. Bleaney in Rare Earth Research, Proceedings of the 3rd Rare Earth Research Conference, K. S. Voores, ed., Vol. 2, p. 499. Gordon and Breach, New York (1964).

41. B. Leon and W. E. Wallace, J. Less-Common Met. 22, 1 (1970).

42. S. K. Malik and W. E. Wallace, Solid State Commun. 24, 283 (1977).

43. See ref. 9.

44. S. K. Malik, W. E. Wallace and T. Takeshita, Solid State Commun. 28, 977 (1978).

45. E. R. Bauminger, D. Davidov, I. Felner, I. Nowik, S. Ofer and D. Shaltiel, Physica 86-88B, 201 (1977).

46. K. H. J. Buschow and R. C. Sherwood, J. Appl. Phys. 49, 1480 (1978).

47. H. Oesterreicher and H. Bittner, J. Mag. Magn. Mat. 15-18, 1264 (1980).

48. S. A. Marei, R. S. Craig, W. E. Wallace and T. Tsuchida, J. Less-Common Met. 13, 391 (1967).

49. See ref. 24, pp. 187-194.

50. H. R. Kirchmayr and W. Steiner, J. Phys. Suppl. 32, C1-665 (1977).

51. C. A. Bechman, K. S. V. L. Narasimhan, W. E. Wallace, R. S. Craig and R. A. Butera, J. Phys. Chem. Solids 37, 245 (1976).

52. See ref. 14.

53. See ref. 13.

54. F. Pourarian, E. B. Boltich, W. E. Wallace and S. K. Malik, J. Less-Common Metals, in press.

55. F. Pourarian, E. B. Boltich, W. E. Wallace, R. S. Craig and S. K. Malik, J. Mag. Magn. Mat., in press.

56. A. Delapalma, J. Déportes, R. Lemaire, K. Hardman and W. J. James, J. Appl. Phys. 50, 1987 (1979).

57. K. Hardman, Ph.D. Thesis, University of Missouri, Rolla, 1979.

58. G. Hilscher and H. Rais, J. Phys. F. Metal Phys. 8, 511 (1978).

59. S. K. Malik, F. J. Arlinghaus and W. E. Wallace, Phys. Rev. B 16, 1242 (1977).

60. S. K. Malik, F. J. Arlinghaus and W. E. Wallace, to be published.

ORDER-DISORDER PHENOMENA IN METAL HYDRIDES

Makoto Hirabayashi and Hajime Asano*

The Research Institute for Iron, Steel and Other
Metals, Tohoku University, Sendai 980, Japan
* Institute of Materials Science, University of
Tsukuba, Ibaraki 305, Japan

In the last decade, considerable progress has
been made in the study of order-disorder phenomena
in metal hydrides. The purpose of this paper is
to review the recent investigations on the order-
disorder transitions in hydrides of the group V
transition metals (V, Nb and Ta) and of Pd. The
emphasis is put upon the features of structural
changes in the equilibrium states on the basis of
experimental data.

INTRODUCTION

In recent years, order-disorder phenomena in metal hydrides
have attracted much interest and activity of numerous workers. To
our knowledge, the occurrence of order-disorder phenomena in a
metal-hydrogen system has first been suggested in 1940 by Kelley[1]
He has found an anomalous heat absorption in the calorimetric study
of Ta-H alloys containing small amounts of hydrogen (2-9 at%), and
attributed it to the disordering of hydrogen distribution in the
bcc metal lattice. This result, which is now interpreted to be due
to the dissolution of precipitated ordered phase into the matrix,
has stimulated Wallace and his colleagues to start their pioneer
work[2-4] on the order-disorder transition of Ta-H system.

The order-disorder transition in metal hydrides is a prototype
of the order-disorder phenomena of interstitial atoms dissolved in
metal crystals. As to the order-disorder phenomena in interstitial
alloys, Honda[5] suggested as early as in 1928 a possibility of the

53

disordering of carbon atoms in the initial stage of tempering of
Fe-C martensite. In 1946, Zener[6] has presented an important theory
concerning the order-disorder phenomena in Fe-C martensite, and
demonstrated that the stress-induced interaction between iron and
carbon atoms is responsible for the ordering phenomena. Accordingly,
such phenomena are named as stress-induced ordering or stress
ordering. A similar theory of elastic interaction has been proposed
independently by Sato[7]. However, experimental verification of the
order-disorder transition in Fe-C martensite has not been provided
because the martensite is metastable and the solid solution decomposes at lower temperatures. Many examples of the order-disorder
transition are recently found in the interstitial alloys based on
the transition metals of the group IV and V[8].

Hydrogen atoms are usually located at either tetrahedral (T-)
or octahedral (O-) interstitial sites in the host metal lattice.
The number of these sites is larger than the number of hydrogen
atoms; for example the numbers of T- and O-sites in the bcc metal
lattice are 12 and 6 per unit cell, respectively, whereas that of
the hydrogen atoms dissolved in the bcc metals is usually less than
2 per unit cell. Therefore, there are many possible ordered
configurations of hydrogen atoms. In fact, we know that various
ordered structures are formed at the compositions of M_2H, M_3H_2, M_4H_3,
MH and others, and transform from one to another stepwise with
lowering temperature at a certain composition. This type of hydrogen
ordering has been interpreted in terms of a branching scheme by
Somenkov and his colleagues[9,10]; the hydrogen ordering proceeds as
a result of successive branching of energy levels for the interstitial sites with changing composition and temperature, which become
all equivant in the disordered solid solution.

In the lattice-gas model, a metal-hydrogen system can be
regarded as a quasi-monocomponent system of hydrogen atoms
dissolved in the host metal lattice[11]. In this model, the hydrogen
ordering from a random solid solution corresponds to the solidifica-
tion of 'lattice liquid', and a successive ordering at lower
temperatures corresponds to a polymorphic transition in the 'solid'
state. Here, the metal lattice is assumed to take no part in these
transitions. However, this is not the case but generally the
elastic interaction between metal and hydrogen atoms plays a
dominant role in the hydrogen ordering. The host metal lattice
distorts locally around the hydrogen atoms, and the elastic
displacement of metal atoms results in such a manner that the loss
of free energy due to the distortion, which is called the self-
trapping distortion[12], turns out to be minimal. In an ordered
phase, the periodic displacement of metal atoms takes place
reflecting the ordered distribution of hydrogen atoms, and the cubic
host lattice distorts to tetragonal or monoclinic.

Various kinds of the ordered structures have been known to

exist in metal-hydrogen systems as will be presented in the
following sections. From theoretical points of view, Khachaturyan[13]
has predicted the ordered structures on the basis of elastic theory
taking account of anisotropy and discrete structure of host metals.
A similar theory has been developed by Cook and de Fontaine[14] for
substitutional alloys; ordered structures are described in terms of
static concentration waves in the wave number space using a Fourier
transform method. Recently, Moriya and Ino[15] have presented the
ordered structures of interstitial solutes in bcc metal lattice on
the basis of Kanamori's theory[16,17]. In his theory, the ground
state structures are deduced in real space from the pairwise interaction model.

In this paper, we confine our attention to the order-disorder
phenomena in hydrides of the group V transition metals V, Nb and
Ta, and of Pd, because these hydrides are studied most extensively
so far.

ORDER-DISORDER PHENOMENA IN HYDRIDES OF V, Nb AND Ta

The phase diagrams of V-H, V-D, Nb-H and Ta-H systems are shown
in Fig. 1. The existence of many ordered phases is proposed at
temperatures below the random solid solution α, but contradictions
do remain in part on the constitutional relations. Comprehensive
surveys regarding the phase equilibria and the crystal structures
of these systems have been presented in recent publications[10,18,19].

Hydrogen atoms occupy randomly the T-sites of the bcc metal
lattice in the disordered phase α of these systems. The T-sites
are occupied regularly in the ordered phases except the β_1, β_2 and
δ phases of V-H system and the β phase of V-D system. In the
following, the features of order-disorder transitions in these
metal hydrides are reviewed. Because of the dramatic isotope
effect, the phase transitions of V-H and V-D systems are described
separately.

V-D System

β-α transition of V_2D. The order(β)$\rightarrow$disorder(α) transition of
V_2D is the first-order transition accompanied by the change in
hydrogen occupation from the O- to T-sites and the metal lattice
transition from monoclinic (pseudotetragonal) to cubic. The
subject to be considered first is the occupation probabilities of
hydrogen as a function of temperature.

At room temperature, the hydrogen atoms in β-V_2D occupy
preferentially a specific set of the O-sites named O_{z_1}-sites, as
shown in the nuclear density map on the (001) plane* (Fig. 2)[20-22].

---

* The indices refer to the host metal lattice.

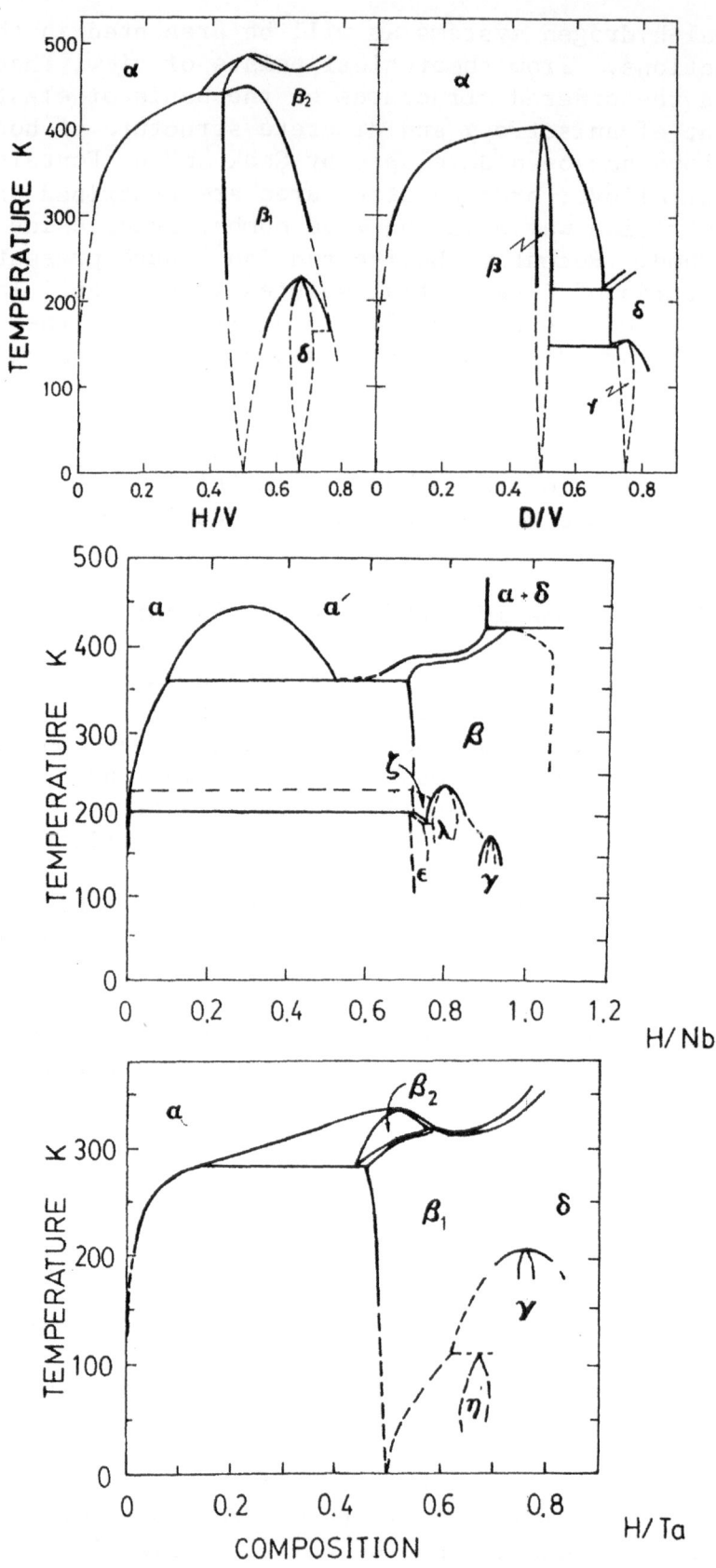

Fig. 1 Phase diagrams of V-H, V-D, Nb-H and Ta-H systems.

The deuterium density shows a main maximum at the O_{z_1} site, a
subsidual maximum at the O_{z_2} site and small hills at the O_{x_1} sites.
We note that the maxima at O_{z_1} and O_{z_2} have skirts covering the
four neighboring T-sites (4T-sites), and the contours extend along
the row of O_{x_1}-T-O_{z_1}-T-O_{x_1} parallel to the Y axis. It is interesting to compare this result with the wave function of deuterium in
vanadium calculated by Fukai and Sugimoto[12]. According to their
calculation, a continuous change between the O- and 4T-occupation
occurs as the self-trapping distortion becomes larger, and the O-

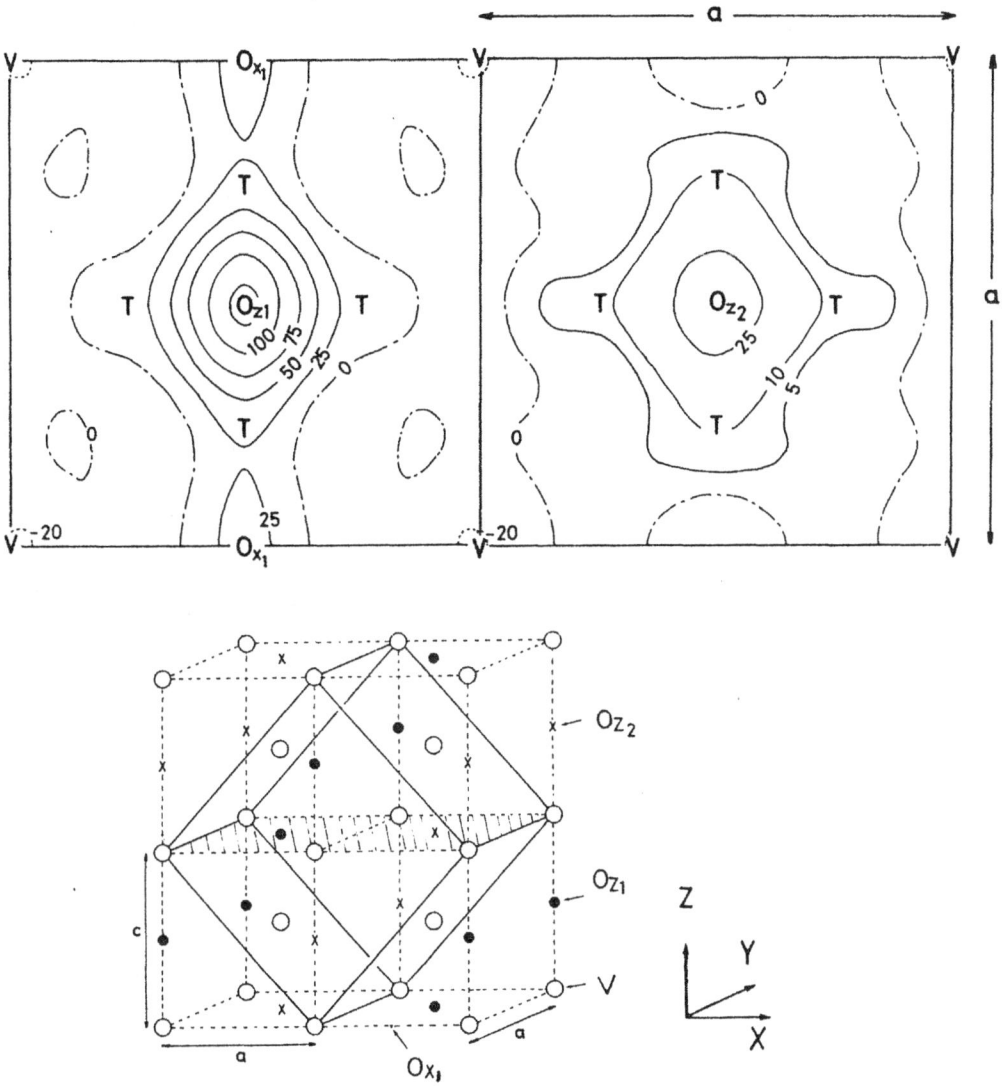

Fig. 2 Nuclear density map of β-VD$_{0.51}$, showing a section of
        three-dimensional Fourier transform parallel to the (001)
        plane shaded in the structure model, where O_{z_1} and O_{z_2}-sites
        are indicated by full circles and crosses, respectively.
        Contours are drawn at different intervals in the left- and
        right-hand sides.

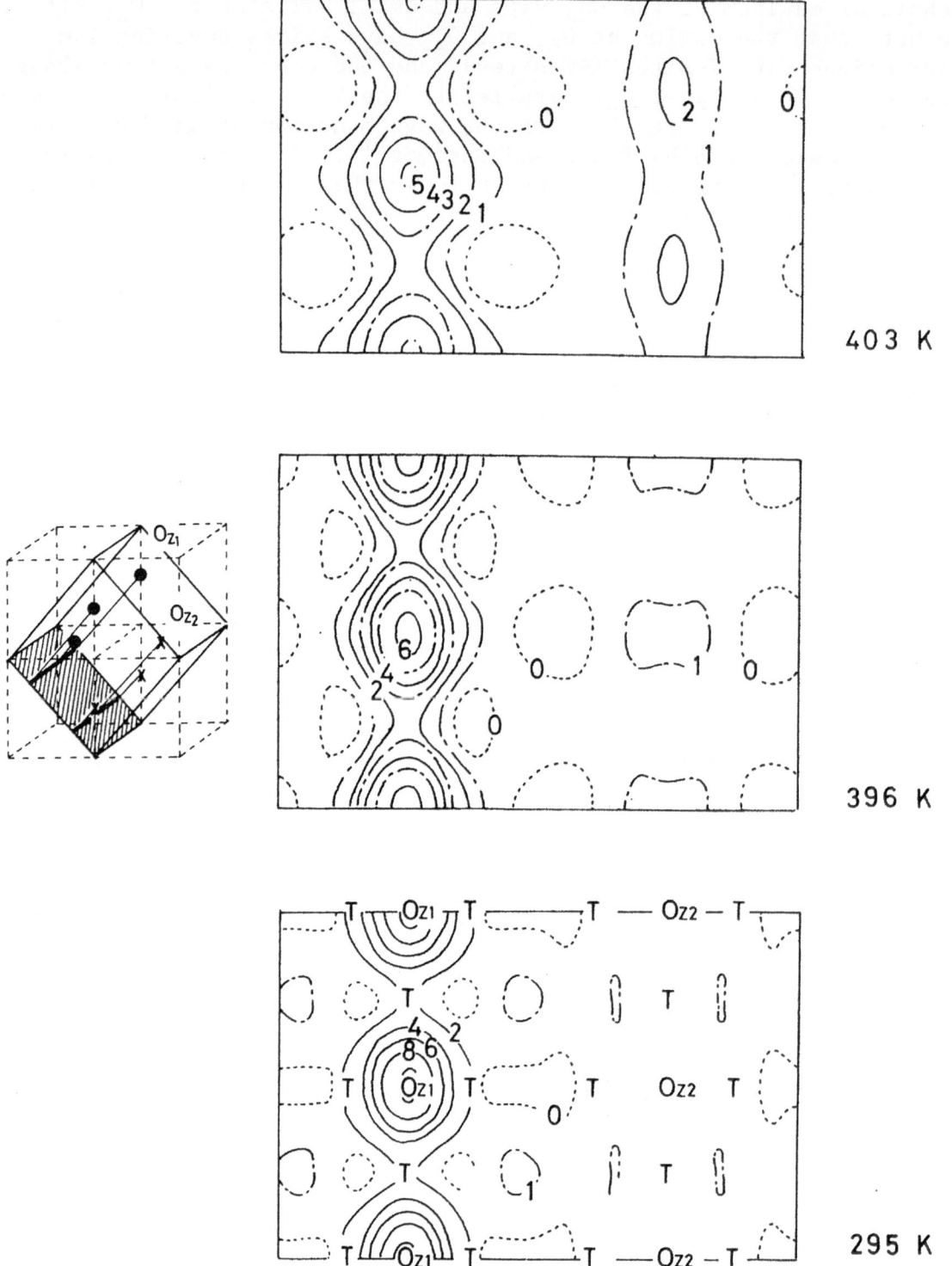

Fig. 3 Change in Fourier projections on the (101) plane of β-VD$_{0.51}$
at 295, 396 and 403 K.

or T-sites as geometrical points almost lose their meaning. They
interpreted also the fact that the β-V$_2$D phase is much narrower
than the phases (β_1, β_2, δ) of O-site occupancy in the V-H system.

Figure 3 shows the nuclear density maps of V$_2$D observed at
295, 396 and 403 K[22]. Note that the density maxima at the O$_{z_1}$-
sites become lower with increasing temperature, and a dramatic
change occurs between 396 and 403 K. At 403 K, just below the
transition temperature (T$_c$ = 406 K), a maximum appears at the
T-sites neighboring the O$_{z_2}$-sites, and an appreciable deuterium
distribution is seen in the row of O$_{z_2}$-T-O$_{z_2}$. The result indicates
a gradual change in the deuterium occupation from the O$_{z_1}$- and O$_{z_2}$-
sites to the neighboring T-sites. The occupation fractions of
deuterium at the O$_{z_1}$, O$_{z_2}$ and T-sites are plotted against temperature
in Fig. 4; approximately 70 % of the deuterium atoms occupy the
T-sites just below the transition temperature. Above the transition
temperature, a fractional O-site occupancy of deuterium was proposed
from the analysis of X-ray diffuse scattering in the α phase
region by Jo and Moss[23].

The metal lattice of β-V$_2$D is pseudotetragonal with the
approximate axial ratio c/a $\simeq$ 1.1 at room temperature. The
tetragonality is a function of the occupation fraction of deuterium

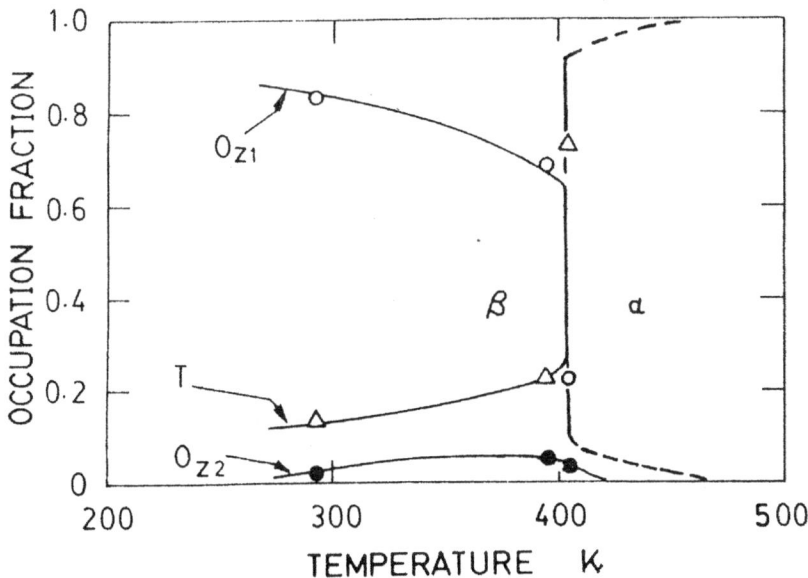

Fig. 4 Occupation fractions of deuterium at the O$_{z_1}$, O$_{z_2}$ and T-
        sites in the β-α transition of VD$_{0.51}$, determined by the
        least squares fitting of neutron diffraction intensity.

in the O_z- and T-sites, and disappears at the transition temperature discontinuously. The lattice parameters are expressed in the form[24]

$$c = a_o (1 + 0.195 P_z + 0.0548 P_T)$$

$$a = a_o (1 - 0.0256 P_z + 0.0548 P_T),$$
(1)

where a_o is the lattice constant of vanadium (3.03 Å), and P_z and P_T are the occupation fractions of deuterium at the O_z- and T-sites, respectively. If the occupation fractions at the O_{z1}- and O_{z2}-sites are designated as P_1 and P_2, respectively, $P_z = (P_1 + P_2)/2$. Similar relations to eq. (1) have been obtained from the X-ray diffraction work by Metzger et al.[25], who have taken into account the thermal expansion for completeness. As plotted in Fig. 5, the values calculated with the occupation fractions P_1, P_2 and P_T observed at three temperatures agree well with the X-ray data[26]. In Fig. 5, the experimental data are compared with a calculated curve, which was obtained from the intensities of superlattice reflections 001 and 110, assuming $P_z = P_1/2$ and $P_2 = 0$[24]. Slight deviation just below the transition temperature implies that P_2 is not negligible as compared with P_1, since the structure factor of these reflections is proportional to $(P_1 - P_2)$.

In Fig. 6, the population difference $(P_1 - P_2)$ is plotted against temperature[27], and compared with a long-range order (LRO) parameter, S, estimated from the electrical resistivity measurement by Bambakidis et al.[28] Sharp increase in the resistivity has been observed at the $\beta \rightarrow \alpha$ transition. They have analyzed the excess resistivity by assuming that the deuterium atoms rearrange from O_z- to T_z-sites, of which the available number is 4 per unit cell of β-$V_2 D$ because of the blocking effect. The excess resistivity $\Delta\rho$ is derived as

$$\Delta\rho = K \left(\frac{4 - S^2}{12}\right),$$
(2)

where K is a constant. As seen in Fig. 6, the curve of S thus evaluated is higher than the neutron diffraction data of $(P_1 - P_2)$. This is possibly because S is proportional to $(P_1 + P_2)$. Actually the normalized value of $(P_1 + P_2)$ at 396 K is very close to the above phenomenological curve. The drop of $(P_1 - P_2)$ at the transition point is much steeper for the data of the single crystal study[27] than those of the powder specimen work[24].

Transition of $V_4 D_3$. The order(δ)-disorder(α) transition above $VD_{0.65}$ is a deuterium rearrangement from local T- to random T-occupations[29,30]. The deuterium atoms in δ-VD occupy a specific set of the T-sites aligned along the [110] direction with the nearest neighbor distance $a/\sqrt{2}$; the deuterium lattice is orthorhombic with $A = \sqrt{2}a$, $B = a/\sqrt{2}$ and $C = a$, being isomorphic with β-NbD (see Fig. 1 of reference 21). At the non-stoichiometric composition of δ-VD_{1-x},

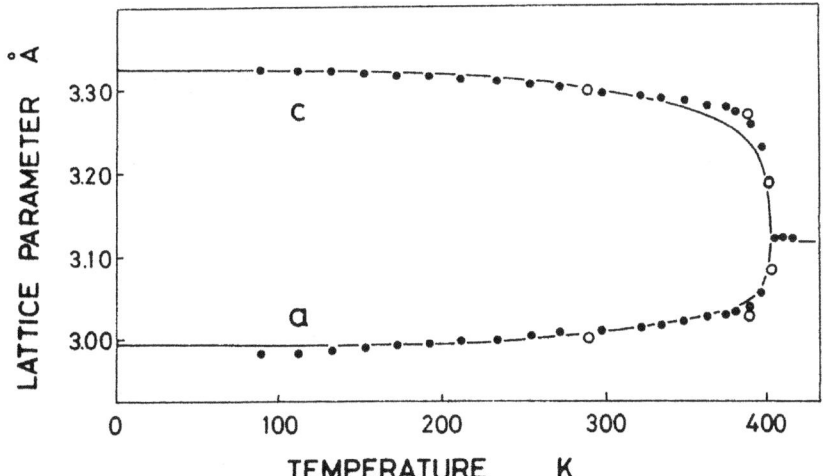

Fig. 5 Temperature dependence of lattice parameters of V_2D. Open
circles are calculated from P_1, P_2 and P_T using eq. (1).
Full circles are determined by X-rays[26], and solid curves
are calculated from powder neutron diffraction data[24].

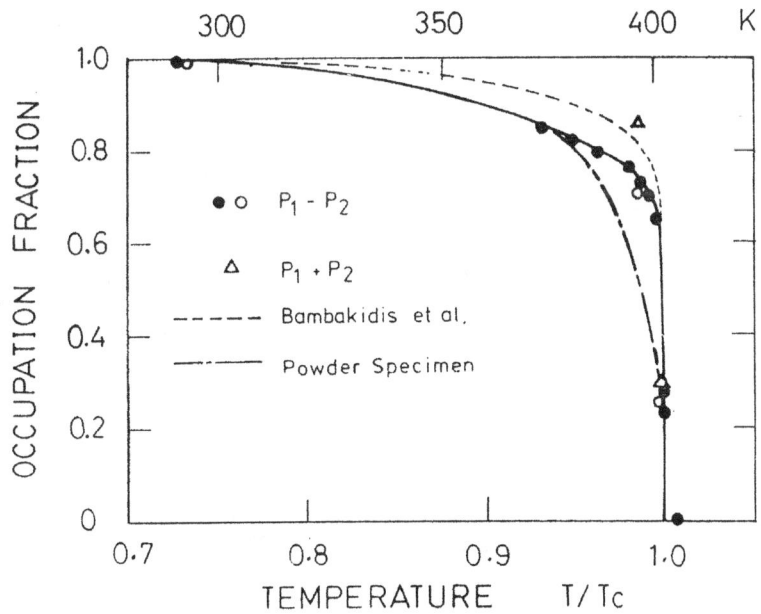

Fig. 6 Occupation fraction of deuterium and a LRO parameter in
V_2D, normalized to the value at room temperature. Full
circles are (P_1-P_2) determined from the 001 and 110 superlattice reflections of a single crystal[27]. Open circles
(P_1-P_2) and triangles (P_1+P_2) are obtained from the data of
Fig. 4. Dotted curve indicates the parameter S obtained by
Bambakidis et al.[28] Broken curve (P_1-P_2) is determined
from powder specimen[24].

the deuterium occupation is statistical. Below 150 K, the deuterium
ordering proceeds further to form the γ-V_4D_3 phase. This transition
is of the 'solid'-'solid' transition, but its feature is not fully
clarified yet, as mentioned below.

The ordered structure of γ-V_4D_3 shown in Fig. 7 has an
orthorhombic deuterium lattice of A = $\sqrt{2}a$, B = $a/\sqrt{2}$ and C = 2a, which
is doubled along the C axis of δ-VD_{1-x}. Two deuterium atom planes
parallel to (001) are followed by one vacant plane; the δ-γ
transition may be described in terms of the freezing of a deuterium
concentration wave having the wave vector parallel to [001].
However, the structure of γ-V_4D_3 is considered to be metastable[10];
it changes into a stable γ'-V_4D_3 after long annealing at 80 K which
has the dimension doubled along the A axis of the γ-V_4D_3 cell. Two
models of the deuterium arrangement are proposed but not distinguishable by the powder neutron diffraction pattern. Another open
question is the existence of an intermediate phase ζ. This phase
has been proposed by Schober and his colleagues[18,31] in a narrow
temperature range between the α and δ phases near V_4D_3.

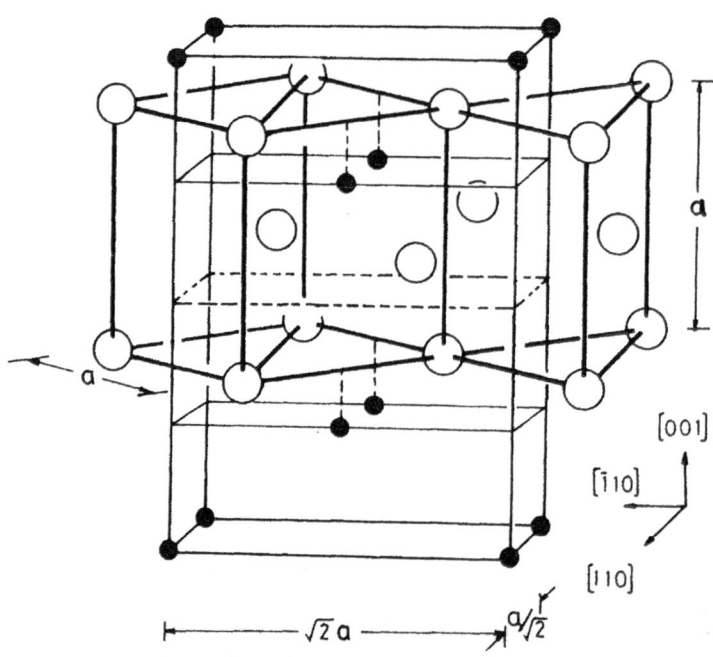

Fig. 7 Structure of γ-V_4D_3. Open and full circles correspond to
        V and D atoms, respectively.

V-H System

β_1-β_2-α transition of V_2H. The order-disorder transition of V_2H occurs stepwise as $\beta_1 \to \beta_2 \to \alpha$ with increasing temperature. The structure of β_1-V_2H is isomorphic with that of β-V_2D; the hydrogen nuclear density distribution is similar to that in β-V_2D (see Fig. 11 of ref. 21). A gradual change in the hydrogen occupation in the β_1-β_2-α transition is shown in Fig. 8[27]. In the $\beta_1 \to \beta_2$ transition, the O_{z_1}- and O_{z_2}-sites become discernible; the hydrogen atoms distribute equally in both sites in the β_2 phase. It is also noted that the occupation fraction at the T-sites becomes almost equal to those at the O_{z_1}- and O_{z_2}-sites. This result is consistent with the changes in the localized vibration mode observed in the β_1-β_2-α transition of V_2H[22].

The lower transition β_1-β_2 is likely to be the second-order; the occupation fractions change continuously, and thermal hysteresis is hardly detected on the heating and cooling measurements of electrical resistivity[28,32-34] and internal friction[22]. The upper transition β_2-α is the first-order transition accompanying with the change in hydrogen occupation from O_z- to T-sites and with the change in metal lattice from tetragonal to cubic.

The diffusion of hydrogen in the β_1-β_2 transition of V_2H has been studied by Fukai and Kazama using a pulsed-NMR technique[35]. They have observed anomalous increase in the mean jump frequency of

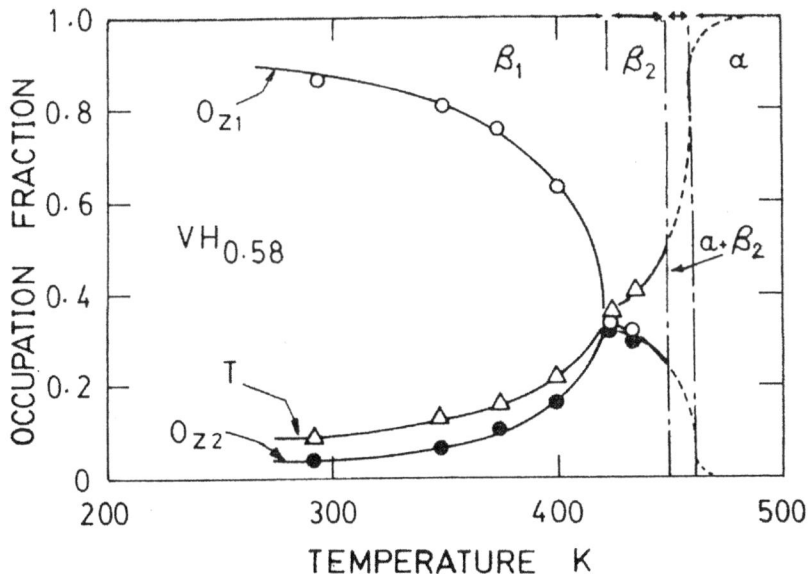

Fig. 8. Occupation fractions of hydrogen at the O_{z_1}, O_{z_2} and T-sites in the β_1-β_2-α transition of $VH_{0.58}$. Two-phase region exists between β_2 and α near 450 K.

protons near the transition temperature T_c, and interpreted to be
due to that the activation energy tends to be smaller in approaching
T_c. Based on this observation, they calculated the temperature
dependence of energy difference (e_1-e_2) between O_{z1}- and O_{z2}-sites
by assuming that the hydrogen population in the O_{z1}- and O_{z2}-states
depends linearly on the energy states e_1 and e_2 for these sites,
respectively. As shown in Fig. 9, the calculated curve reproduces
well the continuous drop in the population difference at T_c.
However if we compare with the experimental data, the calculation
is slightly lower than the values of $(P_1 - P_2)$, but better fitting
is obtained for the values of $(P_1-P_2-P_T)$. This result supports the
fact that the energy level of 4T-sites is very close to that of
O_{z2}-sites.

To throw light on the feature of the order-disorder transition
of metal hydrides, quantitative measurements of the entropy changes
are very valuable. The experimental values of the V-H and V-D
systems have been given already elsewhere in comparison with those
of the Nb- and Ta-H systems[19]. It is worth noting that the molar
entropy change in the β_2-α transition (0.40 cal/deg) of V_2H is less
than half of that in the β_1-β_2 transition (0.85 cal/deg)[26]. This is
attributed to negative change in the vibrational entropy on the
$\beta_2 \rightarrow \alpha$ transition; the vibrational entropy of the β_1 and β_2 phases is
larger than that of the α phase since the localized vibration energy
of hydrogen at the O-site is much lower than that at the T-site.

The β_1-β_2-α transition of $V_2'H$ is sensitive to external stress
as well as internal stress. Takano et al.[34] have studied the

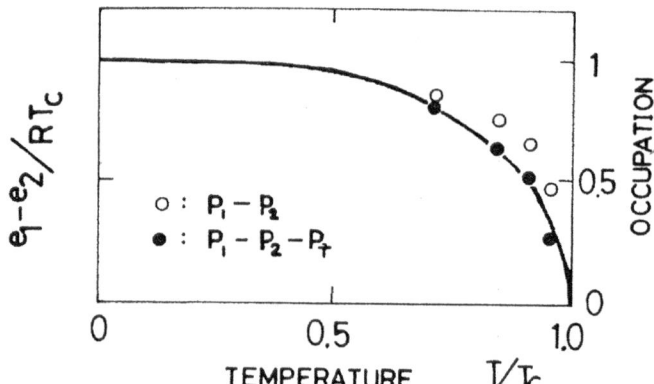

Fig. 9   Comparison of energy difference and occupation fraction of
the O_{z1}- and O_{z2}-sites in the β_1-β_2 transition of V_2H.
Full line is calculated by Fukai and Kazama[35]. Open and
full circles are the experimental data (P_1-P_2) and
$(P_1-P_2-P_T)$ of $VH_{0.58}$, respectively.

external stress effect using single crystals of V_2H. They have found that the transition temperature rises with increasing stress and the transition occurs directly from α to β_1 under stress-cooling instead of passing through the β_2 region. By the application of uniaxial tension on cooling, they have succeeded in growth of V_2H crystals composed of a single variant of the monoclinic β_1-hydride.

Transition of V_3H_2. When an alloy V_3H_2 is cooled down from 500 to 200 K, the stepwise ordering proceeds in the sequence of $\alpha \to \beta_2 \to \beta_1 \to \delta$. The hydrogen arrangements in these phases are schematically illustrated in Fig. 10. As mentioned before, hydrogen atoms occupy statistically the O_{Z_1}- and O_{Z_2}-sites in the β_2 phase. In the β_1 phase, the O_{Z_1}-sites are fully occupied whereas the O_{Z_2}-sites randomly; the hydrogen atoms occupy the O_Z-sites in an alternating scheme on every other $(0\bar{1}1)$ plane. In the δ-V_3H_2 structure, every third layer of the $(0\bar{1}1)$ planes is vacant, while other two layers are fully occupied. Consequently, the successive ordering of V_3H_2 may be described as the freezing of hydrogen concentration waves having the wave vector parallel to the $\langle 011 \rangle$ direction; the structures of β_1-V_2H_{1+x} and δ-V_3H_2 are distinguishable from each other by the period of the concentration wave. Both phases are considered to be separated by the two phase region ($\beta_1 + \delta$) (see Fig. 1).

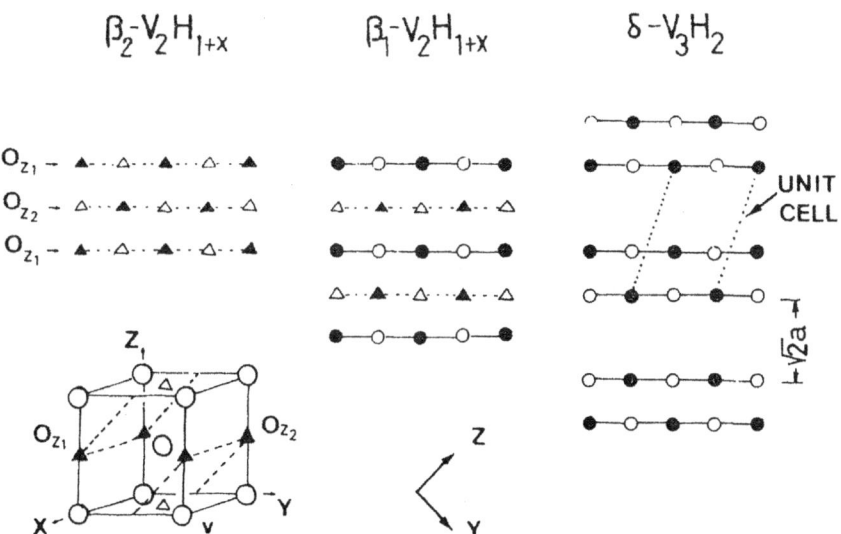

Fig. 10 Hydrogen arrangements in β_2-V_2H_{1+x}, β_1-V_2H_{1+x} and δ-V_3H_2, projected on (100) plane. Full and open triangles indicate partly filled O_Z-sites at $X = 0$ and $1/2$, respectively. Full and open small circles correspond to hydrogen atoms located at the O_Z-sites of $X = 0$ and $1/2$, respectively.

The freezing of the <011> concentration wave takes place in the $\alpha \rightarrow \beta$ transition of V_2D as well as in the $\beta_2 \rightarrow \beta_1$ transition of V_2H. These transitions were interpreted by elastic interaction theory; considering the interaction between vanadium atoms up to the 3rd nearest neighbors, the elastic energy minimum is attained for the freezing of the <011> concentration wave[36].

Nb-H (D) System

The order(β)-disorder(α) transition of NbD_{1-x} has attracted considerable interest of many workers as a typical example of the 'solid'-'liquid' transition in the lattice gas model[11], which is the deuterium rearrangement from local T- to random T-occupations. This transition is the first-order; discontinuous changes are observed in the electrical resistivity and the lattice parameters [10,18]. The metal lattice of β-NbD_{1-x} is orthorhombic and the deuterium arrangement is partly ordered[37], being isomorphic with δ-VD_{1-x}. The metal lattice transition from cubic (α) to orthorhombic (β) owing to the hydrogen ordering is interpreted by central forces for the atomic pairs of the Nb-H neighbors and the H-H neighbors[38].

At low temperatures, the existence of four ordered phases ζ, ε, λ and γ is reported, but the phase relation is not well established yet. Near Nb_4H_3, the stepwise ordering $\alpha \rightarrow \beta \rightarrow \zeta \rightarrow \varepsilon$ takes place on cooling, but an alternative sequence $\varepsilon \rightarrow \eta \rightarrow \zeta \rightarrow \beta \rightarrow \alpha$ is proposed upon heating postulating the existence of an intermediate phase η between ε and ζ[39]. The fully ordered structure of ε-Nb_4H_3 is schematically illustrated in Fig. 11 in comparison with the partly ordered structures of β-NbH_{1-x} and ζ-Nb_2H_{1+x}. In the β-NbH_{1-x} structure, the hydrogen lattice is orthorhombic ($A = \sqrt{2}a$, $B = a/\sqrt{2}$, $C = a$). In the ζ-Nb_2H_{1+x} structure, the B axis is doubled ($A \cup B = \sqrt{2}a$, $C = a$); the T-sites at every second plane parallel to (110) are orderly occupied by hydrogen. The hydrogen distribution in every other plane becomes ordered in the ε-Nb_4H_3 structure; the $\zeta \rightarrow \varepsilon$ transition is considered to be an intralayer ordering in every second T-site plane parallel to (110). As illustrated in Fig. 11, the ε-Nb_4H_3 structure is composed of an alternating sequence of the pairs of T-site planes parallel to (100); the pairs of occupied (100) planes are followed by the pairs of half-filled planes. Therefore, the $\zeta \rightarrow \varepsilon$ transition is described as the freezing of a hydrogen concentration wave having the wave vector parallel to [100]. Diffuse streaks along [100] were observed in the electron diffraction patterns by Makenas[40], and were interpreted as indicating a tendency toward the ε-type ordering in the ζ phase. According to his results, the behavior of the β-ζ transition is consistent with the second-order transition. The η phase might be a metastable phase in the $\varepsilon \rightarrow \zeta$ disordering process.

Brun et al.[41] have recently proposed an ordered structure of λ-$NbD(H)_{0.78-0.84}$ from neutron and electron diffraction studies.

This is a modulated structure having long periodicities composed of microdomains of the β and ζ type, as seen in Fig. 11. This structure is described as a deuterium concentration wave having the wave vector parallel to [001] direction of the slightly distorted bcc metal. The wavelength increases from 16 to 21 Å with increasing deuterium concentration. The freezing of such concentration waves in the β-λ transition has been explained by means of self-consistent elastic interaction theory by Kajitani et al.[42]

The γ phase having the pseudocubic metal lattice exists below 200 K at $NbH_{0.88-0.92}$[43]. The hydrogen arrangement in this phase has been proposed[44], but not confirmed experimentally. The mechanism of the β-γ transition will be an interesting subject of future work.

Ta-H (D) System

The β_1-α transition near $TaD_{0.60}$ is a first-order transition, which is characterized with a sharp peak of the specific heat and a

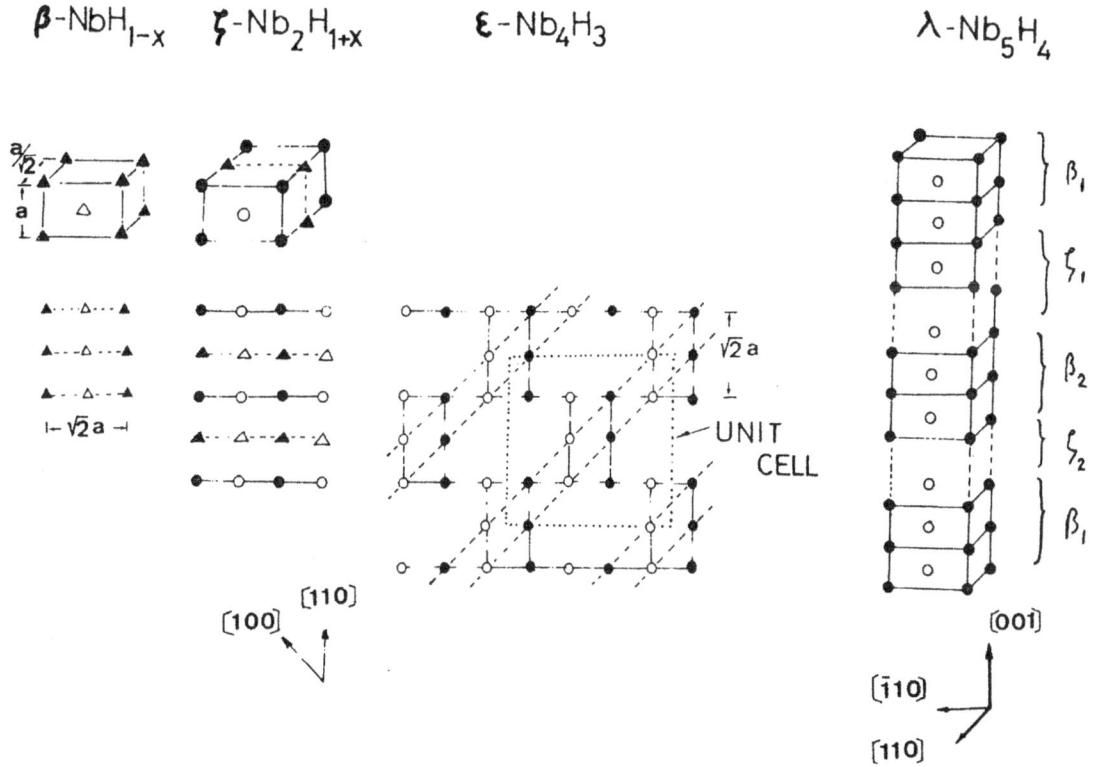

Fig. 11 Hydrogen arrangements in β-NbH_{1-x}, ζ-Nb_2H_{1+x}, ε-Nb_4H_3 and λ-Nb_5H_4. Triangles and circles indicate partly and fully filled T-sites, respectively. Full and open symbols correspond to the sites at Z = 0 and 1/2, respectively. Broken lines in the ε-Nb_4H_3 structure indicate concentration waves parallel to (100).

discontinuous drop of the LRO parameter at the critical point (Fig. 12)[45]. The two-phase region ($\beta_1 + \alpha$) is hardly detectable at this composition. In the α phase, the T-site occupation of hydrogen (deuterium) is not completely random, but the existence of SRO is concluded from the entropy change of the $\beta_1-\alpha$ transition. The experimental entropy change agrees well with the calculation based on the blocking model[46]; the first, second and third neighbor shells of T-sites surrounding a hydrogen atom are prohibited to the occupation by the other hydrogen atoms. This conclusion is consistent with the SRO parameters α_j determined from neutron diffuse scattering of $TaD_{0.75}$[9,10] ($\alpha_1 = \alpha_2 = \alpha_3 = -0.154$); unlike pairs, deuterium atoms–vacant interstices, are preferential in the first three

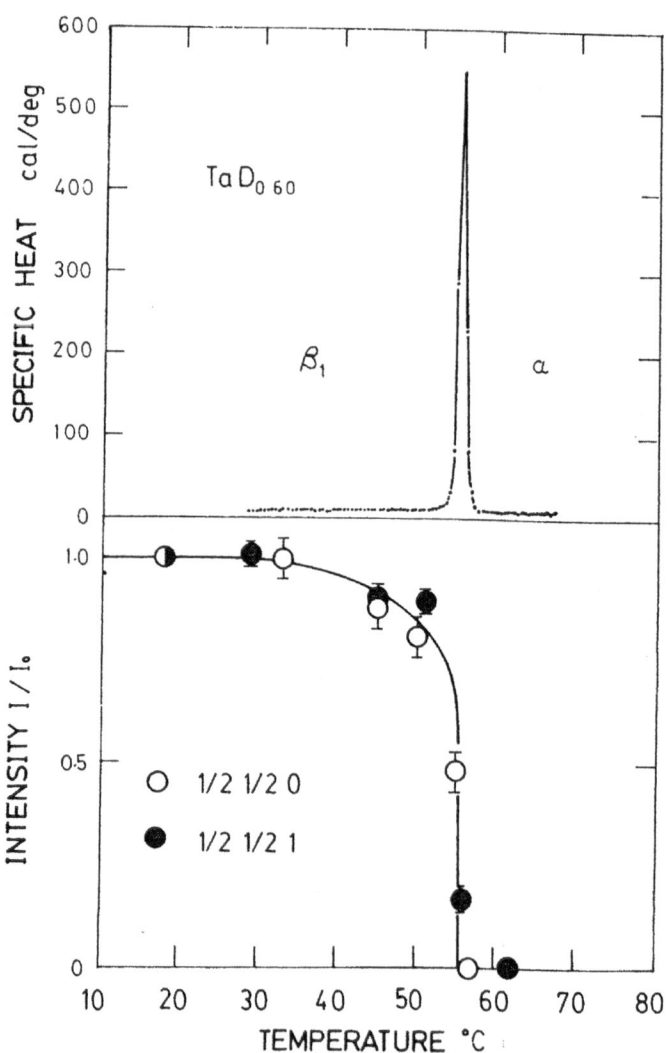

Fig. 12   Specific heat and neutron diffraction intensity of super-lattice reflections, 1/2 1/2 0 and 1/2 1/2 1, in the $\beta_1-\alpha$ transition of $TaD_{0.60}$[45].

neighbor shells.

It is known to occur the double-step transition $\beta_1-\beta_2-\alpha$ at
Ta_2H (Ta_2D), but the features of this transition are undefined
because the hydrogen arrangement of β_2-Ta_2H is still an open
question[21]. Nevertheless the $\beta_1-\beta_2$ and $\beta_2-\alpha$ transitions are of the
first-order accompanying with discontinuous changes in specific
heat[47], electrical resistivity[48,49] and thermoelectric power[49].

The structure of $\beta_1-Ta_2H_{1+x}$ is isomorphic with $\zeta-Nb_2H_{1+x}$ (Fig.
11). The hydrogen arrangements in $\beta_1-Ta_2H_{1+x}$ and $\delta-TaH_{1-x}$ are not
discernible at the non-stoichiometric compositions, because the
hydrogen population in every second layer of the T-site planes
parallel to (110) may change continuously with concentration.
Accordingly, no two-phase regions ($\beta_1 + \delta$) are predicted if the
energy states of these sites get degenerated at a certain composition, as in the case of the β_1 and β_2 phases of V-H. In fact,
experimental evidence for the two-phase region seems to be lacking
[45,47].

The stepwise ordering near Ta_4H_3 (Ta_4D_3) proceeds below 200 K
in the sequence of $\alpha\rightarrow\beta_1(\delta)\rightarrow\gamma$[10,45,47]. The $\gamma-Ta_4H_3$ phase has a fully
ordered structure with a pseudo-orthorhombic cell A = $2\sqrt{2}a$, B
= $\sqrt{2}a$ and C = a, as illustrated in Fig. 13. This is assigned as ζ
in the phase diagram of Schober and Wenzl[18] (Fig. 14). A new
ordered structure with a pseudo-orthorhombic cell A = $4\sqrt{2}a$, B = $3\sqrt{2}a$
and C = a is found near Ta_3H_2 (Ta_3D_2) by low temperature electron
and neutron diffraction work[50]. This phase corresponds to $\eta-Ta_3H_2$
in Fig. 1, and the stepwise transition $\eta-\beta_1-\alpha$ occurs at temperatures
between 50 and 350 K. An interesting correlation is noticed
between the two ordered structures of $\eta-Ta_3H_2$ and $\gamma-Ta_4H_3$. As
shown in Fig. 13, both structures are described as hydrogen concentration waves having the wave vector parallel to $[3\bar{1}0]$, and the
wavelength is $2\sqrt{10}a/3$ ($\simeq$ 7.0 Å) for $\eta-Ta_3H_2$ and $4\sqrt{10}a/9$ ($\simeq$ 4.6
Å) for $\gamma-Ta_4H_3$. Consequently, the $\beta_1\rightarrow\gamma$ and $\beta_1\rightarrow\eta$ transitions may
be identified as the freezing of these concentration waves,
although no theoretical interpretation is given yet.

Many efforts have been devoted to the study of phase relations
of the Ta-H (D) systems[48,50-52], but the complexity at low temperatures remains unsolved. In the diagram of Fig. 14[18], three ordered
phases θ, γ and ℓ are proposed below 250 K, but the structures are
unknown.

ORDER-DISORDER PHENOMENA IN PALLADIUM HYDRIDE

Palladium hydride has been a broad and continual subject of
many investigations for long[53]. Particular interests have been
focused to the peculiar physical properties; the inverse isotope
effect of diffusion coefficient[54], the dispersion modes of optic

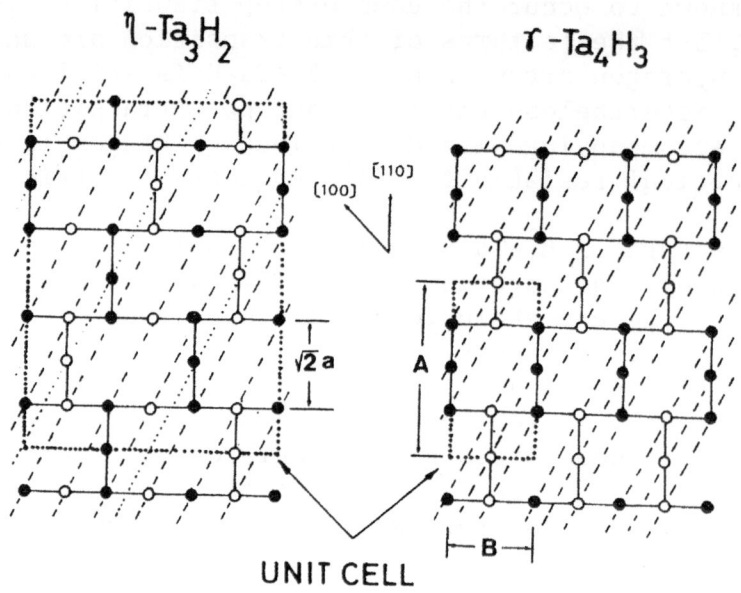

Fig. 13   Hydrogen arrangements in η-Ta$_3$H$_2$ and γ-Ta$_4$H$_3$, projected on
(001) plane. Open and full circles correspond to hydrogen
atoms at the T-sites of Z = 1/4 and 3/4, respectively.
Broken lines indicate concentration waves parallel to ($3\bar{1}0$).

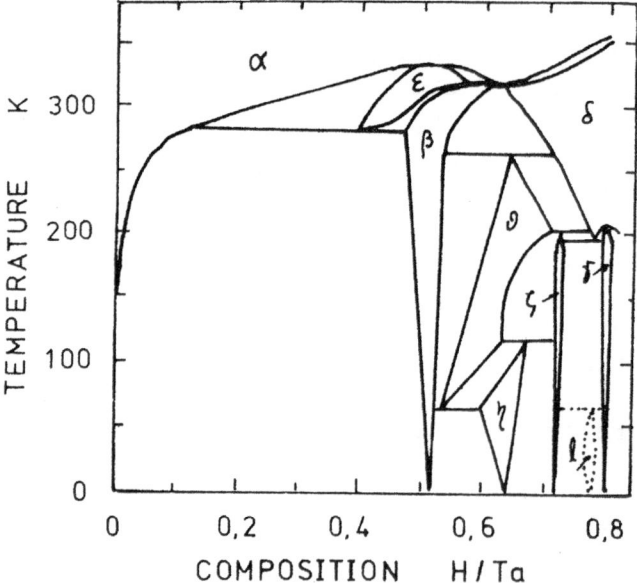

Fig. 14   Phase diagram of Ta-H proposed by Schober and Wenzel[18].

and acoustic phonons[55], the superconductivity of PdH_{1-x}[56] and the so-called '50 K anomaly.' The 50 K anomaly was a matter of speculation since the first discovery by Nace and Aston in 1957[57]. Early efforts poured into this phenomenon are cited in a literature of Jacobs and Manchester[58]. As a result of current studies by numerous workers, the origin of this anomaly is now attributed to an ordering of hydrogen atoms in the O-sites of the fcc metal cell of PdH_{1-x}. The structure at the stoichiometric composition PdH is the NaCl type.

Two superstructures have been proposed from the appearance of two types of superlattice reflections in the composition range of $PdD_{0.64}$-$PdD_{0.76}$. Anderson et al.[59,60] first succeeded to observe a weak superlattice reflection at 1 1/2 0 reciprocal lattice point by neutron diffraction at 46 K using a single crystal $PdD_{0.64}$. Their observation was supported by γ-ray and neutron diffraction of $PdH_{0.73}$ by Blaschko et al.[61] This result is consistent with theoretical predictions of Khachaturyan[62] and Goldberg[63]. On the contrary, Ellis et al.[64] reported the existence of 4/5 2/5 0 superlattice reflection, instead of 1 1/2 0 (Fig. 15), by neutron diffraction of $PdD_{0.76}$ annealed for 1 week at 70 K. Then the question arises if the two observed superlattice reflections of 1 1/2 0 and 4/5 2/5 0 are due to different concentrations or different thermal treatments. Blaschko et al.[65] investigated $PdD_{0.73}$ annealed at different temperatures and showed again the existence of the 1 1/2 0 reflection but no indication of the 4/5 2/5 0 reflection after long annealing up to 200 hr at 53 K. Neither the 1 1/2 0 nor 4/5 2/5 0 reflection could be observed after annealing at 77 K for 120 hr and at 65 K for 110 hr.

Based on the observation of the 1 1/2 0 reflection, Anderson et al.[66] proposed a superstructure model, which is tetragonal (A = a, C = 2a), belonging to the space group $I4_1/amd$. We call it model A for convenience. Another model B which is also tetragonal (A = $\sqrt{5}$a, C = a), I4/m, was proposed by Mueller et al.[67] to interpret the existence of the 4/5 2/5 0 reflection. Both models are schematically illustrated in Fig. 16. As the model A is based on the

Fig. 15 Superlattice reflections of PdD_{1-x}. Open and full circles correspond to the reflections of 1 1/2 0 and 4/5 2/5 0 types, respectively.

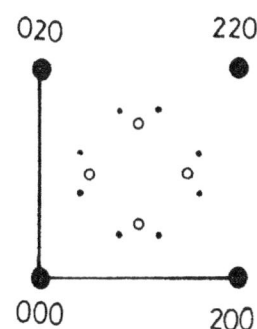

composition $PdD_{0.5}$, the vacant O-sites are considered to be filled statistically with the excess deuterium atoms. The arrangement of deuterium in the model B is analogous with that of nickel in the Ni_4Mo structure, and the vacant sites correspond to molybdenum positions. The composition of this model corresponds to Pd_5D_4. In fact, the structure of model B was found at $PdD_{0.76}$ [64,67] while the model A was proposed for the alloys containing less deuterium: $PdD_{0.64}$ [59,60], $PdD_{0.67}$ [66] and $PdD(H)_{0.73}$ [61,65]. A tentative phase diagram is presented in Fig. 17, where the structures of model A and B appear in the γ and δ phases, respectively.

These two superstructures are closely and consistently correlated with each other. In Fig. 16, a series of parallel $(0\bar{1}2)$

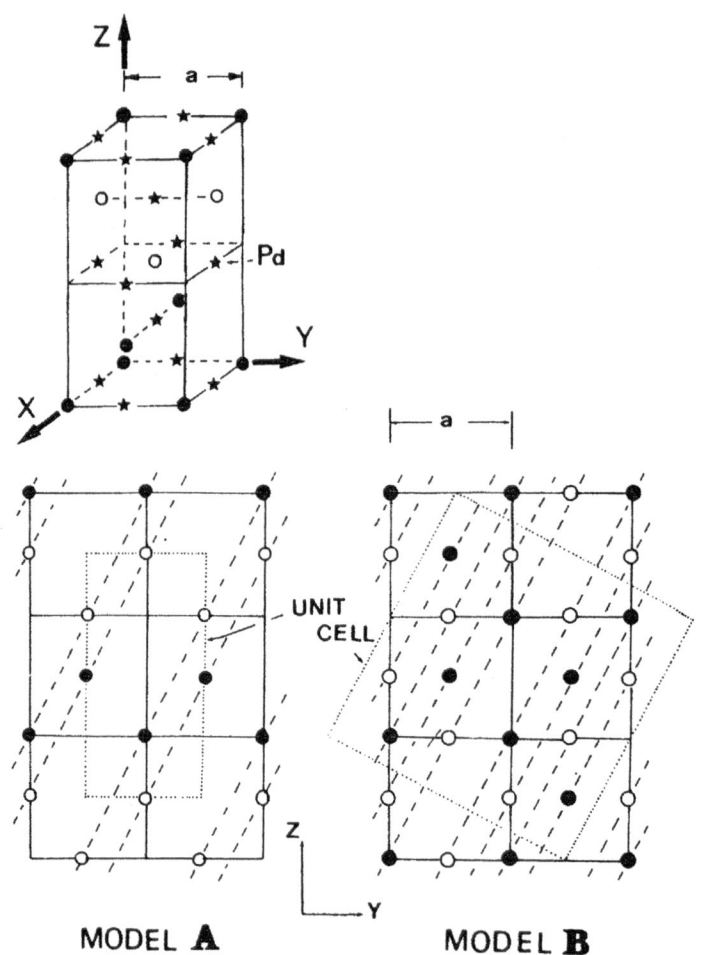

Fig. 16   Models A and B of the ordered structures of PdD_{1-x}. Full and open circles indicate deuterium atoms at the O-sites of X = 0 and 1/2, respectively. Concentration waves parallel to $(0\bar{1}2)$ are shown by broken lines. Palladium atoms are neglected in the projections of the models.

planes represent a modulated deuterium concentration wave. In the
model B, the concentration waves consist of four fully occupied
($0\bar{1}2$) planes followed by one vacant plane. In the model A, the
waves consist of two fully occupied planes followed by two vacant
planes. Consequently, the $\alpha' \to \gamma$ and $\alpha' \to \delta$ transitions are the 'liquid'
$\to$'solid' transitions, and are described in terms of the freezing of
concentration waves having the wave vectors parallel to the <012>.

Elastic free energy calculations[68] based on the Kanzaki
approach[69] show that the energy minima for the concentration wave
is attained if the wave vectors appear in the vicinity of 4/5 2/5 0
in good agreement with the experimental result.

The formation of both structures of the models A and B is
discussed by Goldberg and Manchester[70] within the mean-field free
energy theory formulated in reciprocal space originally presented
by de Fontaine[71]. Based on the Ising model, Anderson et al.[66] have
estimated the range and relative magnitude of pairwise interaction
potentials for the first three neighbors, V_1, V_2 and V_3. Following
the Clapp-Moss theory[72], they showed that maxima in the scattering
intensities should occur at the 1 1/2 0 position in reciprocal
space, when $V_2/V_1 = 0.25$ and $V_1 > 0$.

Satterwaite et al.[73] found the rise of the transition temperature with increasing hydrogen composition in the range $PdH_{0.5}$–$PdH_{0.82}$

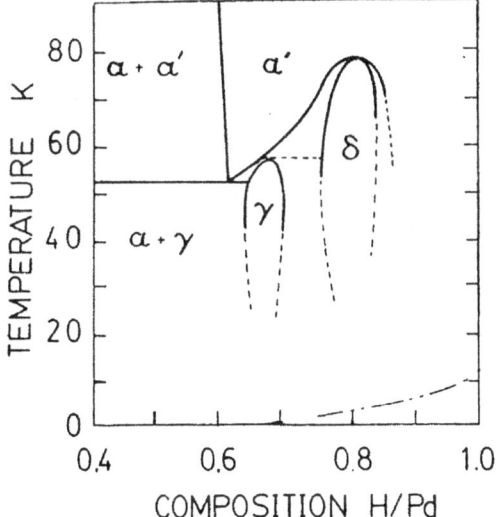

Fig. 17 Tentative phase diagram of Pd-H(D).

from the resistivity measurements. The transition temperature
curve between α' and γ or δ phases in Fig. 17 is based on their
results. They found also an initial increase and a subsequent
sluggish decrease of the resistivity of $PdH_{0.795}$ with aging time
below the transition temperature. This behavior was interpreted in
terms of the nucleation and growth processes of ordered domains of
the structure model B.

In Fig. 17, the transition temperature of superconductivity is
indicated using the data of Miller and Satterthwaite[74]. To our
knowledge, however, no work has been reported on the effect of
hydrogen ordering on the superconductivity.

SUMMARY

As described in the preceeding sections, considerable amount
of experimental results has been accumulated on the order-disorder
phenomena in the hydrides of V, Nb, Ta and Pd. The ordered
structures and the mechanism of phase transitions in these hydrides
are understandable, in principle, from the viewpoint of elastic
interaction.

The ordered phases in the hydrides of V, Nb and Ta are
classified for the sake of convenience into two categories, high-
temperature and low-temperature phases, as listed in Table 1. The
former is formed directly from the disordered solid solution by the
'liquid'-'solid' transition, while the latter is formed from the
high-temperature phase through the 'solid'-'solid' transition. Most
of these transitions are of the first-order, but some of the latter
are likely to be the second-order.

The 'liquid'-'solid' transition temperatures in the individual
systems differ from one another depending primarily on the
magnitudes of the stress-induced interaction. It is interesting to
compare the transition temperatures at a certain composition.
Choosing M_4H_3, the transition temperatures of $\beta_2-VH_{0.75}$, $\delta-VD_{0.75}$,
$\beta-NbH_{0.75}$, $\delta-TaH_{0.75}$ and $\delta-TaD_{0.75}$ are plotted as a function of the
metal lattice constant in Fig. 18 together with that of $PdH_{0.75}$.
The transition temperature decreases with the increase of lattice
constant in consistence with the elastic theory. The highest
temperature is of $VH_{0.75}$; the interaction energy of hydrogen is
strong for the O-site occupation in the distorted bct metal as
compared with the exceptionally low value of $VD_{0.75}$ with the T-site
occupation in the bcc metal. The lowest temperature is of $PdH_{0.75}$;
the hydrogen induced distortion is very small for the O-site
occupation in the fcc metal.

The hydrogen arrangements in the high-temperature ordered
phases are simple, being interpretable taking account of short
range interactions. For example, the stability of the ordered

Table 1. Ordered Phases in Hydrides of V, Nb, Ta and Pd

	High Temperature Phase	Low Temperature Phase	Direction of Concentration Wave	Hydrogen Occupation
V	β_2-V_2H, VH_{1-x} β-V_2D δ-VD_{1-x}	β_1-V_2H, δ-V_3H_2 γ-V_4D_3	[011] [011] [001]	O O T
Nb	β-NbH_{1-x}	ζ-Nb_4H_3, ε-Nb_4H_3 λ-$NbH_{0.8}$ γ-$NbH_{0.9}$	[100] [001]	T T T
Ta	β_2-Ta_2H δ-TaH_{1-x}	β_1-Ta_2H η-Ta_3H_2 γ-Ta_4H_3	 [310] [310]	T T T
Pd	γ-$PdH_{0.6}$ δ-Pd_5H_4		[012] [012]	O O

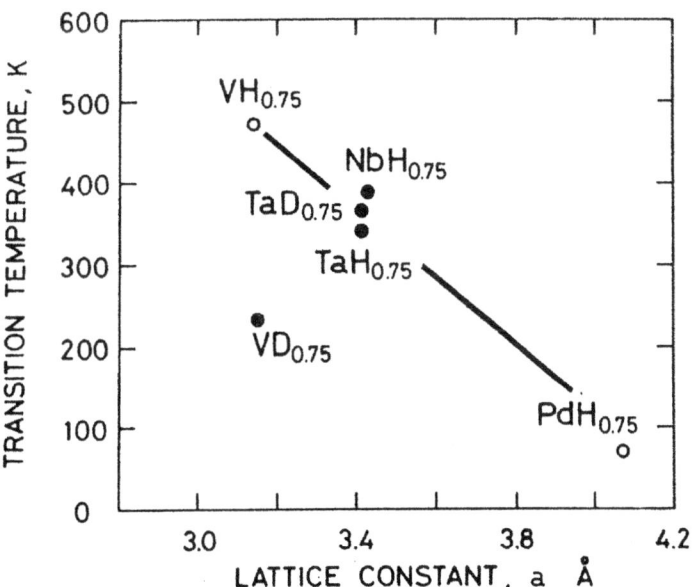

Fig. 18 Transition temperature of M_4H_3 plotted against metal
lattice constant. Full and open circles indicate the T-
and O-site occupation of H(D) in the ordered phases,
respectively.

structure of δ-VD, β-NbH and δ-TaH is interpreted in terms of the repulsive interaction between the metal-hydrogen and hydrogen-hydrogen nearest neighbors[38]. The δ-VD and β_2-V_2H (β-V_2D) structures are proved to be the ground state ordered structures from the Kanamori theory by Moriya and Ino[15] taking account of the pairwise interactions up to the fourth neighbors of hydrogen atoms. These two structures are also predicted from the elastic theory by Khachaturyan[13] as the stable ordered structures for the T- and O-site occupation in the bcc metals, respectively.

More complicated ordered arrangements of hydrogen atoms are stabilized in the low-temperature phases, as described in the preceeding sections. The long range interaction should be considered to interpret the stability of these structures. As summarized in Table 1, these ordered structures are described in terms of the static concentration waves having various kinds of wave vectors. The elastic free energy theory explains successfully the occurrence of the concentration waves in the V-, Nb-, and Pd-H systems as stated before. However, the reason why the wave vectors in the low-temperature phases are so different in the V-, Nb- and Ta-H systems is not clear yet.

The above argument is mostly concerned with the structural changes in equilibrium states of the transition metal hydrides. Our knowledge of the kinetics of hydrogen ordering is far behind that of the ordering in substitutional alloys. Future studies on this subject will give insight into the freezing mechanism of concentration waves.

In this paper, no attention has been given to the order-disorder phenomena in the hydrides of binary alloys and intermetallic compounds, and of the rare-earth and actinide metals. Current progress in these fields has been reported by some workers[9,75,76].

ACKNOWLEDGEMENT

We are grateful to Mr. I. Okada for his important contribution to the experimental study of V_2H and V_2D. We are also indebted to Dr. T. Kajitani for his valuable comments on the manuscript.

REFERENCES

1. K. K. Kelley, Specific heat of Ta at low temperatures and the effect of small amounts of dissolved hydrogen, J. chem. Phys., 8: 316 (1940).
2. T. R. Waite, W. E. Wallace and R. S. Craig, Structures and phase relationships in the Ta-H system between -145 and 70°C, J. chem. Phys., 24: 634 (1956).

3. W. G. Saba, W. E. Wallace, H. Sandmo and R. S. Craig, Heat capacities and the residual entropy of Ta_2H, J. chem. Phys., 35: 2148 (1961).

4. W. E. Wallace, Neutron diffraction data for Ta_2D and the probable arrangement of the deuteriums in two of its polymorphic varieties, J. chem. Phys., 35: 2156 (1961).

5. K. Honda, Die Theorie der Stahlhärtung, Archir. Eisenhüttenwesen, Gruppe E, 8: 527 (1928).

6. C. Zener, Kinetics of the decomposition of austenite, Trans. AIME, 167: 550 (1946). Details are discussed in: Theory of strain interaction of solute atoms, Phys. Rev., 74: 639 (1948).

7. H. Sato, On the ordered arrangement of interstitial atoms in interstitial solid solution, J. Jpn. Inst. Met., 17: 601 (1954). (in Japanese).

8. M. Hirabayashi, S. Yamaguchi, H. Asano and K. Hiraga, Order-disorder transformations of interstitial solutes in transition metals of IV and V groups, in: "Order-Disorder Transitions in Alloys," H. Warlimont, ed., Springer-Verlag, Berlin, 266 (1974).

9. V. A. Somenkov and S. S. Shil'stein, Structural behaviour of hydrogen in metals and intermetallic compounds, Z. Phys. Chem. (N.F.), 117: 125 (1979).

10. V. A. Somenkov and S. S. Shil'stein, Phase transitions of hydrogen in metals, Prog. Mat. Sci., 24: 267 (1980).

11. G. Alefeld, Wasserstoff in Metallen als Beispiel für ein Gittergas mit Phasenumwandlungen, phys. stat. sol., 32: 67 (1969).

12. Y. Fukai and H. Sugimoto, Quantum-mechanical approach to the state of hydrogen in bcc metals, Proc. JIMIS-2, Hydrogen in Metals, Suppl. Trans. Jpn. Inst. Met., 21: 41 (1980).

13. A. G. Khachaturyan, Ordering in substitutional and interstitial solid solutions, Prog. Mat. Sci., 22: 1 (1978).

14. H. E. Cook and D. de Fontaine, On the elastic free energy of solid solutions, I. Microscopic theory, Acta Met., 17: 915 (1969).

15. T. Moriya and H. Ino, Ordered structures of interstitial solutes in bcc lattice, J. Phys. Soc. Jpn., 46: 1776 (1979).

16. J. Kanamori, Magnetization process in an Ising spin system, Prog. theor. Phys., 35: 16 (1966).

17. M. Kaburagi and J. Kanamori, A method of determining the ground state of the extended-range classical lattice gas model, Prog. theor. Phys., 54: 30 (1975).

18. T. Schober and H. Wenzl, The systems NbH(D), TaH(D), VH(D); structures, phase diagrams, morphologies, methods of preparation, in: "Hydrogen in Metals" II, G. Alefeld and J. Völkl, ed., Springer-Verlag, Berlin (1978).

19. H. Asano and M. Hirabayashi, Hydrogen ordering in Va transition metal hydrides, Z. Phys. Chem. (N.F.), 114: 1 (1979).

20. I. Okada, H. Asano and M. Hirabayashi, Refinement of the structure of β-V_2D by single crystal neutron diffraction,

Proc. JIMIS-2, Hydrogen in Metals, Suppl. Trans. Jpn. Inst. Met., 21: 89 (1980).

21. H. Asano and M. Hirabayashi, Neutron diffraction studies of metal hydrides, This proceedings.

22. M. Hirabayashi, Structural phase transition of metal hydrides; V-H and V-D, Proc. JIMIS-2, Hydrogen in Metals, Suppl. Trans. Jpn. Inst. Met., 21: 49 (1980).

23. H. S. U. Jo and S. C. Moss, Deuterium-induced vanadium short-range modulation above T_C in single crystal V_2D, Modulated Structures-1979. Proc. AIP Conf. No. 53: 400 New York (1979).

24. H. Asano and M. Hirabayashi, Hydrogen (deuterium) location and tetragonal distortion in $\beta-VH_x$ and $\beta-V_2D$, Proc. 2nd Int. Cong. Hydrogen in Met., Paris, 1D6, (1977).

25. H. Metzger, H. Jo, S. C. Moss and D. G. Westlake, Single crystal X-ray study of the superstructure modulation and long-range order in V_2D, phys. stat. sol., (a) 47: 631 (1978).

26. H. Asano, Y. Abe and M. Hirabayashi, A calorimetric study of the phase transformation of vanadium hydrides $VH_{0.06}-VH_{0.77}$, Acta Met., 24: 95 (1976).

27. I. Okada, Master Thesis, Tohoku University, (1980).

28. G. Bambakidis, M. W. Pershing and R. C. Bowman, Jr., Electrical resistivity studies of order-disorder phenomena in V_2H and V_2D, Scripta Met., 13: 441 (1979).

29. D. G. Westlake, M. H. Mueller and H. W. Knott, Structural transitions at low temperatures in vanadium deuterides, J. appl. Cryst., 6: 206 (1973).

30. H. Asano and M. Hirabayashi, Interstitial superstructures of V-D, phys. stat. sol. (a) 15: 267 (1973).

31. T. Schober and W. Pesch, The systems V-H and V-D, Z. Phys. Chem. (N.F.), 114: 21 (1979).

32. Y. Fukai and S. Kazama, On the constitution of $\beta-VH_x$, Scripta Met., 9: 1073 (1975).

33. D. G. Westlake, A resistometric contribution to the V-H phase diagram, Scripta Met., 11: 887 (1977).

34. S. Takano, H. Kojima, R. Akaba and T. Suzuki, On the martensitic transformation $\alpha-\beta$ of a V_2H single crystal, Proc. JIMIS-2, Hydrogen in Metals, Suppl. Trans. Jpn. Inst. Met., 21: 85 (1980).

35. Y. Fukai and S. Kazama, NMR studies of anomalous diffusion of hydrogen and phase transition in V-H alloys, Acta Met., 25: 59 (1977).

36. T. Kajitani, private communication.

37. V. A. Somenkov, A. V. Gurskaya, M. G. Zemlyanov, M. E. Kost, N. A. Chernoplekov and A. A. Chertkov, Neutron scattering study of structure and phase transitions in niobium hydrides and deuterides, Sov. Phys.-Solid State, 10: 1076 (1968).

38. G. Mair, K. Bichmann and H. Wenzl, Structural transformation correlated with order-disorder transformations of H in NbH crystals, Z. Phys. Chem., (N.F.), 114: 29 (1979).

39. W. Pesch, T. Schober and H. Wenzl, TEM investigation of anisotropic lattice distortions in ordered NbH and NbD alloys, Scripta Met., 12: 815 (1978).

40. B. J. Makenas, Precipitation and ordering in the Nb-H system, Ph. D. Thesis, University of Illinois, (1978).

41. T. O. Brun, T. Kajitani, M. H. Mueller, D. G. Westlake, B. J. Makenas and H. K. Birnbaum, The structure of the λ-phases of Nb-D, Modulated Structures-1979. Proc. AIP Conf. No. 53: 397, New York (1979).

42. T. Kajitani, T. O. Brun, M. H. Mueller, H. K. Birnbaum and B. J. Makenas, Modulated ordering Nb-H alloys, Modulated Structures-1979, Proc. AIP Conf., No. 53: 394, New York (1979).

43. M. A. Pick, Strukturelle Phasenübergänge in NbH System, Tech. Rpt. Jül-951-FF, KFA Jülich (1973).

44. J. Hauck, Ordering of hydrogen in niobium hydride phases, Acta Cryst., A33: 208 (1977).

45. H. Asano, Y. Ishino, R. Yamada and M. Hirabayashi, Interstitial superstructures in the Ta-D system, J. Solid State Chem., 15: 45 (1975).

46. W. A. Oates, J. A. Lambert and P. T. Gallagher, Monte Carlo calculations of configurational entropies in interstitial solid solutions, Trans. AIME, 245: 47 (1969).

47. H. Asano, R. Yamada and M. Hirabayashi, A calorimetric study of the phase transformation in tantalum hydrides, Trans. Jpn. Inst. Met., 18: 155 (1977).

48. F. Ducastelle, R. Caudron and P. Costa, Etude de systeme Ta-H; diagramme d'equilibre et structure electronique, J. Phys. Chem. Solids, 31: 1247 (1970).

49. T. G. Berlincourt and P. W. Bickel, Electrical properties of Ta_2H, Phys. Rev., B2: 4838 (1970).

50. Unpublished work.

51. T. Schober and A. Carl, A revision of the Ta-H phase diagram, Scripta Met., 11: 397 (1977).

52. U. Köbler and T. Schober, Susceptibility and phase diagram of the Ta-H system, J. Less-Comm. Met., 60: 101 (1978).

53. E. Wicke and H. Brodowsky, Hydrogen in palladium and palladium alloys, in: Hydrogen in Metals II, G. Alfeld and J. Völkl ed., Springer-Verlag, Berlin (1978).

54. J. Völkl, This proceedings.

55. T. Springer, Investigation of vibrations in metal hydrides by neutron spectroscopy, in: Hydrogen in Metals I, G. Alefeld and J. Völkl ed., Springer-Verlag, Berlin (1978).

56. E. N. Economou, This proceedings.

57. D. M. Nace and J. G. Aston, Palladium hydride II, The entropy of Pd_2H at 0°K, J. Am. Chem. Soc., 79: 3623 (1957).

58. J. K. Jacobs and F. D. Manchester, Thermal and motional aspects of the 50 K transition in PdH and PdD, J. Less-Comm. Met., 49: 67 (1976).

59. I. S. Anderson, C. J. Carlile and D. K. Ross, The 50 K transition in β-phase Pd deuteride observed by neutron

scattering, J. Phys. C 11: L 381 (1978).

60.   I. S. Anderson, D. K. Ross and C. J. Carlile, The structure of the γ phase of Pd deuteride, Phys. Letters 68A: 249 (1978).

61.   O. Blaschko, R. Klemencic, P. Weinzierl and O. J. Eder, Investigation of the 50 K transition in $PdH_{0.73}$ by γ-ray and neutron diffraction, Solid State Comm., 27: 1149 (1978).

62.   A. G. Khachaturyan, The problem of symmetry in statistical thermodynamics of substitutional and interstitial ordered solid solutions, phys. stat. sol.   (b) 60: 9 (1973).

63.   H. A. Goldberg, Model possibilities for the 50 K transition in $Pd-H_x$, Transition Metals-1977, Proc. AIP Conf.,No. 39: 504, New York (1977).

64.   T. E. Ellis, C. B. Satterthwaite, M. H. Mueller and T. O. Brun, Evidence for H(D) ordering in $PdH_x(PdD_x)$, Phys. Rev. Letters, 42: 457 (1979).

65.   O. Blaschko, R. Klemencic, P. Weinzierl and O. J. Eder, (1 1/2 0) superlattice reflection in $β-PdD_x$, J. Phys. F. 9: L 113 (1979).

66.   I. S. Anderson, C. J. Carlile, D. K. Ross and D. L. T. Wilson, Interstitial interactions in the Pd-H(D) system, Z. Phys. Chem. (N.F.), 115: 165 (1979).

67.   M. H. Mueller, T. O. Brun, R. L. Hitterman, H. W. Knott, C. B. Satterthwaite and T. E. Ellis, Deuteron density modulation of D atoms in PdD_{1-x}, Modulated Structures-1979. Proc. AIP Conf. No. 53: 391, New York (1979).

68.   T. Kajitani, private communication.

69.   H. Kanzaki, Point defects in face-centered cubic lattice, J. Phys. Chem. Solids, 2: 24 (1957).

70.   H. Goldberg and F. D. Manchester, Low temperature ordering in $PdH_x(D_x)$ alloys, Phys. Letters, 68A: 360 (1978).

71.   D. de Fontaine, k-space symmetry rules for order-disorder reactions, Acta Met., 23: 553 (1975).

72.   P. C. Clapp and S. C. Moss, Correlation functions of disordered binary alloys, I, Phys. Rev., 142: 418 (1966). II, III, ibid, 171: (1978) 754, 764.

73.   C. B. Satterthwaite, T. E. Ellis and R. J. Miller, On the nature of the 50 K anomaly in Pd-H, Transition Metals-1977, Proc. AIP Conf., No. 39: 501, New York (1977).

74.   R. J. Miller and C. B. Satterthwaite, Electronic model for the reverse isotope effect in superconducting Pd-H(D), Phys. Rev. Letters, 34: 144 (1975).

75.   A. Fujimori, M. Ishii and N. Tsuda, Infrared and Raman spectra of non-stoichiometric cerium hydrides, phys. stat. sol. (b), 99: 673 (1980).

76.   D. G. de Groot, R. G. Barnes, B. J. Beaudry and D. R. Torgeson, Nuclear magnetic resonance evidence for the occurrence of an ordered structure in the La-H system near $LaH_{2.65}$ at 250 K, Z. Phys. Chem. (N.F.), 114: 83 (1979).

NEUTRON DIFFRACTION STUDIES OF METAL HYDRIDES

Hajime Asano and Makoto Hirabayashi*

Institute of Materials Science, University of Tsukuba
Ibaraki 305, Japan
*The Research Institute for Iron, Steel and Other Metals
Tohoku University, Sendai 980, Japan

Neutron diffraction is the most powerful tool to determine the hydrogen location in metal hydrides, and many investigations have been carried out. This paper reviews recent progress in the structural studies of the Va transition metal – hydrogen system using neutron diffraction.

In ordinary neutron diffraction experiments of metal hydrides, it is common to use powdered deuteride specimens. This is partly because of the difficulties to prepare single crystal specimens and partly because the lighter isotope H has a strong incoherent scattering cross section of neutrons. In some instances, however, such substitution causes a misunderstanding of the problem due to the isotope effect; the examples are given for V-H(D) alloys and β_2-Ta_2H(D). In this connection is pointed out an important role of single crystals, by which we can study directly the hydride sample, and also obtain detailed structural information based on the diffraction data of good quality. The latest results are reported for single crystals of β-V_2D, β_1-$VH_{0.58}$, β_1-Ta_2H, and α-$VO_{0.02}D_x$ (x = 0.02 - 0.04). An example of the neutron diffraction method using the pulsed white neutron beam is also presented.

INTRODUCTION

Neutron diffraction is distinctive from X-ray and electron diffraction in the point that the atomic scattering length, b, is independent of the atomic number (see Table 1). By virtue of this character, we can determine efficiently the location of light atoms, such as H, B, C, N, and O, in the interstitial compounds of heavy elements. Especially, the metal-hydrogen system is the most

Table 1. Neutron Scattering Parameters

Element	b(10^{-13}cm)	σ_s (barns)*	σ_{inc} (barns)**
H	−3.739	81.7	79.9
D	6.672	7.63	2.2
V	−0.38	5.13	5.10
Nb	7.11	6.6	0.6
Ta	6.91	6.3	0.2

*Total scattering cross section ($\sigma_{coh} + \sigma_{inc}$)
with $\sigma_{coh} = 4\pi b^2$. 1 barn = 10^{-24} cm^2.
**Incoherent scattering cross section.

suitable target of neutron diffraction, because hydrogen is hardly visible by means of X-ray and electron diffraction.

Metals which absorb hydrogen and form metal hydrides belong to the transition elements of groups IIIa (Sc, Y, rare earth metals), IVa (Ti, Zr, Hf), and Va (V, Nb, Ta). Among these metal hydrides, the Va transition metal – hydrogen system has been studied most extensively by neutron diffraction in recent years. This is because the bcc lattice of the Va metals contains far more available inter-stitial sites than the number of the dissolved hydrogen atoms, and there appear many varieties of ordered hydrogen configuration depending on the hydrogen composition and temperature. Details of these superstructures have already been described in successive reviews[1-3] of recent years. In this paper, therefore, we will survey historically the recent progress in the structural studies of the Va transition metal hydrides using neutron diffraction.

Hydrogen is known to exist in three isotopes H, D, and T. From the experimental point of view, the light isotope H is unsuitable for neutron diffraction. As shown in Table 1, the majority part (98%) of the scattering cross section of H is the incoherent one, which gives rise to a high background intensity in the diffraction pattern and makes the experiment using powdered samples practiclly impossible. For this reason, it is common to substitute D for H in the usual neutron diffraction studies. In fact, most of the superstructures in the metal-hydrogen system have been determined by powder neutron diffraction experiments using deuteride specimens. In some instances, however, such substitution causes a misunder-standing of the problem due to the isotopic difference in the crystal structures and phase relationships. The examples will be demonstrated in V-H(D) alloys and β_2-Ta$_2$H(D). In this connection is pointed out an important role of the single crystal, by which we can study directly the hydride specimen, and also obtain detailed structural information based on the diffraction data of good quality. The latest results for single crystals β-V_2D, β_1-$VH_{0.58}$, and β_1-Ta_2H will be reported.

It is also seen in Table 1 that the scattering amplitude of vanadium is very small in comparison with other elements. Therefore, neutron diffraction has little merit in obtaining information on the vanadium positions. It will be shown, however, that just the small scattering amplitude of vanadium is skilfully utilized in determining the location of deuterium in dilute α phase alloys of V-D and V-O-D. An example of the neutron diffraction method using the pulsed white neutron beam is also presented.

GENERAL ASPECTS

The Va transition metals dissolve hydrogen in the interstitial sites of the metal matrix and form nonstoichiometric hydrides over a wide composition range. The phase diagram of these M-H systems generally shows a primary solid solution α at high temperatures up to the composition H/M ~ 1 and a low-temperature phase β in the range $MH_{0.5}$ - MH. The α phase keeps the bcc metal lattice, in which hydrogen atoms are distributed randomly in the tetrahedral sites (T-sites). The β phase is distinguished from α by the slightly deformed metal structure, which is body-centered tetragonal (bct) with the axial ratio c/a ~ 1.1 in the V-H system[4] and monoclinic with a = b, c/a = 1.01, and γ ~ 90.5° in the systems Nb-H[5] and Ta-H.[6] The distortion of the metal lattice in the β phase strongly suggests the hydrogen ordering in the specific interstitial sites, which has motivated the neutron diffraction studies on the β phase alloys.

EARLY INVESTIGATION

The first attempt of neutron diffraction experiments on the Va metal - hydrogen system was made by Roberts[7] as early as 1955. He studied the vanadium deuteride V_4D_3 by means of X-ray and neutron diffraction and reported that the metal lattice of V_4D_3 was bcc down to the liquid nitrogen temperature and the deuterium ordering took place below -66°C. The unit cell of the superstructure was reported as 2a×2a×2a (a is the bcc vanadium lattice parameter), but the ordered deuterium configuration was not determined.

In 1961, Wallace and his colleagues studied the order-disorder phenomenon of hydrogen (deuterium) in $Ta_2H(D)$ by means of heat capacity measurements[8] and neutron diffraction.[9] The specific heat curve indicated that Ta_2H underwent stepwise phase transformations β_1-β_2-β_3-α at 33, 59, and 60°C. The neutron diffraction patterns for β_1-$TaD_{0.52}$ obtained at 4 K, 78 K, and room temperature were indexed by the superlattice unit cell of $2a_0 \times 2a_0 \times 2a_0$, where a_0 is

the pseudocubic metal lattice parameter.* In order to determine the
deuterium configuration, it is necessary to locate 8 deuterium atoms
among 96 T-sites. The number of manners for such configuration
amounts to 10^{12}, and the laborious structure analysis was unsuccess-
ful. The diffraction pattern for the β_2 form obtained at 53°C
showed absence of the superlattice reflections seen in the β_1 form.
Accordingly, the unit cell of β_2-Ta$_2$D was the same as the tantalum
lattice, and a model of a partially ordered deuterium configuration
was reported.

In 1968, Somenkov et al.[10] determined for the first time
a superstructure of β-NbD shown in Fig. 1. The structure is described by an orthorhombic unit cell of A = B = $\sqrt{2}a_0$ and C = a_0, and
the deuterium atoms occupy the specific T-sites 2(a): 1/4 1/4 1/4,
3/4 3/4 3/4 and 2(b): 3/4 1/4 1/4, 1/4 3/4 3/4 of space group Pnnn.
By further extensive neutron diffraction studies, Somenkov and his
colleagues identified superstructures δ-VD[11] and δ-TaD[12] to be isomorphic with β-NbD, and revealed other ordered structures β-V$_2$D,[13]
ε-Nb$_4$D$_3$,[14] β_1-Ta$_2$D,[15] and γ-Ta$_4$D$_3$.[16] These results have been summarized in a review by Somenkov.[17]

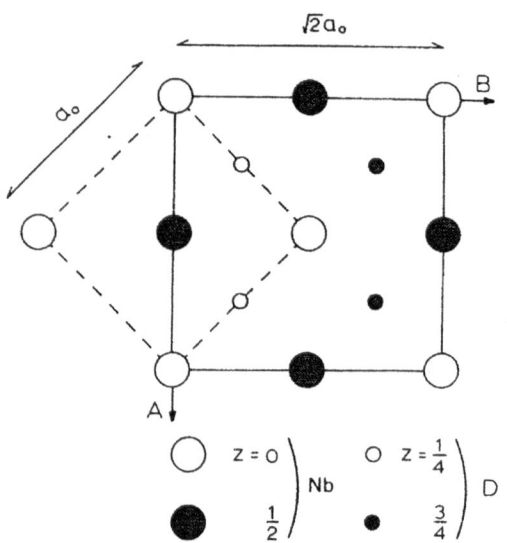

Fig. 1. Superstructure of β-NbD. Dotted lines show the pseudocubic
metal lattice of the parameter a_0. By later investigations,
δ-VD and δ-TaD were shown to be isomorphic with β-NbD.

---

*Although the metal lattice in the β phase of Nb-H and Ta-H alloys
is slightly deformed to monoclinic, limited resolution of powder
neutron diffraction cannot detect the distortion, and the metal
lattice is regarded as pseudocubic with the lattice parameter a_0.

RECENT RESULTS AND DISCUSSION

Interpretation for the Result of Roberts

The observation by Roberts[7] on V_4D_3 was inexplicable by the knowledge about the V-H system[4] in those days. The V-H alloys undergo the phase transformation into β of the bct metal lattice due to the hydrogen ordering below about 200°C, while V_4D_3 keeps the bcc metal lattice down to the liquid nitrogen temperature and the deuterium ordering takes place below -66°C. The problem was solved in 1972-1974 by the discovery[18-21] that V-H and V-D systems show quite different phase relationships as shown in Fig. 2. The observation by Roberts is fully consistent with the V-D phase diagram of Fig. 2. The metal lattice of V_4D_3 is bcc in the whole temperature range of α, δ, and γ phases, and the deuterium ordering observed at -66°C corresponds to the transformation from α to δ. The deuterium arrangement in δ-VD[11,20] has been clarified to be isomorphic with β-NbD of Fig. 1, where the T-sites, 2(a) and 2(b), are occupied by deuterium with the probability of 0.75 at the composition of V_4D_3.

The V-D phase diagram also shows that the nonstoichiometric δ phase undergoes further ordering into γ at lower temperatures. The neutron diffraction pattern[20] on the γ phase alloy shows additional

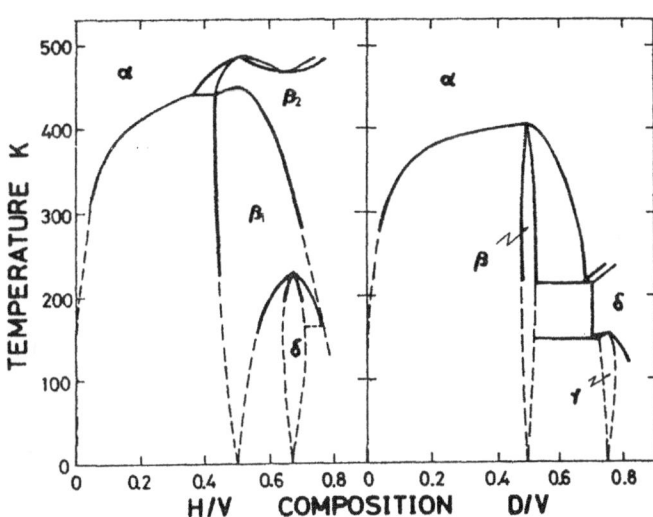

Fig. 2.   Comparison of the phase diagrams in the V-H and V-D systems.[22] The V-D system contains three ordered phases β-V_2D, γ-V_4D_3, and δ-VD. The metal lattice is bcc in the α, γ, and δ phases, and bct in the β phase. The V-H phase diagram is a revised one, in which the so-called β phase is divided into three ordered phases β_1, β_2, and δ. Each ordered phase possesses the bct metal lattice and is distinguished only by the hydrogen configuration in the interstitial sites.

superlattice reflections besides the reflections of the δ-VD structure, indicating an occurrence of full ordering among deuterium atoms
and vacancies in the 2(a) and 2(b) sites. The analysis of the diffraction pattern leads to a superstructure γ-V_4D_3 shown in Fig. 3.
The superlattice unit cell has a lattice parameter twice that of
δ-VD along the C-axis and the deuterium atoms are fully ordered in
three quarters of the 2(a) and 2(b) sites of the δ-VD structure.
Similar superlattice formations into ε-Nb_4D_3 and γ-Ta_4D_3 (Fig. 3)
take place in the nonstoichiometric β-NbD and δ-TaD phases at lower
temperatures.

Superstructure of β-V_2D

The structure of β-V_2D shown in Fig. 4 was determined independently by three research groups[13,19,20] by means of powder neutron
diffraction. In contrast to the T-site occupation in δ-VD and
γ-V_4D_3, the deuterium atoms in β-V_2D are ordered in the specific
octahedral sites (Oz_1) between two neighbouring vanadium atoms along
the metal lattice c-axis.

The model of Fig. 4, however, was insufficient to explain the
intensity data[13,20] obtained at room temperature, and some structure
modification seemed necessary. Somenkov et al.[13] located 95% of the
deuterium atoms in the Oz_1-sites and the remainder in the specific
T-sites in an ordered manner. The present authors[20] interpreted the
intensity data in terms of the order parameter in the β-α transformation. It is assumed that the atomic arrangement shown in Fig. 4 is
realized at 0 K. At room temperature (T = 0.73 Tc, where Tc is

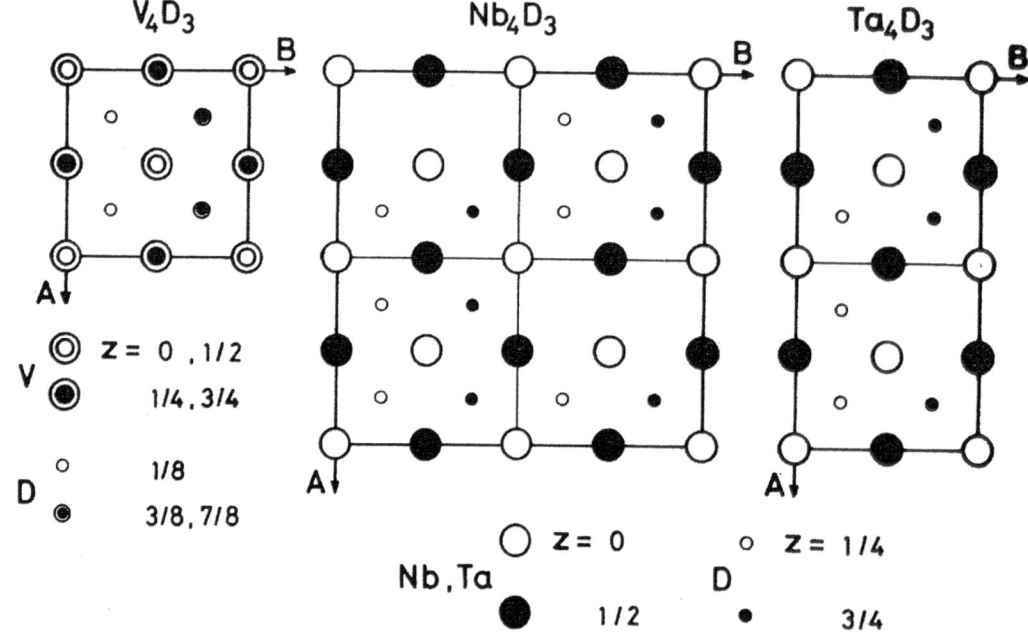

Fig. 3. Superstructures of γ-V_4D_3,[20] ε-Nb_4D_3,[14] and γ-Ta_4D_3.[16,23]

the β–α transition point of 406 K), substantial decrease of the order parameter is expected, and some fractions of the deuterium atoms occupy randomly the T-sites as in the disordered α phase. The observed intensity data were compared with the calculation taking the degree of order, or the occupation probability of deuterium in the Oz_1-site, as an adjustable parameter. Fig. 5 shows a variation of R-factor with the order parameter. The R-factor is minimized at the order parameter of 0.85, indicating that 85% of the deuterium atoms occupy the Oz_1-site and the remainder is distributed randomly in the T-site.

Quite recently, Okada et al.[24,25] carried out a single crystal neutron diffraction study on β-V_2D. The sample was prepared by loading deuterium from the gas phase into a vanadium single crystal. Because of the cubic-tetragonal metal lattice transition from α to β,

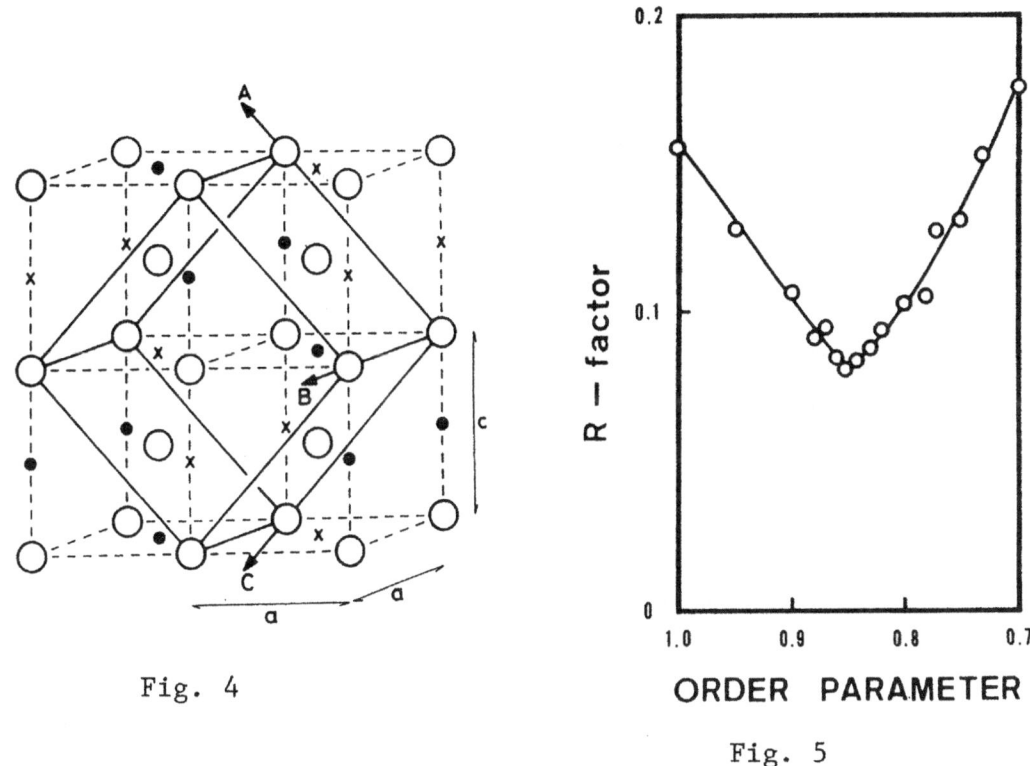

Fig. 4   ORDER PARAMETER

                                          Fig. 5

Fig. 4.   Superstructure of β-V_2D. Vanadium atoms are shown by open circles, forming the bct metal lattice depicted by dotted lines. The z-type octahedral sites are shown by full circles (Oz_1) and crosses (Oz_2), the former being occupied by deuterium in β-V_2D.

Fig. 5.   Variation of R-factor with the order parameter. R-factor is defined as $\Sigma|I_{obs} - I_{cal}|/\Sigma I_{obs}$.

the obtained crystal consisted of multiple β domains, whose metal lattice c-axis was nearly parallel to the cubic axes of the original single crystal. The neutron diffraction data were collected on one dominant domain using a four-circle diffractometer, where irrational reflections from other domains were eliminated by the collimator of good resolution. After the correction for the absorption effect and the Lorentz factor, the intensity data were subjected to the Fourier analysis. The nuclear density distribution $\rho(XYZ)$, which is modified by the atomic scattering amplitude, is formulated by

$$\rho(XYZ) = \frac{1}{V} \sum_{HKL}\sum\sum F_{HKL}\, e^{-2\pi i(HX+KY+LZ)},$$

where V is the volume of the unit cell. Fig. 6 shows a Fourier projection of β-V_2D onto the (010) plane, which is constructed from 31 H0L-type reflections by the relation

$$\rho(XZ) = \frac{1}{S} \sum_{HL}\sum F_{H0L}\, e^{-2\pi i(HX+LZ)},$$

where S is the area of the projected unit cell. It is evident that the density maxima represent the deuterium atoms at the Oz_1-sites,

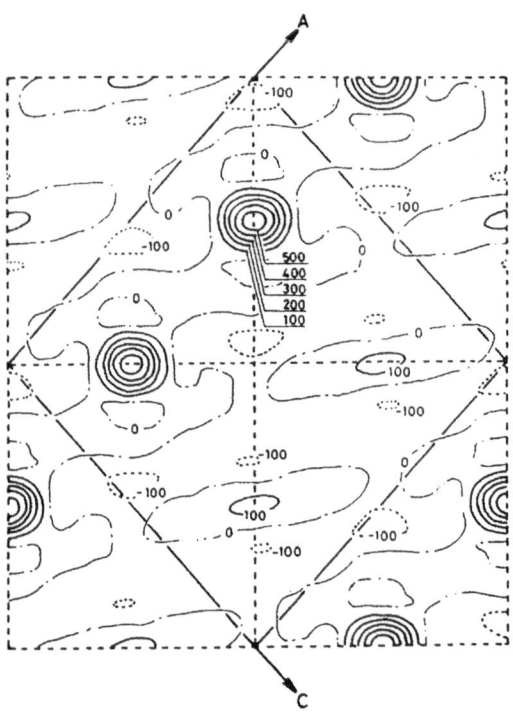

Fig. 6.   Fourier projection of β-V_2D on (010). Contours are drawn
          at equal intervals in arbitrary units; full, dotted, and
          broken lines indicate positive, negative, and zero contours,
          respectively.

which are slightly elongated along the [101] direction. A small
fraction of the deuterium atoms exists also in the Oz_2-sites, while
vanadium atoms do not show any sharp peaks because of the nearly
null scattering amplitude. Fig. 7 shows a section of the three-
dimensional density map, where it is seen that the density maximum
of deuterium at the Oz_1-site is stretched along the B-axis over the
neighbouring T-sites and Ox-sites.

Fig. 8 shows another Fourier section for the plane (a). The
elongation of the deuterium density at the Oz_1-site along the B-axis
is consistent with the result of Fig. 7, and bulges can also be seen
in the perpendicular direction towards the T-sites. On the basis
of these results, we can conclude that the deuterium atoms in β-V_2D
are distributed like a string of beads passing through the interstitial sites Oz_1-T-Ox-T-Oz_1 along the B-axis, and each bead is
slightly compressed in the direction of the metal lattice c-axis and
has bulges along the orthogonal direction.

The above intensity data were analyzed also by the least squares
refinement technique to find out quantitative occupation parameters

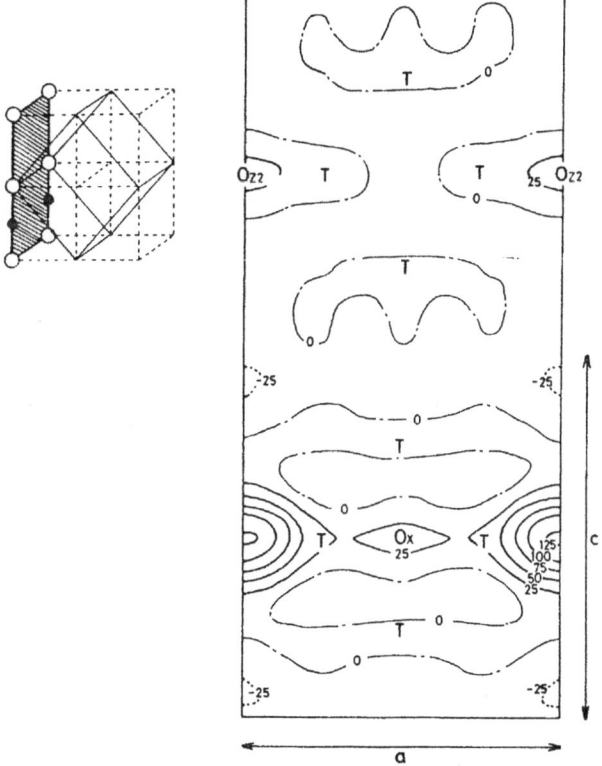

Fig. 7. A section of the three-dimensional Fourier map corresponding
to the hatched plane in the left figure, in which vanadium
and deuterium atoms are indicated by open and full circles,
respectively.

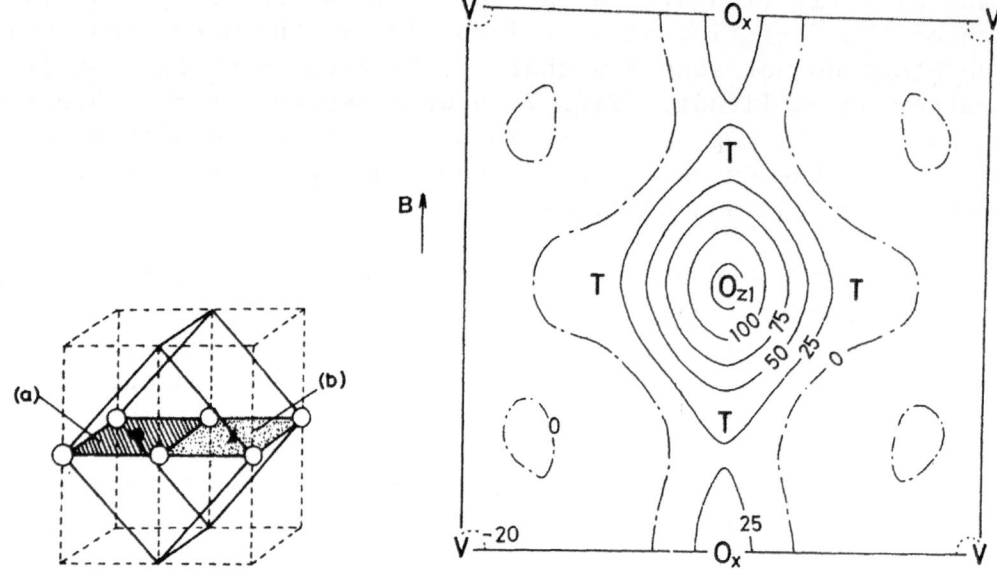

Fig. 8.   A Fourier section for the plane (a) in the left figure, in
which Oz_1- and Oz_2-sites are indicated by the full circle
and the cross, respectively.

of deuterium in the interstitial sites. The following were adopted
as the fitting parameters: the deuterium occupation fraction P_1 for
the Oz_1-site, P_2 for the Oz_2-site, Px for the Ox-site on the line
of Oz_1-T-Ox-T-Oz_1, and P_T for the quartet of T-sites surrounding the
Oz_1- and Oz_2-sites shown in Fig. 8 ($P_1 + P_2 + Px + P_T = 1$).
Anisotropic Debye-Waller parameters for the O-site deuterium, Bx,
By, and Bz, were also treated as fitting parameters, where suffixes
x, y, and z indicate respectively the [101], [010], and [10$\bar{1}$]
directions referred to the principal axes of the bct metal lattice.
Debye-Waller parameters for vanadium (0.8 A^2) and T-site deuterium
(0.9 A^2), and the static vanadium atom displacement[26] $u_z = 0.12$ A
were used as the known parameters. The observed structure factors
$|Fo|$ were compared with the calculations $|Fc|$ by the least squares
fitting method and the minimum R-factor $= \Sigma||Fo| - |Fc||/\Sigma|Fo|$
$= 0.078$ was obtained when $P_1 = 0.84$, $P_2 = 0.03$, Px $= 0.01$, and
$P_T = 0.12$, and Bx $= 4.9$, By $= 6.2$, and Bz $= 1.0$ A^2. The present
result is consistent with the above-described Fourier map,* and also
supports the occupation parameter obtained in Fig. 5.

---

*The Fourier maps of Figs. 6-8 have been constructed from finite
 numbers of experimentally observed structure factors, although the
 true density distribution should be determined by all sets of F_{HKL}.
 Reconstructed density maps by using 388 structure factors
 calculated from the above best-fitted parameters agree with Figs.
 6-8, and also show a more remarkable contraction of the deuterium
 density in the direction of the metal lattice c-axis.[25]

Superstructures in the V-H System

It is evident from Fig. 2 that the results of neutron diffraction on vanadium deuterides are not always applicable to vanadium hydrides, and separate inquiry is necessary. As seen in Table 1, both hydrogen and vanadium scatter neutrons for the most part incoherently, and the neutron diffraction experiment without a single crystal seems impossible. As the first clew to the problem, however, we performed a powder neutron diffraction study[27] directly on a hydride sample V_2H to make sure whether β_1-V_2H and $\beta-V_2D$ are isomorphic or not. The result is given in Fig. 9, and we can barely observe several reflections in spite of the poor signal-to-noise ratio. The reflections 001 and (110, $\bar{1}10$) are the superlattice reflections characteristic of the $\beta-V_2D$ structure. The reflections indexed as (002, 200, $11\bar{1}$, $\bar{1}11$) and (111, $1\bar{1}1$) are the fundamental reflections corresponding to 101 and 110 of the bct metal lattice, respectively. From the extinction rule of reflections and the intensity data, we can conclude that β_1-V_2H is isomorphic with $\beta-V_2D$.

In contrast with the $\beta-V_2D$ phase, the β_1 hydride phase has a wide homogeneity range in both sides of the composition V_2H (see Fig. 2). It has been proposed by the present authors[28] that the occupation site of hydrogen in excess of the composition V_2H is the

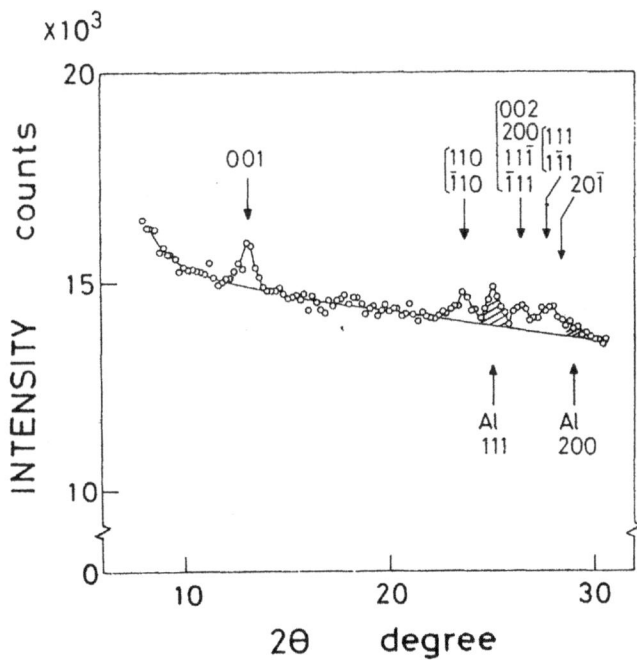

Fig. 9. Powder neutron diffraction pattern of β_1-V_2H. The indexing is given in terms of the monoclinic superlattice cell of $\beta-V_2D$ shown by solid lines in Fig. 4. Hatched reflections come from an aluminum sample holder of 0.1 mm thick.

Oz_2-site. Quite recently, we carried out a neutron diffraction study on a single crystal β_1-$VH_{0.58}$. Although the sample was composed of multiple β_1 domains as in the case of β-V_2D, we chose one dominant domain and gathered intensity data by a four-circle diffractometer. Fig. 10 shows a series of 00L reflections, where even and odd indices of L represent fundamental and superlattice reflections, respectively. First of all, our attention should be paid to an excellent signal-to-noise ratio in comparison with the powder pattern of Fig. 9. The obtained diffraction data were proved to be consistent with the extinction rule for the β-V_2D superstructure. The intensity data were analyzed both by the Fourier synthesis method and by the least squares fitting technique as in the study of β-V_2D. Fig. 11 shows a Fourier projection $\rho(YZ)$ of β_1-$VH_{0.58}$ onto the (100) plane constructed from 22 0KL-type reflections. The result is analogous to the deuterium distribution in β-V_2D, and intense negative peaks correspond to the hydrogen atoms at the Oz_1-sites. The Oz_2-site occupation of the excess hydrogen atoms is not clear in this figure, but the least squares fitting analysis gives the following occupation parameters: $P_1 = 1.00$, $P_2 = 0.05$, and $P_T = 0.11$ ($P_1 + P_2 + P_T = 2\times0.58$). It is concluded from these results that β_1-V_2H_1+x is isomorphic with β-V_2D, and the hydrogen atoms in excess of the V_2H composition occupy the Oz_2-sites in accordance with our previous proposal.[28]

The structure of the β_2 phase was studied in connection with the β_1-β_2 phase transformation. According to the phase diagram, the β_2 phase appears as the high-temperature form of β_1 near the composition of V_2H. The neutron diffraction experiments at elevated

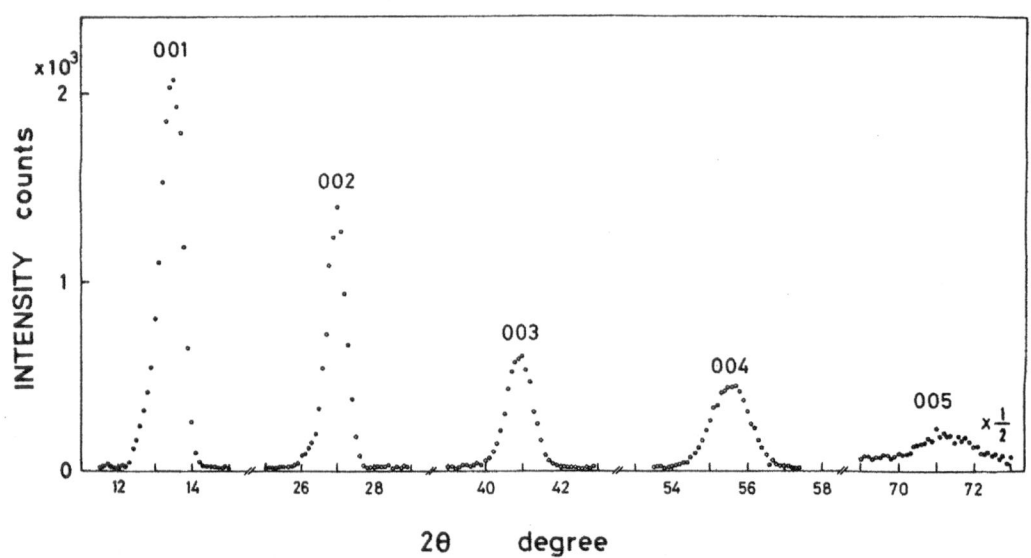

Fig. 10. Diffractometer scanning of the 00L reflections for the single crystal β_1-$VH_{0.58}$.

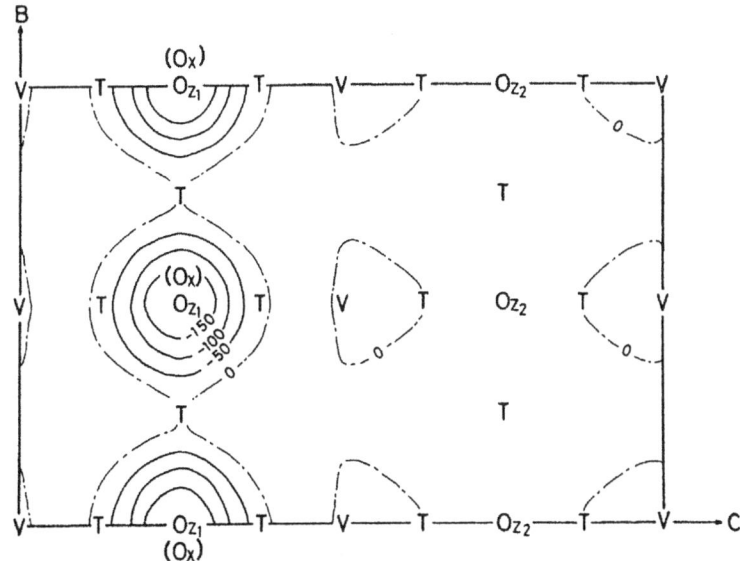

Fig. 11. Fourier projection of β_1-$VH_{0.58}$ on (100).

temperatures on the powdered sample V_2H[27] and the single crystal $VH_{0.58}$ indicate that the intensity of the superlattice reflections characteristic of the β_1 form decreases with increasing temperature and disappears at the β_1-β_2 transition as seen in Fig. 12. The structure factor for these superlattice reflections is written as $b_H(P_1-P_2)$, where b_H is the atomic scattering amplitude of hydrogen, and P_1 and P_2 are occupation probabilities of hydrogen at the Oz_1- and Oz_2-sites, respectively. The decrease of intensity with

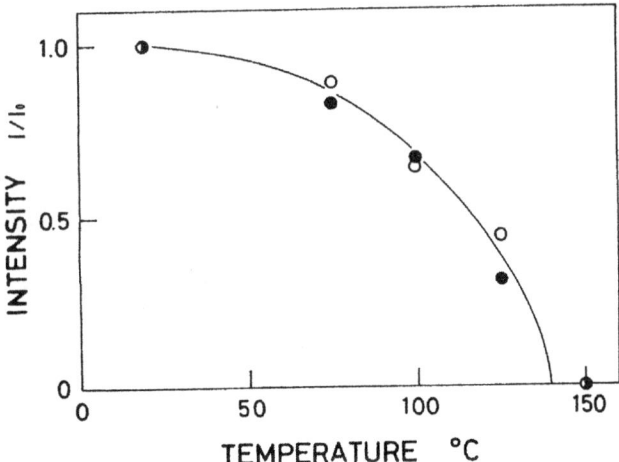

Fig. 12. Temperature dependence of the superlattice reflection intensities for the single crystal $VH_{0.58}$. Open and full circles correspond to the 001 and 003 reflections shown in Fig. 10, respectively, and I_0 is the intensity at 20°C.

temperature indicates the hydrogen rearrangement from the Oz_1-site
to the Oz_2-site, and no superlattice reflection in the β_2 phase means
that Oz_1- and Oz_2-sites are occupied by hydrogen with the equal
probability. It is clear, therefore, that the stoichiometric formula
of the β_2 phase is VH, and all the Oz-sites are occupied by hydrogen
to form a superstructure β_2-VH shown in Fig. 13.

In Fig. 13 is also shown a superstructure δ-V_3H_2 proposed by
the present authors[29] on the basis of the electron diffraction
patterns of Wanagel et al.[30] The unit cell is drawn by thick lines,
in which one third of the Oz-site plane parallel to the (101) metal
lattice plane is regularly missing. The confirmation of the structure by neutron diffraction, however, has not yet been made.

Superstructure of $Ta_2H(D)$

The structure of β_1-Ta_2D remained unknown since the early
investigation by Wallace.[9] From the analogy of the δ-TaD superstructure of Fig. 1, Somenkov et al.[12] proposed a structure model
of β_1-Ta_2D, which was obtained by subtracting two deuterium atoms
at the T-sites (3/4 1/4 1/4, 3/4 3/4 3/4) from the δ-TaD structure.
Calculated intensity data using this model agreed well with the
experimental ones given by Wallace.[9] Subsequent powder neutron
diffraction by the same authors[15] confirmed their proposal,[12] and
also revealed a static displacement of the tantalum atom due to the
deuterium ordering. In Fig. 14, the tantalum atoms are shifted from
the normal positions as shown by arrows, and the displacement Δ has
been estimated from the intensity data as 0.012 and 0.017 at room
and liquid nitrogen temperatures, respectively.

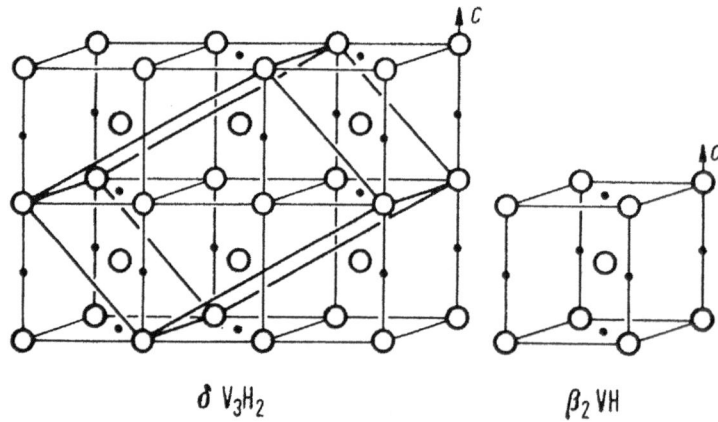

δ V_3H_2 β_2 VH

Fig. 13.   Superstructures of δ-V_3H_2 and β_2-VH. Open and full
           circles represent vanadium and hydrogen atoms, respec-
           tively. The bct metal lattice is shown by thin lines,
           and the c-axis is indicated by arrows.

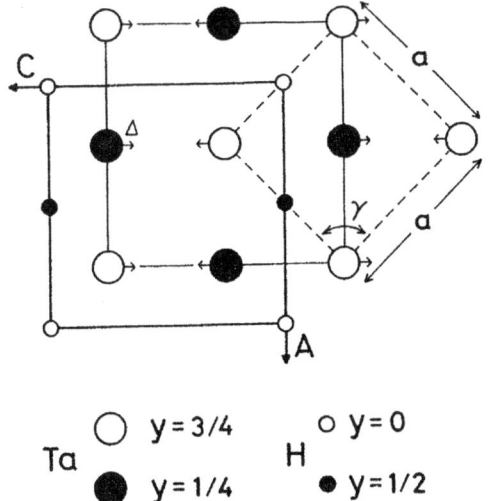

$$Ta \quad \bigcirc \ y = 3/4 \qquad H \quad \circ \ y = 0$$
$$\bullet \ y = 1/4 \qquad \bullet \ y = 1/2$$

Fig. 14.   Superstructure of β_1-$Ta_2H(D)$. The structure was studied
also by electron diffraction[31] and single crystal X-ray
diffraction,[32] which determined the relative orientation
between the superlattice (solid lines) and the monoclinic
metal lattice (dotted lines).

Recently, we reinvestigated the structure of β_1-Ta_2H by neutron
diffraction using a single crystal.[33] The method of sample preparation is the same as in β-V_2D and β_1-$VH_{0.58}$. The specimen has been
decomposed into multiple β_1 domains through the ordering process,
and the B-axis of the superlattice, which is identical with the
c-axis of the monoclinic metal lattice, has six orientations parallel
to the cubic axes of the original single crystal. The obtained
extinction rule of reflections is consistent with the β_1-Ta_2D superstructure, indicating that the hydride and the deuteride are isomorphic. The intensity measurements were made only on superlattice
reflections* from one domain. The structure factor of the HOL-type
superlattice reflections is written as

$$F = 2b_H \pm 4b_{Ta} \sin 2\pi L\Delta,$$

where the second term comes from the tantalum atom displacement,
and the signs $\pm$ depend on the reflection index. In the case of
X-ray diffraction, the second term is dominant, and the superlattice
reflections with high index of L become detectable.[32] In the case
of neutron diffraction, the first and the second terms are in the
same order of magnitude. Depending on the signs $\pm$ in the structure
factor, two terms are summed up or cancelled to give an oscillatory
intensity distribution as shown in Table 2. It is also seen that

---

*Measurements on the fundamental reflections give superposed
 information from domains with six orientations.

Table 2. Comparison of $|Fo|$ and Fc with $\Delta = 0.016$

HKL	F	Fc	$|Fo|$
001	$2b_H - 4b_{Ta}\sin 2\pi\Delta$	-10.1	10.8
201	$2b_H + 4b_{Ta}\sin 2\pi\Delta$	-4.4	5.4
401	$2b_H - 4b_{Ta}\sin 2\pi\Delta$	-8.8	9.1
601	$2b_H + 4b_{Ta}\sin 2\pi\Delta$	-2.5	4.0
003	$2b_H + 4b_{Ta}\sin 2\pi 3\Delta$	1.2	1.3
203	$2b_H - 4b_{Ta}\sin 2\pi 3\Delta$	-14.3	15.2
403	$2b_H + 4b_{Ta}\sin 2\pi 3\Delta$	1.9	1.7
005	$2b_H - 4b_{Ta}\sin 2\pi 5\Delta$	-18.0	18.7
205	$2b_H + 4b_{Ta}\sin 2\pi 5\Delta$	6.7	6.7
405	$2b_H - 4b_{Ta}\sin 2\pi 5\Delta$	-16.4	12.6

the oscillation becomes remarkable with increasing index of L. The observed structure factors $|Fo|$ were compared with the calculated ones $|Fc|$, taking Δ as an adjustable parameter, and the minimum R-factor of 0.11 was obtained when $\Delta = 0.016$.

As for the structure of β_2-Ta$_2$D, Wallace[9] reported that the superlattice reflections seen in the β_1 phase were absent at 53°C in the sample TaD$_{0.52}$. He proposed a model of a partially ordered deuterium configuration in a pseudocubic unit cell of a_0. As indicated in Fig. 15, however, the β_2 phase of Ta$_2$D, and of TaD$_{0.52}$ as well, seems to exist only in a very narrow region, and it is not evident whether the neutron diffraction result of Wallace is really representative of the β_2 form. For this reason, we performed a neutron diffraction experiment directly on a powdered hydride sample Ta$_2$H above room temperature up to 67°C. The room temperature result is consistent with the superstructure of β_1-Ta$_2$H. Fig. 16 shows the result at 50°C, where the superlattice reflections still exist even in the β_2 phase; the reflection 1/2 1/2 0, which corresponds to 001 for the β_1-Ta$_2$H structure, remains unchanged through the β_1-β_2 transformation, while another one seen at $2\theta = 19.5°$ cannot be indexed with an appropriate unit cell. These two superlattice reflections were proved to be intrinsic for the β_2 form, because they disappeared at the β_2-α transition point of 59°C. At the present stage, however, further discussion seems impossible on the basis of the powder pattern with the poor signal-to-noise ratio, and the structure is still an open question. Neutron diffraction using a single crystal is highly desirable.

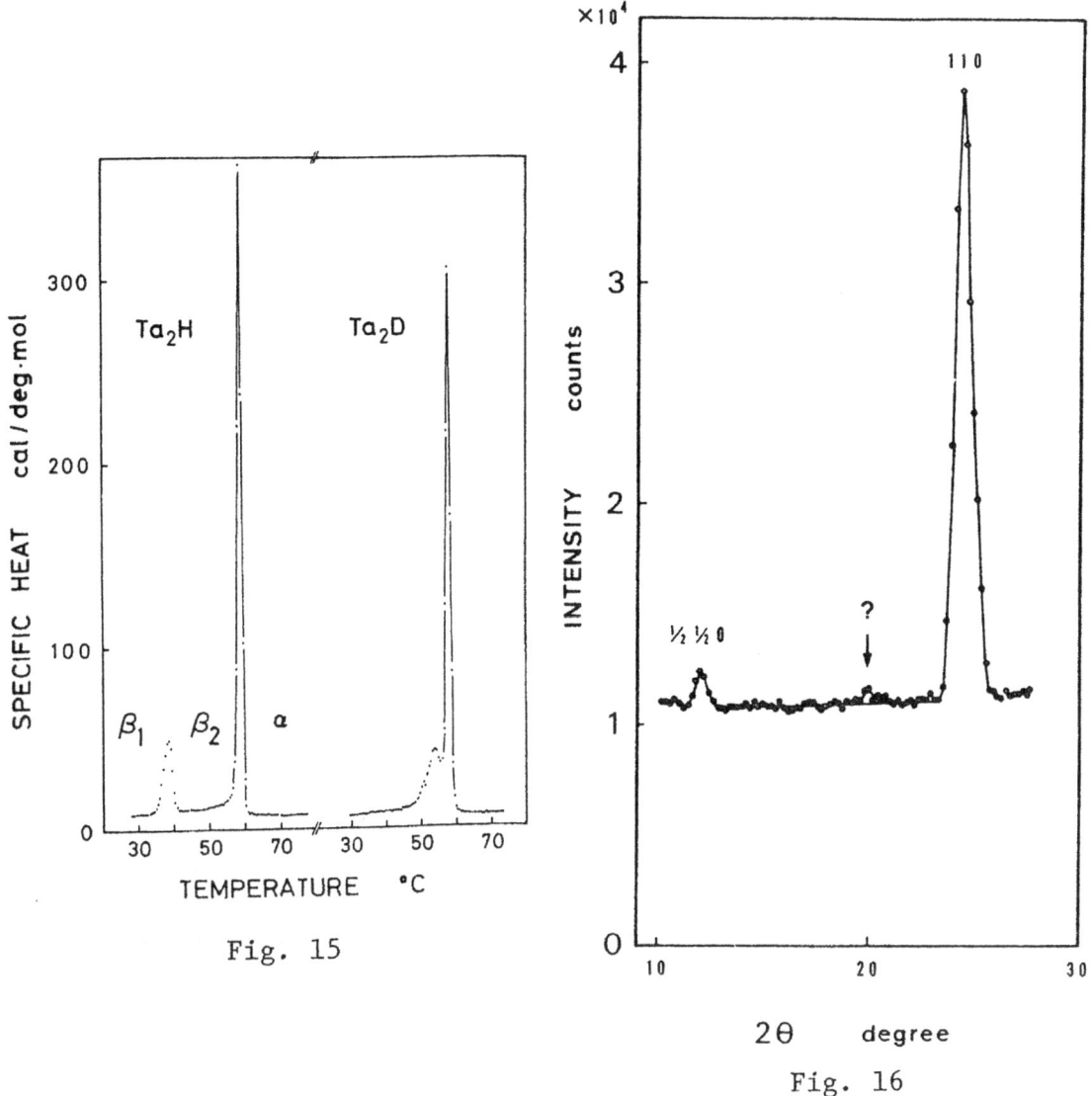

Fig. 15

2θ   degree

Fig. 16

Fig. 15.   Specific heat curves of Ta_2H and Ta_2D to demonstrate
a remarkable isotopic difference in the β_1-β_2-α transformation.

Fig. 16.   Powder neutron diffraction pattern of β_2-Ta_2H at 50°C.
The indexing is based on the pseudocubic metal lattice,
and 110 is the fundamental reflection.

Location of Deuterium in α Phase Alloys of V-D and V-O-D

The lattice location of low concentration interstitial
impurities in metals has been successfully determined by the ion
channeling technique.[34-36] With the aid of the small scattering
length of vanadium, the same information can be obtained also by
neutron diffraction using single crystals. Table 3 shows structure

Table 3. Structure factors of VI_x

hkl	T-site	O-site
110	$b_V - \frac{x}{3} b_I$	$b_V - \frac{x}{3} b_I$
200	$b_V + \frac{x}{3} b_I$	$b_V + x\, b_I$
211	$b_V + \frac{x}{3} b_I$	$b_V - \frac{x}{3} b_I$
220	$b_V - \frac{x}{3} b_I$	$b_V + x\, b_I$
310	$b_V - \frac{x}{3} b_I$	$b_V - \frac{x}{3} b_I$
222	$b_V - x\, b_I$	$b_V + x\, b_I$
321	$b_V + \frac{x}{3} b_I$	$b_V - \frac{x}{3} b_I$
400	$b_V + x\, b_I$	$b_V + x\, b_I$

factors assuming a random occupation of interstitial atoms, I, in the T- or O-sites of bcc vanadium, where b_I is the scattering length of the interstitial atom and x is the concentration. Because b_V and x are small, the first and the second terms in the structure factors are comparable in their order of magnitude. Accordingly, the occupation of T- or O-sites can be clearly distinguished from the intensity of reflections other than 110, 310, and 400. Actually, single crystal neutron diffraction studies clarified the O-site occupation of carbon, nitrogen, and oxygen in vanadium[37] and the T-site occupation of deuterium in $VD_{0.01}$.[38]

The location of deuterium in vanadium containing oxygen impurities was also studied by neutron diffraction using single crystals of $VO_{0.02}D_x$ (x = 0.02, 0.03, and 0.04).[22] The location of the oxygen atoms in $VO_{0.02}Dx$ has been determined in advance to be the O-site by the ion channeling technique on the same crystal.[36] The neutron diffraction intensities do not agree with models of deuterium occupation purely in either T-sites or O-sites, but mixed occupation of both T- and O-sites seems to satisfy the experimental results. Therefore, the observed intensity data are fitted with calculations taking the occupation probability P of the O-sites as an adjustable parameter. Table 4 shows a comparison of the observed structure factors with the calculations for the models of P = 0, 0.5, and 1 in the alloys $VO_{0.02}D_{0.04}$, and the best agreement is obtained when P = 0.5. Similarly, P = 0.1 and 0.2 were determined for the alloys of x = 0.02 and 0.03, respectively. This phenomenon may be

Table 4. Comparison of $|Fo|$ and $|Fc|$ of $VO_{0.02}D_{0.04}$

hkl	$|Fo|$	$P = 0$	$|Fc|$ 0.5	1
110	6.26	6.29	6.29	6.29
200	2.01	2.88	2.01	1.14
211	5.70	4.55	5.42	6.29
220	2.85	4.62	2.88	1.14
222	3.69	6.36	3.75	1.14
400	1.14	1.14	1.14	1.14

connected with the fact that the terminal solubility of deuterium
in vanadium increases with the addition of oxygen, which has been
interpreted in terms of the trapping effect.[39]

Neutron Diffraction Method Using Pulsed White Neutrons

In ordinary neutron diffraction experiment using a nuclear
reactor, we radiate a monochromatic neutron beam of wavelength λ on
a sample and measure scattered neutrons as a function of the
scattering angle 2θ by scanning a detector, which determines the
diffraction intensity of the spacing d in the Bragg's law $2d\sin\theta = \lambda$.
An alternative is also possible when we radiate white neutrons on
the sample and analyze the wavelength of scattered neutrons by
a counter of fixed angle. A facility for the latter method was
installed by Kimura and his colleagues[40] at the Tohoku University
Electron LINAC. Pulsed white neutrons are generated by the bombard-
ment of the electron beam from 250 MeV LINAC on the tungsten target,
run through a flight tube of about 9 m long and then hit the sample.
The scattered neutrons are counted by a detector of fixed angle
connected with a multichannel time-analyser and the neutron wavelength is determined by the time of flight (TOF) from the neutron
source to the detector. Fig. 17 shows an example of the diffraction
pattern by the TOF method on the sample β-$NbD_{0.89}$, where the
scattering angle 2θ is fixed at 158°. The peak thermal neutron flux
is of the order of 10^{12} neutrons/sec cm^2 and the exposure time is
about 8 hours. The abscissa is convertible to the wavelength of
scattered neutrons and high channel numbers correspond to the low-
angle side in the ordinary neutron diffraction pattern. The
indexing is based on the superlattice cell of β-NbD, and unmixed
and mixed indices represent fundamental and superlattice reflections, respectively. Similar patterns were taken also by the
detectors of $2\theta = 90°$ and 32°. The diffraction intensity in the
TOF method is given by the formula

$$I = ki(\lambda)\lambda^4 j|F|^2 e^{-2W}A(\lambda),$$

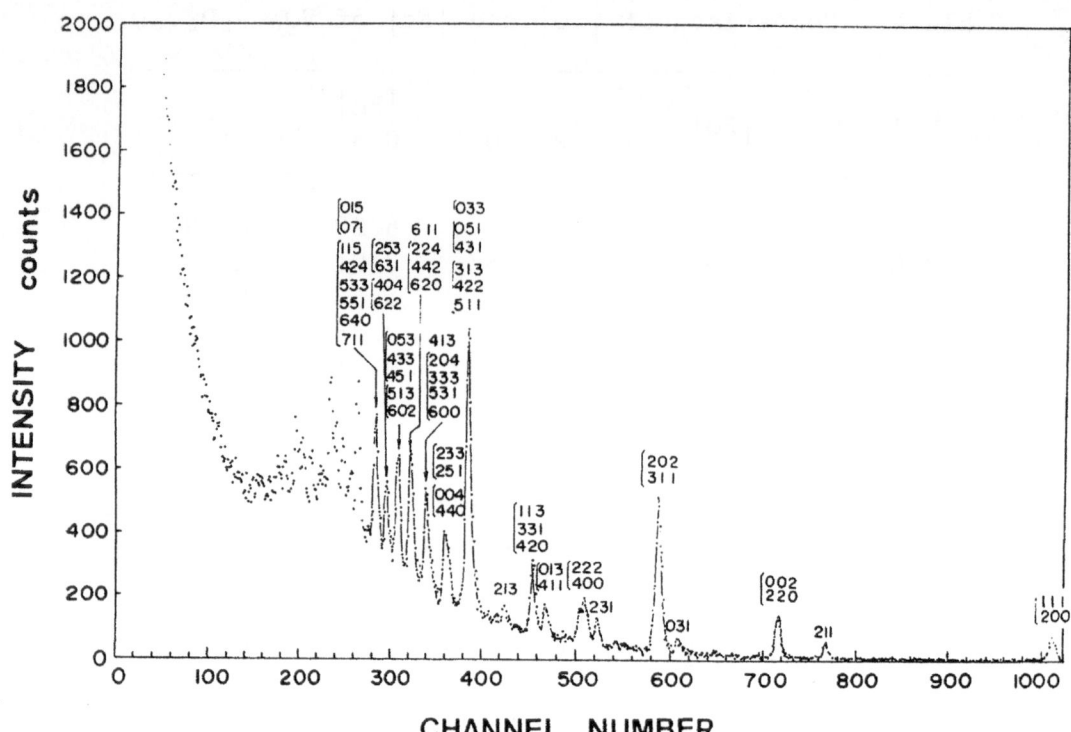

Fig. 17. Neutron diffraction pattern of β–NbD$_{0.89}$ by the TOF method.

where k is the scale factor, i(λ) the intensity spectrum of the incident neutrons depending on λ, j the multiplicity factor, e^{-2W} the temperature factor, and A(λ) the absorption factor. The incident neutron spectrum i(λ) has been determined in advance by the incoherent neutron scattering of a vanadium sample. The observed intensity data agreed well with the calculation based on the superstructure of β–NbD using the above intensity expression.

ACKNOWLEDGMENTS

     We would like to thank Mr. I. Okada for his help in the single crystal neutron diffraction work on V$_2$D and VH$_{0.58}$, Dr. S. Yamaguchi for the collaboration in the work on α–V(O$_{0.02}$)Dx, Dr. N. Watanabe for his help in the TOF neutron diffraction experiment, Dr. T. Kajitani for his critical comments on the manuscript, and Mr. M. Terada for the technical assistance. This work was supported in part by the Basic Chemical Research Fund from the Japan Society for the Promotion of Science.

REFERENCES

1. T. Schober and H. Wenzl, The systems NbH(D), TaH(D), VH(D): structures, phase diagrams, morphologies, methods of preparation, in: "Hydrogen in Metals II," G. Alefeld and J. Völkl, ed.,

Springer-Verlag, Berlin-Heidelberg-New York (1978).

2. H. Asano and M. Hirabayashi, Hydrogen ordering in Va transition metal hydrides, Z. Phys. Chem. Neue Folge 114:1 (1979).

3. V. A. Somenkov and S. S. Shil'stein, Phase transitions of hydrogen in metals, Prog. Mat. Sci. 24:267 (1980).

4. A. J. Maeland, Investigation of the vanadium-hydrogen system by X-ray diffraction techniques, J. phys. Chem. 68:2197 (1964).

5. M. A. Pick and R. Bausch, The determination of the force-dipole tensor of hydrogen in niobium, J. Phys. F: Metal Phys. 6:1751 (1976).

6. B. Stalinski, X-ray analysis and magnetic susceptibilities of tantalum hydrides, Bull. Acad. Polon. Sci. Cl. III 2:245 (1954).

7. B. W. Roberts, A neutron diffraction study of VD_x, Phys. Rev. 100:1257 (1955).

8. W. G. Saba, W. E. Wallace, H. Sandmo, and R. S. Craig, Heat capacities and the residual entropy of Ta_2H, J. chem. Phys. 35:2148 (1961).

9. W. E. Wallace, Neutron diffraction data for Ta_2D and the probable arrangement of the deuteriums in two of its polymorphic varieties, J. chem. Phys. 35:2156 (1961).

10. V. A. Somenkov, A. V. Gurskaya, M. G. Zemlyanov, M. E. Kost, N. A. Chernoplekov, and A. A. Chertkov, Neutron scattering study of structures and phase transitions in niobium hydrides and deuterides, Soviet Phys. - Solid State 10:1076 (1968).

11. A. Yu. Chervyakov, I. R. Entin, V. A. Somenkov, S. Sh. Shil'shtein, and A. A. Chertkov, Order-disorder transition in $VD_{0.8}$, Soviet Phys. - Solid State 13:2172 (1972).

12. V. A. Somenkov, A. V. Gurskaya, M. G. Zemlyanov, M. E. Kost, N. A. Chernoplekov, and A. A. Chertkov, Neutron-diffraction study of the structure and phase transformations of tantalum hydrides and deuterides, Soviet Phys. - Solid State 10:2123 (1969).

13. V. A. Somenkov, I. R. Entin, A. Yu. Chervyakov, S. Sh. Shil'shtein, and A. A. Chertkov, Abnormal phase transition in the vanadium deuteride, V_2D, Soviet Phys. - Solid State 13:2178 (1972).

14. V. A. Somenkov, V. F. Petrunin, S. Sh. Shil'shtein, and A. A. Chertkov, Neutron-diffraction study of the decomposition and ordering in the niobium-hydrogen system, Soviet Phys. - Cryst. 14:522 (1970).

15. V. F. Petrunin, V. A. Somenkov, S. Sh. Shil'shtein, and A. A. Chertkov, The structure of Ta_2D, Soviet Phys. - Cryst. 15:137 (1970).

16. V. A. Somenkov, A. Yu. Chervyakov, S. Sh. Shil'shtein, and A. A. Chertkov, Stepwise phase transitions in Ta_4D_3, Soviet Phys. - Cryst. 17:274 (1972).

17. V. A. Somenkov, Structure of hydrides, Ber. Bunsenges. Phys. Chem. 76:733 (1972).

18. K. I. Hardcastle and T. R. P. Gibb, Jr., An X-ray diffraction

investigation of the vanadium-deuterium system, J. phys. Chem. 76:927 (1972).

19. D. G. Westlake, M. H. Mueller, and H. W. Knott, Structural transitions at low temperatures in vanadium deuterides, J. appl. Cryst. 6:206 (1973).

20. H. Asano and M. Hirabayashi, Interstitial superstructures of vanadium deuterides, phys. stat. sol. (a) 15:267 (1973).

21. R. R. Arons, H. G. Bohn, and H. Lütgemeier, Comparison of the V-H and V-D phase diagrams by means of NMR, J. Phys. Chem. Solids 35:207 (1974).

22. M. Hirabayashi, Structural phase transitions of metal hydrides; V-H and V-D, Proc. JIMIS-2 Hydrogen in Metals, Suppl. Trans. Jpn. Inst. Met. 21:49 (1980).

23. H. Asano, Y. Ishino, R. Yamada, and M. Hirabayashi, Interstitial superstructures in the Ta-D system, J. Solid State Chem. 15:45 (1975).

24. I. Okada, H. Asano, and M. Hirabayashi, Refinement of the structure of β-V_2D by single crystal neutron diffraction, Proc. JIMIS-2 Hydrogen in Metals, Suppl. Trans. Jpn. Inst. Met. 21:89 (1980).

25. I. Okada, Structures and phase transitions in the V-H(D) systems (in Japanese), Master Thesis, Tohoku University (1980).

26. H. Metzger, H. Jo, S. C. Moss, and D. G. Westlake, Single crystal X-ray study of the superstructure modulation and long-range order in V_2D, phys. stat. sol. (a) 47:631 (1978).

27. H. Asano, Y. Abe, and M. Hirabayashi, A neutron diffraction study of V_2H, J. Phys. Soc. Jpn. 41:974 (1976).

28. H. Asano, Y. Abe, and M. Hirabayashi, A calorimetric study of the phase transformation of vanadium hydrides $VH_{0.06}$-$VH_{0.77}$, Acta Met. 24:95 (1976).

29. H. Asano and M. Hirabayashi, Low-temperature phase transition near V_3H_2, phys. stat. sol. (a) 16:69 (1973).

30. J. Wanagel, S. L. Sass, and B. W. Batterman, Low-temperature phase transformation in the vanadium-hydrogen system, phys. stat. sol. (a) 11:767 (1972).

31. T. Schober and H. Wenzl, An electron microscope study of the lattice distortions in ordered β-Ta_2H alloys, Scripta Met. 10:819 (1976).

32. H. Asano, Y. Ishikawa, and M. Hirabayashi, Single-crystal X-ray diffraction study on the hydrogen ordering in Ta_2H, J. appl. Cryst. 11:681 (1978).

33. H. Asano, K. Kishi, and M. Hirabayashi, Static displacement of Ta in Ta_2H studied by single crystal X-ray and neutron diffraction, Proc. JIMIS-2 Hydrogen in Metals, Suppl. Trans. Jpn. Inst. Met. 21:93 (1980).

34. M. Antonini and H. D. Carstanjen, Location of interstitial deuterium in $TaD_{0.067}$ by channeling, phys. stat. sol. (a) 34:K153 (1976).

35. J. Takahashi, M. Koiwa, M. Hirabayashi, S. Yamaguchi,

Y. Fujino, K. Ozawa, and K. Doi, A lattice location study of oxygen in vanadium by 1-MeV deuteron channeling, J. Phys. Soc. Jpn. 45:1690 (1978).

36. K. Ozawa, S. Yamaguchi, Y. Fujino, O. Yoshinari, M. Koiwa, and M. Hirabayashi, Channeling studies on the trapping of deuterium in vanadium by oxygen interstitials, Nucl. Instr. and Meth. 149:405 (1978).

37. K. Hiraga, T. Onozuka, and M. Hirabayashi, Location of oxygen, nitrogen and carbon atoms in vanadium determined by neutron diffraction, Mat. Sci. and Eng. 27:35 (1977).

38. S. Yamaguchi, private communication.

39. O. Yoshinari, M. Koiwa, H. Asano, and M. Hirabayashi, Low frequency internal friction study on vanadium-deuterium alloys, Trans. Jpn. Inst. Met. 19:171 (1978).

40. M. Kimura, M. Sugawara, M. Oyamada, Y. Yamada, S. Tomiyoshi, T. Suzuki, N. Watanabe, and S. Takeda, Neutron Debye-Scherrer diffraction works using a linear electron accelerator, Nucl. Instr. and Meth. 71:102 (1969).

DIFFUSION OF HYDROGEN IN METALS

Johann Völkl

Physik-Department
der Technischen Universität München
D-8046 Garching, W.-Germany

ABSTRACT

Many experimental methods have been applied to determine diffusion coefficients of H in metals[1]. The following methods are briefly reviewed: Permeation methods, mechanical relaxation methods (Snoek effect, Gorsky effect), electrochemical methods, magnetic disaccommodation[2], resistivity relaxation, tracer method, x-ray method, gravimetric method, nuclear magnetic resonance (NMR)[2], quasielastic neutron scattering (QNS)[2], Mössbauer effect[3], nuclear acoustic resonance[4] and a nuclear physical method[5].

The results for the diffusion coefficient of H at small concentrations are discussed for some representative metals following Ref.6. The activation energies U and preexponential factors D_O are tabulated below:

	U (eV)	D_O (cm^2/s)	
Palladium:	0.230	$2.90 \cdot 10^{-3}$	
Nickel:	0.420	$6.9 \cdot 10^{-3}$	T > Curie temp.
	0.408	$4.8 \cdot 10^{-3}$	T < Curie temp.
α-Iron:	0.044-0.105	$2.3 \cdot 10^{-3} - 4.15 \cdot 10^{-4}$	

Niobium:

Hydrogen	0.106	$5.0 \cdot 10^{-4}$	$T > 0^{\circ}$ C
	0.068	$0.9 \cdot 10^{-4}$	$T < -50^{\circ}$ C
Deuterium	0.127	$5.2 \cdot 10^{-4}$	
Tritium[7]	0.135	$4.5 \cdot 10^{-4}$	

Tantalum:

Hydrogen	0.140	$4.4 \cdot 10^{-4}$	$T > -20^{\circ}$ C
	0.040	$2.0 \cdot 10^{-6}$	$T < -75^{\circ}$ C
Deuterium	0.160	$4.6 \cdot 10^{-4}$	$T > -20^{\circ}$ C
	0.039	$1.2 \cdot 10^{-6}$	$T < -75^{\circ}$ C

Vanadium:

Hydrogen	0.045	$3.1 \cdot 10^{-4}$
Deuterium	0.073	$3.8 \cdot 10^{-4}$

The strong non-classical behavior of the hydrogen diffusion coefficient in the bcc metals V, Nb and Ta and the fcc metal Pd is stressed which manifests itself in the isotope dependence as well as in deviations from the Arrhenius relation observed in Nb and Ta. A classical behavior cannot be expected for these metals because the thermal energies are much lower than the relevant phonon energies, especially than the energies of the local hydrogen modes which are known from inelastic neutron scattering experiments[2]. Additionally, in the bcc metals V, Nb and Ta the local mode energies are in the same order or even larger than the activation energies for diffusion ruling out the classical overbarrier-jump model.

Furthermore, the influence of traps[1,8] and the influence of the structure of the host lattice on the diffusion coefficient[6] are discussed.

In the second part results are reviewed for the Fick's or chemical diffusion coefficient in Nb and Ta at large H concentrations obtained with the Gorsky-effect technique[6,8,9]. It is demonstrated how drastic interaction effects described by the so-called thermodynamic factor influence the Fick's diffusion coefficient leading, e.g., to critical slowing down. In Gorsky-effect experiments the thermodynamic factor can be determined simultaneously with the diffusion coefficient and, therefore, be eliminated easily from the measured (Fick's) diffusion coefficient; one obtains the mobility or tracer-diffusion coefficient, respectively, which is directly comparable to NMR- and QNS-results. The experimental results for the temperature and concentration dependence of the thermodynamic factor are presented and described in a simple theoretical model[8,9]. The temperature and concentration dependence of the tracer-diffusion coefficient in Nb and Ta and the concentration dependence of the activation energies and preexponential factors[6,9] are shown.

As an unique feature of H-metal systems the Fick's or chemical diffusion coefficient is dependent on the geometry of the sample[6,8,10]. It is shown that the reason for this behavior is a shape-dependence of the thermodynamic factor which can be understood within the scope of the theory of elastic interaction[2,8,10]. The elimination of this shape-dependent thermodynamic factor leads - as it must be - to a shape-independent hydrogen mobility.

In concluding remarks the theoretical situation is touched and attention is drawn to a promising theoretical approach[11].

Details about the above-mentioned topics can be found in recent review articles[1,6]. The Gorsky-effect results are summarized in Ref.8.

REFERENCES

1. J. Völkl, G. Alefeld, in:"Diffusion in Solids, Recent Developments" (A.S. Nowick, J.J. Burton, eds.), Academic Press, New York 1975, p. 231.
2. For review articles by H. Wagner (elastic interaction, p. 5), T. Springer (inelastic neutron scattering, p. 75), R.M. Cotts (NMR, p. 227), K. Sköld (QNS, p. 267), and H. Kronmüller (magnetic disaccommodation, p. 289) see "Hydrogen in Metals I, Topics in Applied Physics", Vol. 28 (G. Alefeld and J. Völkl, eds.), Springer-Verlag, Berlin - Heidelberg - New York 1978.
3. A. Heidemann, G. Kaindl, D. Salomon, H. Wipf, G. Wortmann, Phys. Rev. Lett. 36:213 (1976); A. Heidemann, H.Wipf, G. Wortmann, Hyperfine Interactions 4:844 (1978); H. Wipf, A. Heidemann, J. Phys. C: Solid State Phys., in press (1980); A. Heidemann, H. Wipf, G. Wortmann, to be published (1980).
4. B. Ströbel, Thesis, Universität Konstanz (1979); B. Ströbel, K. Läuger, E.H. Bömmel, Appl. Phys. 9:39 (1976).

5. R. Dörr, E. Brauer, R. Gruner, F. Rauch, Z. Physik.
    Chem. N.F. 116:271 (1979).
6. J. Völkl, G. Alefeld, in:" Hydrogen in Metals I,
    Topics in Applied Physics", Vol.28 (G. Alefeld and
    J. Völkl, eds.), Springer-Verlag, Berlin - Heidelberg-
    New York 1978, p. 321.
7. G. Matusiewicz, H.K. Birnbaum, J. Phys. F: Metal
    Phys. 7:2285 (1977); H.K. Birnbaum, G. Matusiewicz,
    C.G. Chen, P. Zapp, Proc. 2nd Int. Congr. on Hydrogen
    in Metals, Paris 1977.
8. J. Völkl, G. Alefeld, Z. Physik. Chem. N.F. 114:123
    (1979).
9. H.C. Bauer, J. Völkl, J. Tretkowski, G. Alefeld,
    Z. Phys. B29:17 (1978).
10. J. Tretkowski, J. Völkl, G. Alefeld, Z. Phys.
    B28:259 (1977).
11. D. Emin, M.I. Baskes, W.D. Wilson, Hyperfine Inter-
    actions 6:255 (1979); Phys. Rev. Lett. 42:791 (1979);
    Z. Physik. Chem. N.F. 114:231 (1979).

HYDROGEN MOBILITY AT HIGH CONCENTRATIONS

R. C. Bowman, Jr.*

Monsanto Research Corporation
Mound Facility**
Miamisburg, Ohio 45342

ABSTRACT

The roles of crystal structure, hydrogen site occupancy, phase transition, and isotope substitution on the hydrogen diffusion behavior will be considered. After brief discussions of general diffusion concepts and the experimental methods that have been most often used to determine the hydrogen mobility in the high concentration metal hydride phases, the nuclear magnetic resonance techniques that have been extensively applied in hydride diffusion studies will be reviewed in some detail. The emphasis will be on relating measured nuclear relaxation times with diffusion parameters as well as NMR techniques to directly measure the hydrogen diffusion constants. The diversity of diffusion behavior will be illustrated with several specific metal hydride systems including: PDH_x, γ-TiH_x and structurally related TiCuH, and VH_x. Plausible models of the microscopic diffusion mechanisms will also be presented and compared with the observed diffusion behavior in these systems.

---

* Present temporary address is Division of Chemistry and Chemical Engineering, California Inst. of Technology, Pasadena, CA 91125.

**Operated by Monsanto Research Corp. for the U.S. Department of Energy under Contract No. DE-AC04-76-DP00053.

INTRODUCTION

The mobility of hydrogen isotopes in those metal hydride phases containing large hydrogen contents will be described with emphasis on methods used to identify the microscopic diffusion processes. The roles of host crystal structure, hydrogen site occupancy, phase transitions, alloying, and isotope substitution will be considered for several representative metal hydrides. Although knowledge of diffusion behavior has often been related to the kinetics of the hydriding or dehydriding reactions, the diffusion parameters can also provide unique insights on the structure and thermodynamic properties of metal hydrides. The intent of the present paper is to illustrate the close relationship between structure and diffusion processes in the interstitial metal hydrides.

From a microscopic viewpoint the diffusion constant D can be represented by atomic jumps between equilibrium positions through

$$D = f_T \frac{a^2}{6\tau_d} \tag{1}$$

where a is the mean jump distance, τ_d is the mean time between the atomic jumps, and f_T is the tracer correlation factor. Although f_T is structure sensitive and usually ranges between 0.5 and 1.0, explicit calculation of f_T is generally difficult--especially, for low symmetry structures or for large amounts of disorder. Consequently, f_T is normally assumed to be unity unless highly accurate comparisons between D and τ_d are made to determine a. Figure 1a illustrates the idealized diffusion process involving

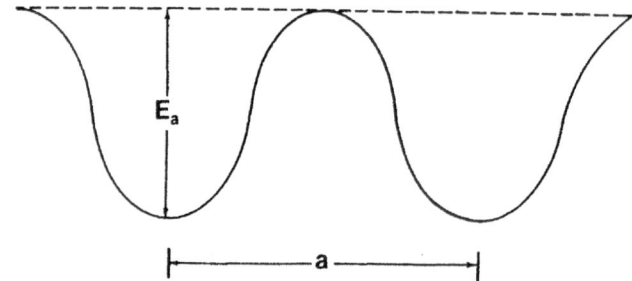

Fig. 1a. Schematic representation of potential well for hydrogen isotopes that is involved in diffusion processes: a simple activated hopping process.

the direct jumps between potential well minima. Here, E_a is the diffusion activation energy that corresponds to the Arrhenius relation

$$D = D_0 \exp(-E_a/k_B T) \qquad (2)$$

where the preexponential factor contains the hydrogen vibrational and jump attempt frequencies, k_B is the Boltzmann's constant, and T is the absolute temperature. However, there is so much evidence that hydrogen mobility in many metal hydrides is actually far more complicated either because several structually distinct sities are occupied by the hydrogen atoms or because the jump process involves one or more metastable intermediate sites. Figure 1b shows a possible intermediate relatively high energy site. In this case the relationship between experimentally observed E_a and the depths of the potential wells is much less straightforward.

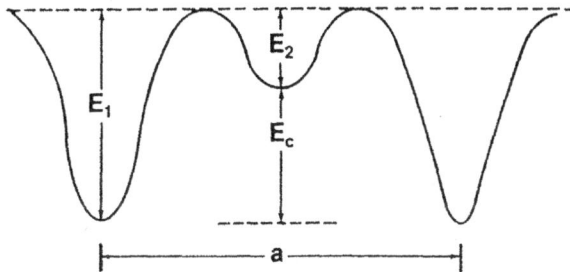

Fig. 1b.   Schematic representation of potential well for hydrogen
           isotopes that is involved in diffusion processes: jumps
           involving intermediate site.

There have been many experimental studies involving numerous techniques of diffusion parameters in the hydride phases of both pure metals and intermetallic compounds. Much of this data is summarized in the reviews of Völkl and Alefeld[1,2] and Cotts[3,4]. Compared to other interstitial atoms such as N, O, or C the mobility of H-atoms in metal hydride phases is quite rapid with observed room temperature D values that cover the rather wide range between 10^{-16} cm^2/s and 10^{-6}cm^2/s. Among the techniques that have been used to obtain diffusion parameters in metal hydrides, four merit special attention.

1. Nuclear Magnetic Resonance (NMR) has probably been applied to more hydrides than any other technique. As will be described in the next section, there are various NMR techniques that are suitable to characterize diffusion parameters between $\sim 10^{-12}$cm^2/s and $\sim 10^{-4}$cm^2/s. Furthermore, NMR has relatively few requirements on sample preparation or properties while the analysis and interpretation of the NMR data is well established and usually provide diffusion parameters without much supplemental information.

2. Quasielectric Neutron Scattering (QNS) -- The broadening of the neutron diffraction peaks when $D \geq 10^{-7}$cm^2/s provides a very powerful technique for studying hydrogen diffusion processes as discussed, for example, in the review by Sköld[5]. In favorable cases, detailed information on the actual atomic jump processes can be extracted from the momentum dependence of the quasielastic peak widths. However, since most metal hydride phases have $D \lesssim 10^{-8}$cm^2/s at temperatures of interest, QNS measurements have been restricted to relatively few high mobility systems such as PdH$_x$ or the b.c.c. α-phases of VH$_x$, NbH$_x$, and TaH$_x$.

3. Gorsky Effect -- The use of thse mechanical relaxation techniques to study diffusion in metal hydrides has been extensively reviewed by Völkl and Alefeld[1,2]. For high concentration metal hydride phases a very large thermodynamic correction factor is required to deduce the diffusion constant defined by Equation 1 from the Gorsky data. The Gorsky effect measurements have been mainly applied to the Group V metal hydrides.

4. Permeation of hydrogen isotopes has been utilized several times, but great precautions must be taken to avoid surface sensitive domination of the permeation rate.

CONCEPTS AND METHODS OF NMR

Because of the versatility and wide application of NMR techniques to characterize diffusion behavior in metal hydrides a brief outline of NMR foundations and experimental approaches will be presented. For more complete descriptions of NMR (with respect to both general background and specific application to metal hydrides) the excellent reviews by Cotts[3,4] are strongly recommended.

The phenomena of nuclear resonance occurs when nuclei with non-zero spin angular momentum (quantum number given by I) are placed in an external magnetic field (H$_0$) and irradiated with electromagnetic waves that satisfy

$$\omega_0 = 2\pi\nu_0 = \gamma_I H_0 \quad \text{(rad/sec)} \tag{3}$$

Here, ν_0 is the resonance frequency in MHz since typical $H_0 \sim 1T$ ($T \equiv$ Tesla $= 10^4$ gauss) and the nuclear gyromagnetic constant (γ_I) is typically between $10^3 - 3 \times 10^4$ rad/gauss. There are two common methods for observing nuclear resonance: 1) apply weak ($\lesssim$ mWatt) continuous rf in the steady-state method (the technique of conventional high-resolution NMR) or 2) apply short bursts of intense ($\sim$ Kwatt) rf to induce a time-dependent signal in the pulse NMR method. It can be shown[6] that the NMR signal (known as the Free Induction Decay (FID)) following the rf-pulse is the Fourier transform of the steady-state spectra. The three hydrogen isotopes have nonzero nuclear moments and their spin properties are summarized in Table 1. Both protons (H) and tritons (T) have $I = 1/2$ and their nuclear spin interactions occur only through their megnetic dipole moments. However, deuteron (D) has electric quadrupolar interactions in addition to the dipolar terms since $I = 1$. In fact, because of its small dipolar moment, γ_I, deuteron NMR properties are often dominated[3,4] by the quadrupolar contributions.

When the diffusion rate is sufficiently slow that the magnetic and/or quadrupolar interactions between the nuclei are time independent, the NMR lineshape will be approximately gaussian with a half-width of ~10-20 gauss. When the diffusion rate becomes rapid enough to begin averaging the magnetic dipolar interactions, the NMR lineshape measured by the steady-state technique will narrow to ultimately approach a linewidth defined by the inhomogenities ($\lesssim$ 1 gauss) of the magnet or the sample itself. From the temperature dependence of the nuclear linewidths the diffusion parameters E_a and the diffusion correlation time τ_c can be determined.[3] In fact, the diffusion behavior in many metal hydrides has been investigated using the steady-state NMR technique. However, because various extraneous magnetic contributions (i.e., ferromagnetic impurities or large paramagnetic susceptibilities) occur for many of these hydrides, the resulting diffusion parameters can be unreliable. Consequently, the alternative pulse NMR methods for studying diffusion will be emphasized in the present paper. The interesting effects[3,4] of diffusion on the deuteron quadrupolar spectra is also beyond the scope of this paper.

At thermodynamic equilibrium the nuclear spin system can be represented by a magnetization vector precessing parallel to the applied field H_0. When an rf-pulse satisfying Equation 3 is applied perpendicular to H_0, this magnetization is rotated away from its equilibrium direction by the angle θ according to

$$\theta = \gamma_I H_1 t_p \tag{4}$$

where H_1 is the magnetic field of the rf in gauss and t_p is rf pulse width in sec. The two most important θ values are $\pi/2$ ($90°$) which gives the maximum NMR signal and $\pi(180°)$ which inverts the

TABLE 1. NUCLEAR SPIN PROPERTIES OF HYDROGEN ISOTOPES

Isotope	Spin I	Gyromagnetic Ratio γ_I (rad/gauss)	NMR Frequency at $H_0 = 1.4092T$ (MHz)	Relative Sensitivity for Equal Number of Spins at Constant Frequency	Quadrupole Moment eQ (10^{-24} cm^2)
H	1/2	26751.0	60.000	1.000	--
D	1	4106.4	9.210	0.409	2.77×10^{-3}
T	1/2	28533.5	63.998	1.067	--

magnetization. After the rf-pulses are over, spins will interact
both among themselves and with external degrees of freedom
(commonly, known as the "lattice") to ultimately return to equilibrium conditions. The time constants for these various processes
are collectively called the nuclear relaxation times and diffusion
mechanisms dominate over substantial ranges. Briefly, pulse
techniques can provide for the nuclear magnetization to decay with
the following relaxation times:

1) Spin-Lattice in the laboratory H_0 field (T_1) describes the
return of the nuclei to thermodynamic equilibrium.

2) Spin-Lattice in the rotating H_1 field ($T_{1\rho}$) is recovery of the
magnetization with the H_1 field remaining parallel to the magnetization (i.e., "spin-locking" condition).

3) Spin-Lattice in the local dipolar magnetic field of the nuclei
(T_{1D}) is very sensitive to the very slow diffusion rates before
the onset of motional narrowing of the NMR lineshape.

4) Spin-Spin (T_2) processes correspond to the loss of spin coher-
ence without loss of the excitation energy from the spin system.
The time-constant of the FID is denoted T_2^* and contains both
homogeneous (i.e., lifetime of spin states and dipolar coupling
interactions) and many potential inhomogeneous (i.e., spatially
variate) contributions. Often, the application of the two pulse
$[90°-\tau-180°]$ spin-echo[3] sequence produces refocused NMR signals
where these echo maxima decay with a constant (T_2') that contains
only the homogeneous terms. When a rapid diffusion rate causes
the nuclei to experience different magnetic field gradients during
T_2 relaxation process, the multiple pulse Carr-Purcell-Meiboom-
Gill (CPMG) sequence[7] yields T_{2m} with minimal inhomogeneous
contribution. The benefit of the CPMG sequence of T_2 measurements
in metal hydrides was originally demonstrated by Zamir and Cotts[8].

The general relationships between nuclear relaxation rates
(i.e., reciprocals of relaxation times) and diffusion modulation
of dipole-dipole and quadrupolar interactions are well
established[3,4,6]. To illustrate this relationship consider a
nuclear spin system with one type of magnetic nuclei which are
mobile and have I = 1/2. Thus, only the magnetic dipolar interactions between identical nuclei need to be included. The dipolar
nuclear relaxation rates in this case are given by expressions[3,4]

$$\frac{1}{T_{1d}} = \frac{3}{2} \gamma_I^4 \hbar^2 I (I+1) [J^{(1)}(\omega_0) + J^{(2)}(2\omega_0)] \qquad (5)$$

$$\frac{1}{T_{1\rho}} = \frac{3}{8} \gamma_I^4 \hbar^2 I (I+1) [J^{(0)}(2\omega_i) + 10 J^{(1)}(\omega_0) + J^{(2)}(2\omega_0)] \qquad (6)$$

$$\frac{1}{T_2} = \frac{3}{8} \gamma_I^4 \hbar^2 I(I+1)[J^{(0)}(0)+10J^{(1)}(\omega_0)+J^{(2)}(2\omega_0)] \qquad (7)$$

where the power spectra of the randomly varying dipolar fields $J^{(q)}(\omega)$ are the Fourier transforms of correlation functions $G^{(q)}(t)$ as

$$J^{(q)}(\omega) = \int_{-\infty}^{\infty} G^{(q)}(t)e^{-i\omega t}dt, \qquad (8)$$

$$G^{(q)}(T) = \sum_j <F_{ij}^{(q)}(t') F_{ij}^{(q)*}(t'+t)>, \qquad (9)$$

$$F_{ij}^{(q)}(t) = dq\ Y_{2q}(\theta_{ij},\phi_{ij})/r_{ij}^3 \qquad (10)$$

$$d_0^2 = \frac{16\pi}{5},\ d_1^2 = \frac{8\pi}{15},\ d_2^2 = \frac{32\pi}{5}. \qquad (11)$$

Here, Y_{2q} are normalized spherical harmonies that are time dependent because of the diffusion of the resonant nuclei. The simplest assumption is that the dipolar field correlations have exponential time dependences as $\exp(-t/\tau_c)$, where τ_c is the diffusion correlation time. Thus,

$$J^{(q)}(\omega) = \frac{G^{(q)}(0)2\tau_c}{1+\omega^2\tau_c^2} \qquad (12)$$

and the relaxation rates in Equations 5-7 lead directly to τ_c. For mobile nuclei $\tau_c = \tau_d/2$ where τ_d is from Equation 1 and is proportional to D. This is known as the BPP model and has been widely used to interpret nuclear relaxation time data in the metal hydrides. Figure 2 shows typical dependences[4] of T_1, $T_{1\rho}$, and T_2 upon τ_c in terms of the BPP model. Although the absolute τ_c values from the BPP model can be in error by as much as 50%, the corresponding E_a values are usually accurate within about 10%. More accurate calculation procedures for $G^{(q)}(t)$ have been developed[4], but the complexities of the resulting numerical analyses have restricted these methods to high symmetry structures. Consequently, unless very accurate estimates of D or a in Equation 1 are desired, the BPP model is generally adequate to determine diffusion parameters from the nuclear relaxation times in the motionally narrowed regime given by $\gamma_I M_{2d}^{1/2}\tau_c \ll 1$ where the second moment M_{2d} defines the strength of the dipolar interactions when the nuclei are immobile. When dipolar interactions between mobile hydrogen isotope and the fixed metal nuclei with $I \neq 0$ are dominant, expressions similar to Equations 5-11 are obtained except $\tau_c = \tau_d$. The electric quadrupole moment of deuterons obey analogous expressions in terms of eQ but are less easily evaluated since the electric field gradients cannot be completely specified.

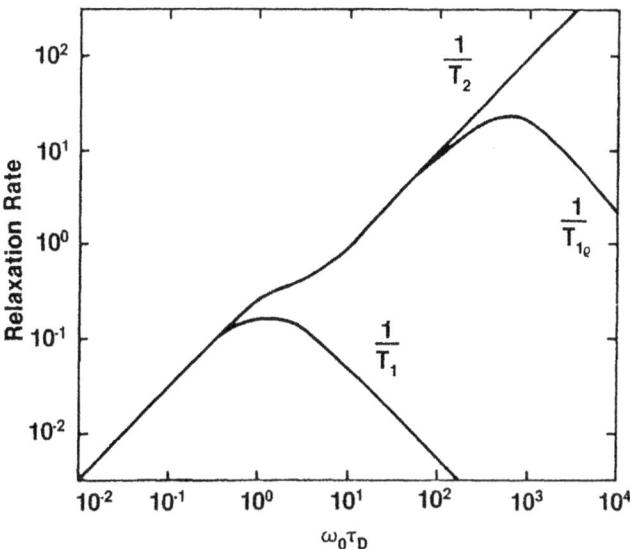

Fig. 2. Relaxation rate, in arbitrary units, as a function of $\omega_0\tau_d$ for T_1^{-1}, $T_{1\rho}^{-1}$, T_2^{-1} as calculated with the BPP model for the correlation function and for $\omega_0 = 500\omega_1$ (after Cotts[4]).

The nuclear relaxation times often have contributions from processes other than diffusion. In general, T_1 for most metal hydrides can be separated into three distinct terms

$$T_1^{-1} = T_{1e}^{-1} + T_{1p}^{-1} + T_{1d}^{-1} \tag{13}$$

where T_{1e} arises from interactions with conduction electrons and obeys $T_{1e}T = K$ with K being the well-known Korringa constant[3,4], T_{1p}^{-1} corresponds to interactions with magnetic atoms (this term is often negligible in metal hydrides when they are not highly paramagnetic or contain large concentrations of magnetic impurities), and T_{1d} is the diffusion term represented by Equation 5 or similar expressions. Since T_{1d} depends exponentially upon temperature as well as having the characteristic T_1^{-1} maximum for $\omega_0\tau_c \approx 1$ shown in Figure 2, it is experimentally possible to separate the T_{1e} and T_{1d} terms providing T_1 data is collected over a large temperature range. The results of Korn and Zamir[9] for γ-TiH$_{1.70}$ given in Figure 3 clearly show the relative contributions of T_{1d} and T_{1e}. Near the T_1 minimum $T_1 \approx T_{1d}$ as expected since T_{1d} now makes its largest contribution.

118   R.C. BOWMAN, JR.

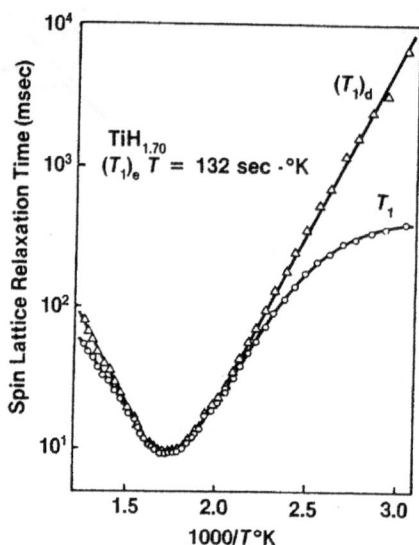

Fig. 3.   An example of extracting T_{1d} from the experimental proton
T_1 data for $TiH_{1.70}$ (after Korn and Zamir[9]).

When $D \gtrsim 10^{-8} cm^2/s$, the spin-echo NMR experiments provide the means for direct measurement of D. Stejskal and Tanner[10] introduced the pulsed field gradient (PFG) technique with the 90°–180° echo sequence shown in Figure 4. The echo amplitude at time 2τ may be expressed as[10]:

$$\ln\left|\frac{A(2\tau)}{A_0(2\tau)}\right| = -\gamma^2 D\left[\frac{2}{3}G_0^2\tau^3 + \delta^2 G^2(\Delta-\delta/3)\right.$$
$$\left. - \delta\underset{\sim}{G}\cdot\underset{\sim}{G}\left\{(t_1^2+t_2^2)+\delta(t_1+t_2)+(2/3)\delta^2-2\tau^2\right\}\right] \qquad (14)$$

where

$A_0(t)$ = Echo amplitude at time t that excludes effects of diffusion, but includes effects of spin-spin relaxation.

γ = magnetogyric ratio of observed species

D = diffusion coefficient

$\underset{\sim}{G}_0$ = background (static) magnetic field gradient

$\underset{\sim}{G}$ = pulsed magnetic field gradient
t_1 = time between 90 rf pulse and field gradient pulse
δ = length of time the field gradient pulse is applied
Δ = time between start of first field gradient pulse and the start of the second field gradient pulse
$t_2 = 2\tau - (\delta + \Delta + t_1)$.

When $G^2 \delta^2 \Delta \gg G_0 \tau^3$, the above simplifies to

$$\ln \left| \frac{A(2\tau)}{A_0(2\tau)} \right| = -\gamma^2 D \delta^2 G^2 (\Delta - \delta/3) \tag{15}$$

and the determination of D is straightforward. Although this condition is often satisfied in gaseous, liquid, or diamagnetic solids (where the main source of G_0 is the H_0 inhomogeneity over the sample dimensions), it is usually not met in many metal hydrides. Because of rf skin depth effects[3,4] powdered hydride samples are normally required for NMR studies. In these powdered samples the bulk magnetic susceptibility can lead to quite inhomogeneous magnetization and large G_0 values. In order to minimize these detrimental effects, multiple-pulse extensions[11,12] of the PFG sequence have been recently developed and successfully applied to

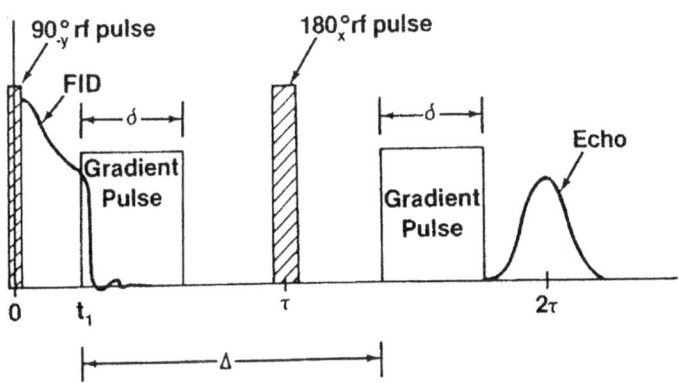

Fig. 4.   The Stejskal-Tanner pulsed gradient spin-echo sequence for direct measurement of D.

some metal hydrides such as $\alpha-NbH_x$[11] and $\beta-LaNi_5H_x$[12]. It appears likely these improved techniques will ve valuable in measuring D in many metal hydrides.

Since NMR can provide independent measurements of both D (via PFG or its modifications) and τ_d (via relaxation times), it is, in principle, possible to experimentally determine the diffusion jump distance a using Equation 1. Knowledge of a gives unique insights into the microscopic diffusion processes by identifying allowed jump paths. The application of the pulse NMR methods to studying hydrogen diffusion in several representative metal hydrides will be described in the following sections.

DIFFUSION IN FCC METAL HYDRIDES

Many metals react with hydrogen to form hydride phases where the metal atoms (M) assume the fcc crystal structure shown in Figure 5. The hydrogen atoms can occupy either the octahedral or

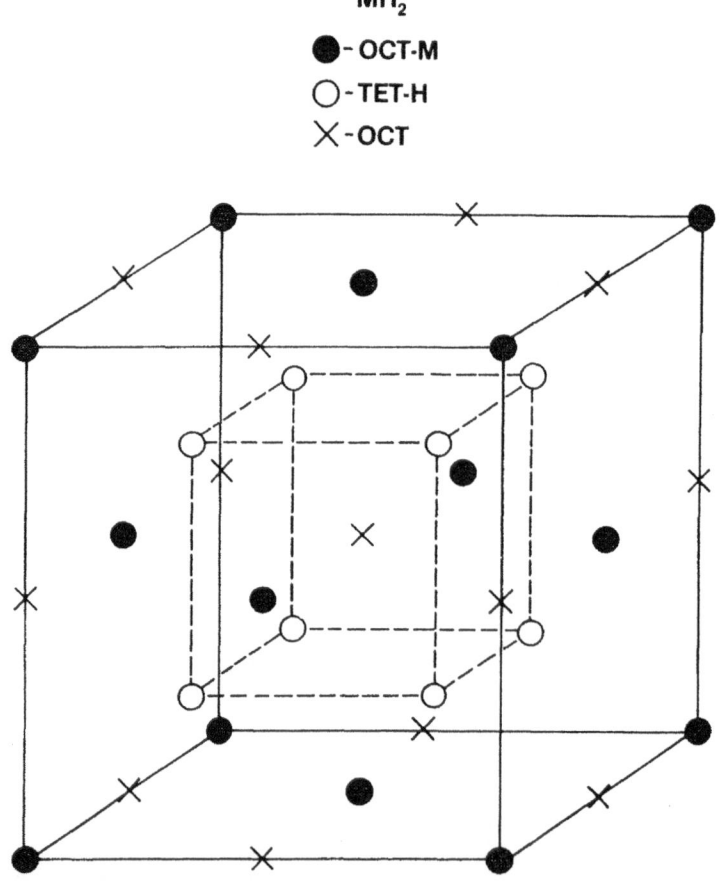

Fig. 5. Unit cell of the fcc metal hydride structure.

tetrahedral interstitial sites of the fcc metal sublattice. When
only the octahedral sites are occupied, the limiting composition
is MH with the NaCl structure. Occupancy of only tetrahedral sites
gives MH_2 with the familiar fluorite (CaF_2) structure. There are
also examples such as LaH_x ($2 \leq x \leq 3$) where both types of interstitial sites are occupied. It is a general characteristic that
metal hydrides are nonstoichiometric with variable numbers of the
hydrogen sites vacant. This can significantly influence the
diffusion behavior.

As a first example of diffusion in metal hydrides, consider
the nonstoichiometric fcc α'-PdH_x system with hydrogen in the octahedral sites. Although there have been many experimental studies[1,2]
of diffusion in PdH_x, two NMR studies (that are consistent with
most other reliable studies) will be emphasized. First, Schoep,
et al.[13] have measured proton T_1 and T_2 at several resonance frequencies. Some of their data for $PdH_{0.76}$ is presented in Figure
6a while Figure 6b plots the temperature dependence of τ_c obtained
with the BPP model. The diffusion activation energy $Q = E_a = 0.22 \pm$
0.02 eV. The T_1 minima show the predicted behavior as ν varies.
Davies et al.[14] have conducted both T_1 and PFG measurements on
PdH_x and some of their data is reproduced in Figure 7. The PFG
technique gave $E_a = 0.228 \pm 0.006$ eV for α'-$PdH_{0.70}$. It is
particularly noteworthy that PFG gave D values in good agreement
with estimates from T_1 results as well as other completely independent techniques. The concentration dependence of D is well
described by the factor (1-X) which corresponds to the number of
vacant octahedral hydrogen sites. Hence, diffusion in α'-PdH_x seems
to correspond to the simple diffusion process of Figure 1a where the
hydrogen atoms jump directly between their octahdral equilibrium
positions.

The prototype fcc MH_2 system is probably γ-TiH_x ($1.6 \lesssim x \leq 2$).
Korn and Zamir[9] performed extensive NMR measurements of proton T_1
on several TiH_x samples and used the BPP model to extract the τ_c
data which are summarized in Figure 8 as ν (i.e., τ_c^{-1}). Korn and
Zamir found the composition dependence of τ_c^{-1} in γ-TiH_x to obey[9]

$$\tau_c^{-1} = (2-x)\tau_0 \exp[-(0.507 \pm 0.010 \text{ eV})/k_B T] \tag{16}$$

where τ_0 is the diffusion attempt frequency and (2-x) is the hydrogen vacancy concentration. Subsequent NMR measurements of proton
T_1 in γ-TiH_x by Schmolz and Noack[15] verified the general correctness
of this relation, which implies a vacancy diffusion mechanism.
However, Figure 9 illustrates three alternative jump-paths for
hydrogen in γ-TiH_x. Korn and Zamir[9] and Schmotz and Noack[15]
propose jumps along the <111>-direction through the vacant octahedral sites to the third nearest-neighbor position. Bustard
et al.[16] recently conducted simultaneous NMR measurements of D by

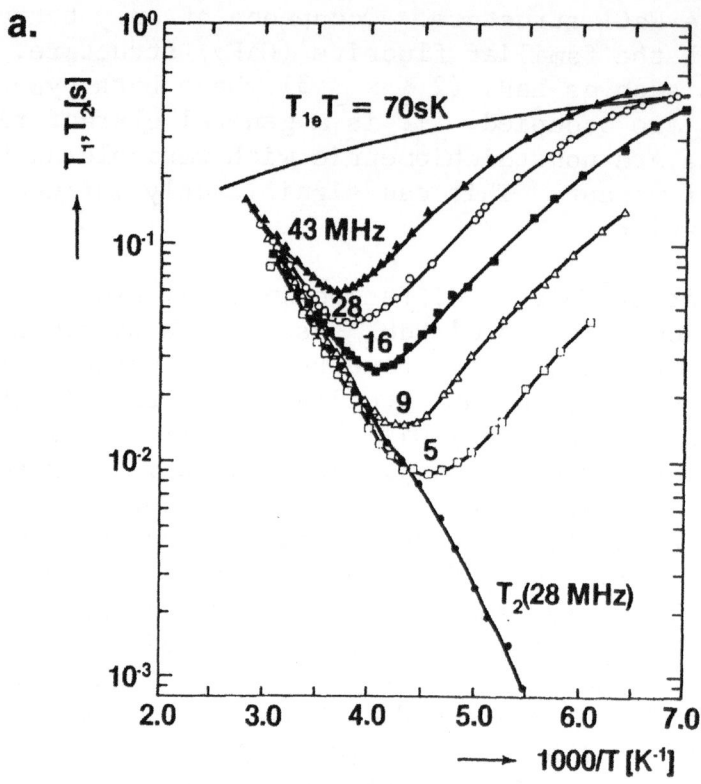

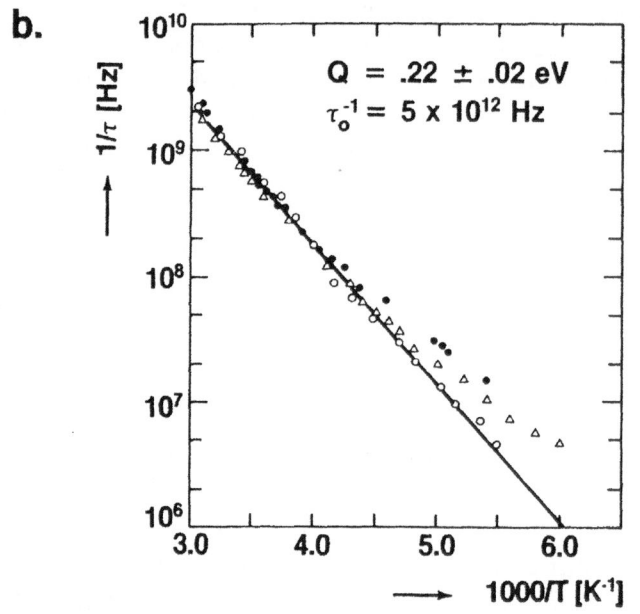

Fig. 6a. T_1 vs T^{-1} for protons in PdH$_{0.76}$ at several frequencies.
          The full curve denoted T_{1e} as determined below 150K.

Fig. 6b. τ_c^{-1} vs T^{-1} for protons in PdH$_{0.76}$. $\triangle$, T_1 at 5 Mhz; $\bullet$,
          T_1 at 28 MHz; O T_2 (CPMG) at 28 MHz. (After Schoep et
          al.[13]).

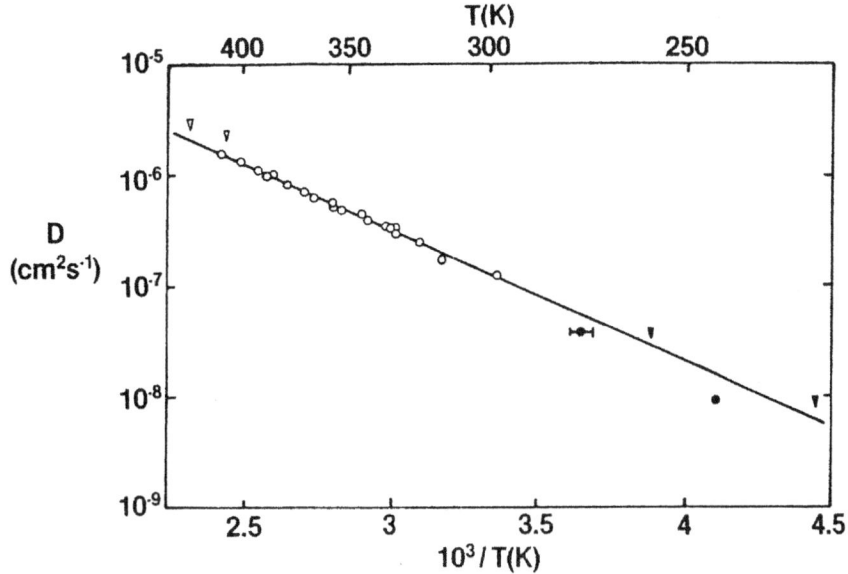

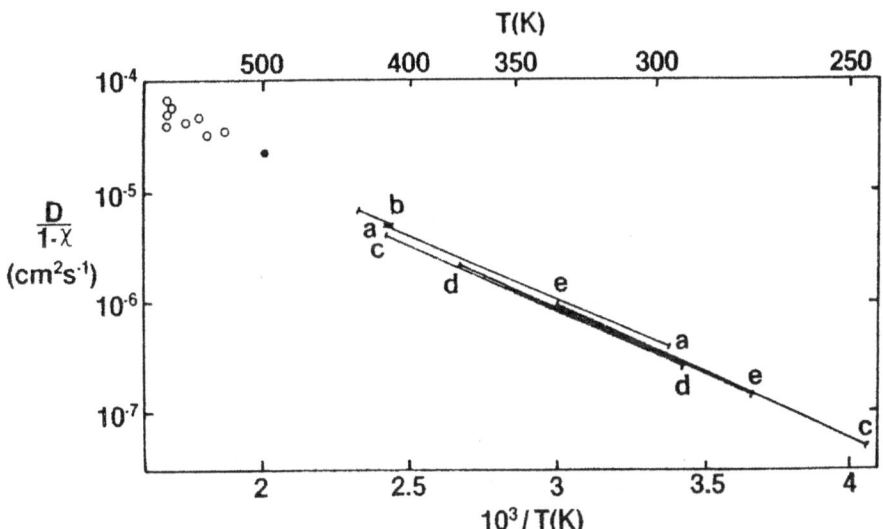

Fig. 7a.   D values measured by PFG method: O, PdH$_{0.70}$; ∇, PdH$_{0.54}$.
The straight line is the best fit to data for PdH$_{0.70}$.
Also shown are D values deduced from temperatures of T$_1$
minima: ●, PdH$_{0.70}$; ▼, PdH$_{0.54}$.

Fig. 7b.   Comparison of measured D for α'-PdH$_x$, corrected for site-
blocking (1-x). aa (PFG) PdH$_{0.76}$; bb (PFG) PdH$_{0.54}$; cc (T$_1$)
PdH$_{0.69}$; dd (permeation) PdH$_{0.74}$; ee (permeation)
PdH$_{0.69}$; O (QNS), PdH$_{0.2}$, PdH$_{0.4}$, PdH$_{0.48}$; ● (QNS)
PdH$_{0.55}$ (after Davis et al.[14]).

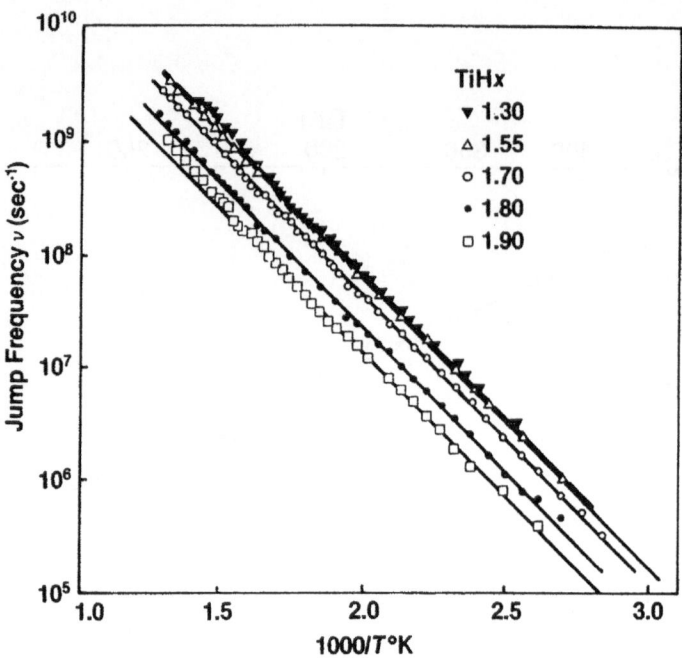

Fig. 8.   Proton jump frequency $\nu = \tau_c^{-1}$ in γ-TiH$_x$ from BPP model
          analysis of T$_1$ data (after Korn and Zamir[9]).

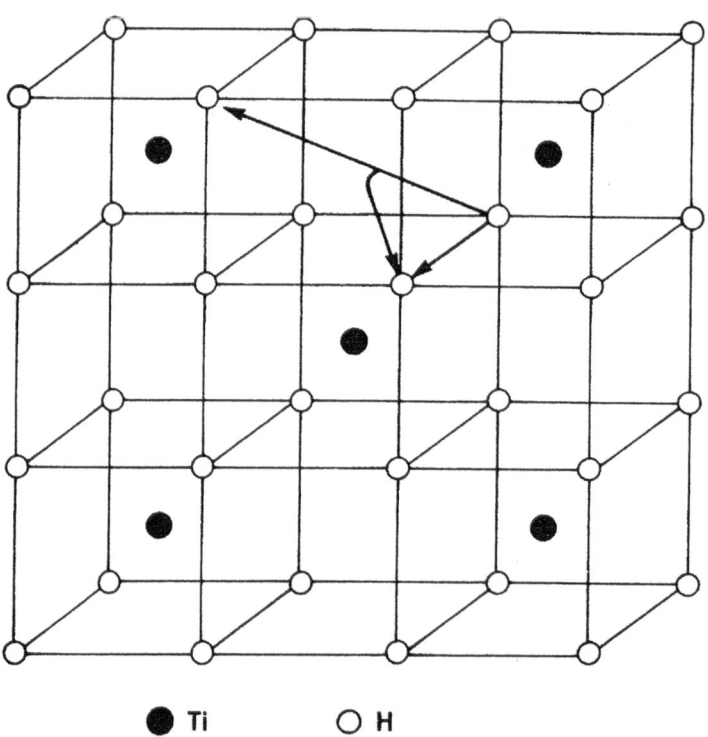

● Ti   ○ H

Fig. 9.   Jump paths of hydrogen atoms to first and third nearest
          neighbors in fcc γ-TiH$_x$.

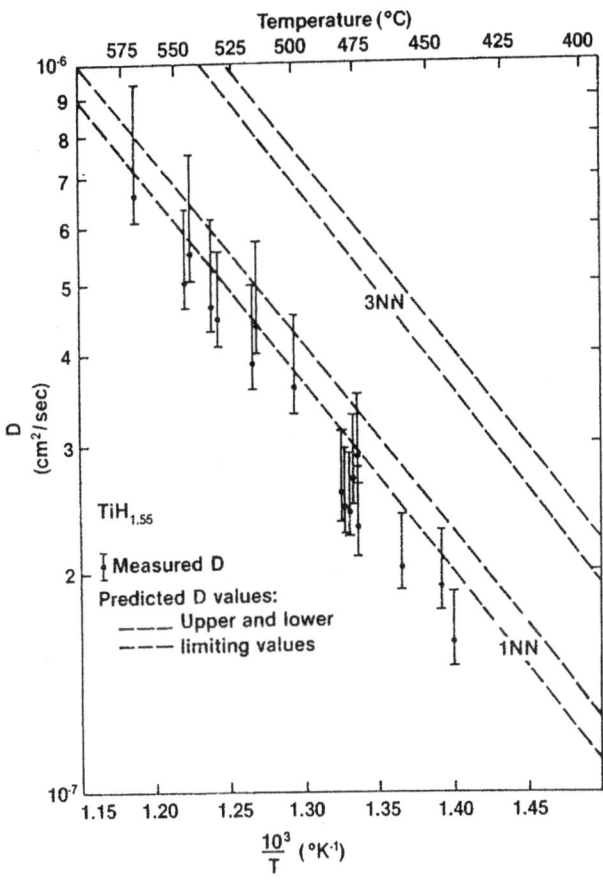

Fig. 10. Measured and predicted values of the diffusion coefficients for TiH$_{1.55}$ (after Bustard et al.[16]).

PFG technique and T_1 for $TiH_{1.55}$ and $TiH_{1.71}$. Figure 10 compares their experimental D values with calculated D values for first and third nearest neighbors using Equation 1 and τ_c from the T_1 data on the same $TiH_{1.55}$ sample. They obtained similar results for $TiH_{1.71}$ and concluded that jump distance a for $\gamma-TiH_x$ corresponds to nearest neighbor hopping. However, theoretical studies[17,18] strongly suggest that hydrogen atom hops through the vacant octahedral sites yield much lower E_a values than the direct tetrahedral-tetrahedral jump between the nearest neighbors.

Additional insight into the diffusion mechanism of $\gamma-TiH_x$ is provided from relaxation time measurements[19] on $TiCuH_x$. The crystal structure of TiCuD from neutron diffraction measurements[20] are compared with the fcc TiD_2 structure in Figure 11. The double layer of titanium atoms in TiCuD gives a two-dimensional analog of the

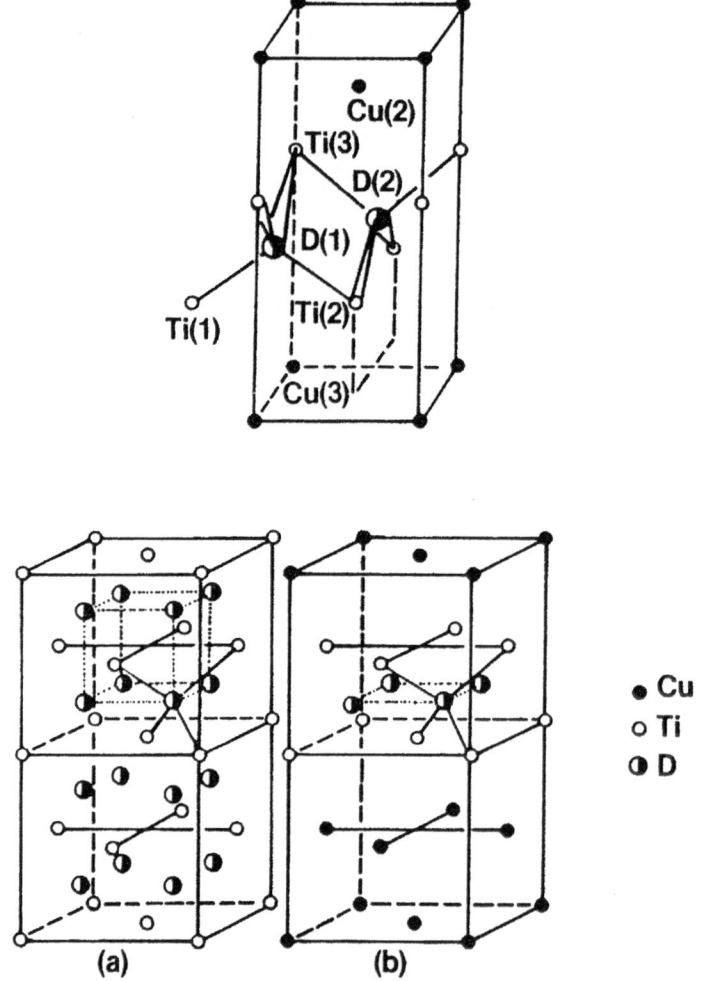

Fig. 11. Comparison between structures of (a) TiD_2, and (b) TiCuD where for clarity the unit cell (upper figure) of TeCuD is doubled (after Santoro et al.[20]).

TiD$_2$ structure. The local symmetry of the deuterium (or hydrogen) as well as Ti-Ti, Ti-D, and D-D interatomic distances in TiCuD$_{20}$ are nearly identical to TiD$_2$. However, the diffusion process in TiCuH$_x$ is restricted to jumps directly between tetrahedral sites within the plane of H-atoms. The proton relaxation times T$_2$ and T$_{1D}$ for TiCuH$_{0.94}$ presented in Figure 12 yield E$_a$ = 0.80 eV, larger by

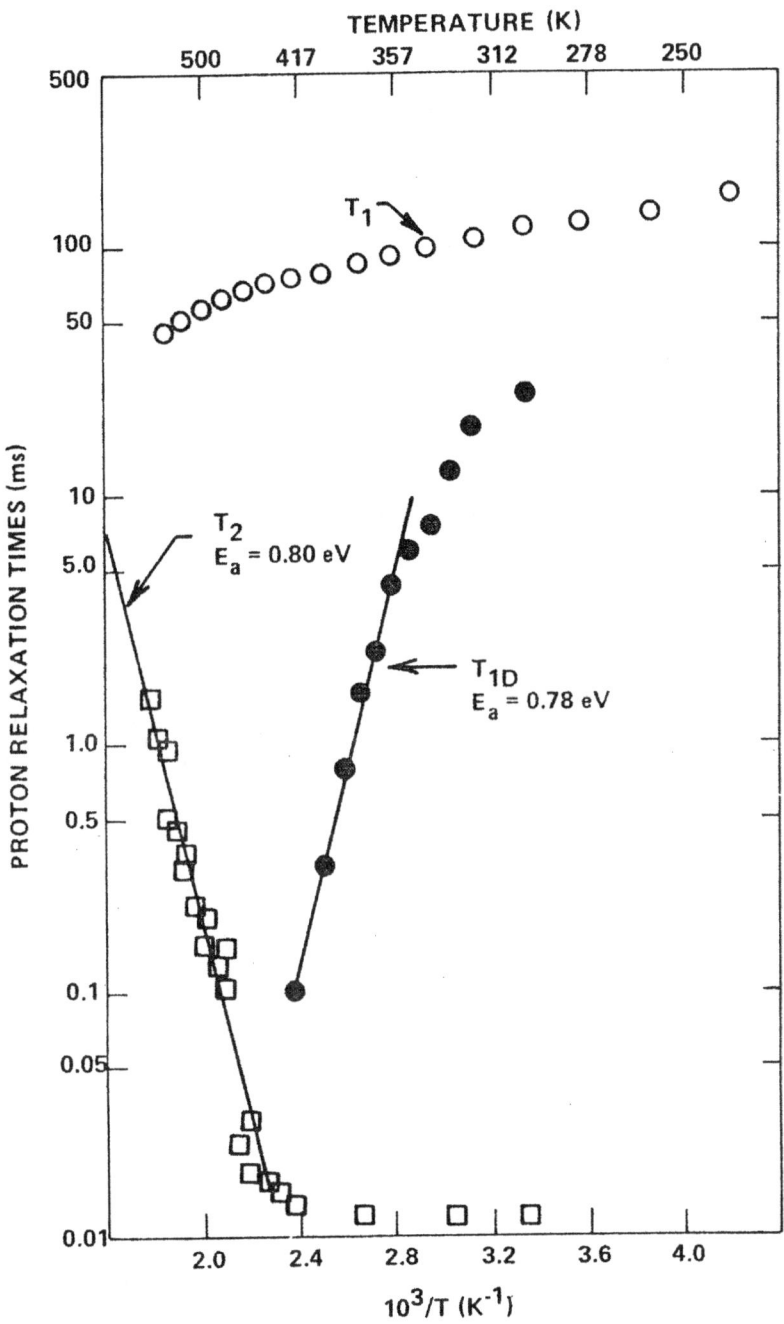

Fig. 12. Temperature dependence of proton relaxation times for TiCuH$_{0.94}$. (after Bowman et al.[19])

0.3 eV than the E_a obtained for γ-TiH$_x$[9,15,16]. Melius and Upton[18] have calculated a 0.2 eV larger E_a for direct tetrahedral-tetrahedral jumps relative to tetrahedral-octahedral-tetrahedral jumps in TiH$_x$. Considering the observations of Bustard et al.[16], the TiCuH$_{0.94}$ results strongly support the curved path through the octahedral site between nearest neighbor tetrahedral sites as indicated in Figure 9. Hence, the actual hydrogen diffusion path in the relatively ideal case of fcc γ-TiH$_x$ is apparently more involved than the naively expected simple hop to a nearest empty site.

DIFFUSION IN VANADIUM HYDRIDES

When metals with b.c.c. crystal structures (i.e., V, Nb, Ta) absorb hydrogen, a large number of interstitial sites can be potentially occupied as illustrated in Figure 13. If all these sites would be filled with hydrogen atoms, a limiting composition of MH$_9$ would occur. However, because minimal H-H separation distances of ~2.1Å are empirically observed for all known metal hydrides, occupancy of one of the sites in Figure 13 prevents any neighbors from containing a hydrogen atom. Furthermore, hydrogen ordering in MH$_x$ also occurs to a very large extent when x < 1.0 and have yielded complex and interesting phase diagrams which have been recently reviewed by Schober and H. Wenzl[21]. The diffusion of hydrogen isotopes in the various hydride phases of the b.c.c. metal have been extensively studied[1,2,3,4]. Much of this interest is due to the often rapid diffusion rates (i.e., D $\geq 10^{-6}$cm^2/s) and non-Arrhenius temperature dependences. There has been much theoretical effort[22] in attempting to develop models for the diffusion mechanism in very dilute hydrogen solutions. However, quantitative theories for diffusion mechanisms in the high hydrogen content phases are not currently available.

Some of the unusual diffusion behavior in hydride phases of the b.c.c. metals will be illustrated with the VH$_x$ system. Figure 14 gives the current VH$_x$ phase diagram that is a revision of the one given by Schober and Wenzl[21] to be more consistent with several recent studies[23,24]. The various phases of VH$_x$ and VD$_x$ are summarized in Table 2 and indicate some very interesting (and not currently completely understood) isotope effects. Reflecting the complexities of the VH$_x$ phase diagram, the diffusion behavior in the VH$_x$ phases exhibits a variety of anomalous features. Although the mobility in the α-phase is very rapid (D $\geq 10^{-6}$cm^2/s)[1,2], the diffusion rate decreases by 2-3 orders of magnitude upon the transition to the ordered β_1 or β_2 phase. The crystal structure[25] of β-VH$_x$ is given in Figure 15 and shows the occupancy of specific subsets of the octahedral sites described in Figure 13. The diffusion rates in the β-VH$_x$ phase are particularly amenable[26,27] to NMR nuclear relaxation time measurements. Consequently, recent unpublished NMR studies by Bowman et al. will be emphasized to point out some of the more unusual aspects of diffusion in β-VH$_x$ phases.

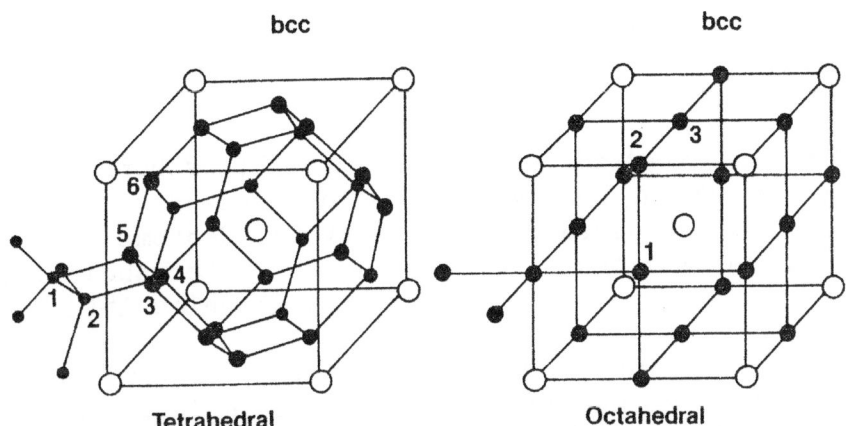

Fig. 13.   Tetrahedral and octahedral interstitial sites in bcc metal,
           where open circles are the metal atoms and the full circles
           are the interstitial sites. The non-equivalent interstitial
           sites are labeled with 1 to 6 (tetrahedral) and 1, 2, 3
           (octahedral).

TABLE 2. SUMMARY OF THE VH_x AND VD_x PHASES
(FROM SCHOBER AND WENZL, REF. 21

Vanadium Hydrides		Vanadium Deuterides	
α_H	Disordered bcc solution	α_D	Disordered bcc solution, ~90% D-atoms occupy tetrahedral sites
$\beta_H(\beta_1)$	V_2H; monoclinic, H-atoms predominantly ordered on octahedral O_{z1} sites, $c_o/a_o \simeq 1.1$ (pseudo-tetragonal) $\beta_H \rightarrow \varepsilon_H$ at 446K	β_D	V_2D; monoclinic, D-atoms ordered on O_{z1} sites, $c_o/a_o \simeq 1.1$, isomorphic with β_H, $\beta_D \rightarrow \alpha_D$ occurs at 407K
γ_H	VH_2; fcc dihydride phase, $a_0 \simeq 0.424$ nm, H-atoms presumably ordered on tetrahedral sites	γ_D	VD_{1-x}; orthorhombic, fully ordered phase with composition V_4D_3, D-atoms predominately occupy tetrahedral sites, forms below 151K
δ_H	V_3H_2; Monoclinic structure, H-atoms presumably ordered on octahedral sites, exists below ~222K	δ_D	VD_{1-x}; orthorhombic phase existing between ~151K and ~222K, partially ordered with D-atoms on tetrahedral sites, composition V_4D_3
$\varepsilon_H(\beta_2)$	V_2H; tetragonal phase exists between 446K and 470K, H-atoms randomly distributed between octahedral O_{z1} and O_{z2} sites. No similar deuteride observed	ε_D	VD_2; a fcc dideuteride phase, presumably isomorphic with γ_H
ζ_H	VH_{1-x} (x=0.1-0.2): similar to ε_H except at higher concentrations	ζ_D	$\approx V_4D_3$, partially ordered phase existing in narrow (5-10K) region between δ_D and α_D phases

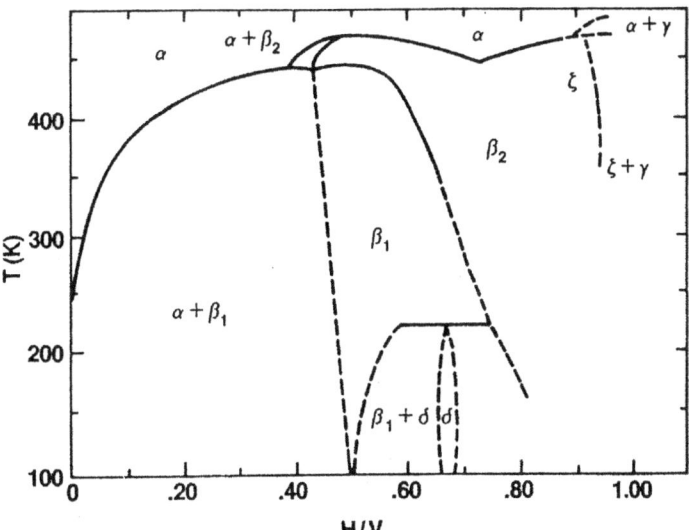

Fig. 14. Phase diagram of VH_x.

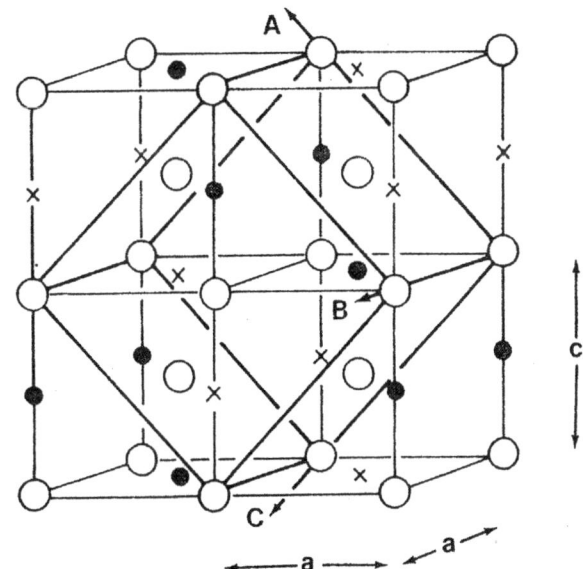

Fig. 15. Crystal structure of βVH_x phases, showing the unit cell
          of the β_1-V_2H superstructure. Open circles représent the
          vanadium atoms, forming the pseudotetragonal lattice by
          thin lines. The two types of octahedral H-sites are O_{z1}
          (full circles) and $O_{z2}(x)$. The unit cell of the β_1-V_2H
          superstructure is drawn by thick lines (after Asano and
          Hirabayashi[25]).

Complete occupancy of the octahedral O_{z1} sites leads to the stoichiometric superstructure β_1-V_2H at low temperature (i.e., $\lesssim$ 300K). However, it was recently shown[23] that β_1-V_2H undergoes a second-order phase transition to the β_2-phase with a critical temperature (T_c) of 445K. The main consequence of this phase transition is the disordering of the H-atoms to equal statistical occupancy of both O_{z1} and O_{z2} sites in the β_2-phase. When· x > 0.50, the additional H-atoms in β_1-VH$_x$ are assumed to randomly occupy the O_{z2}. This reduces T_c as shown in Figure 14 until x $\approx$ 0.70 when the β_1 and β_2 become identical. The effect of hydrogen content on the room temperature (301K) proton relaxation times T_1 and T_2 (CPMG) for 0.39 $\leq$ x $\leq$ 0.85 is shown in Figure 16. It is apparent that T_2

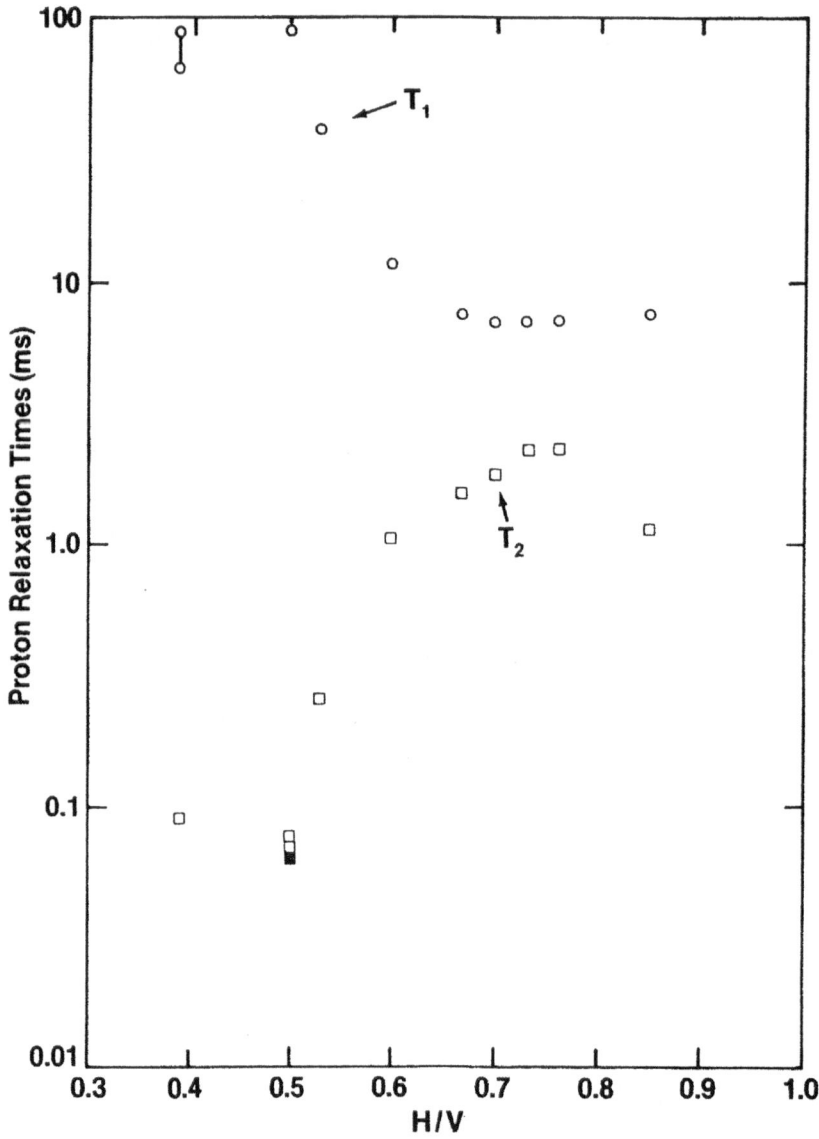

Fig. 16.   Proton relaxation times in β-VH$_x$ at room temperature (301K) for resonance frequency of 34.5 MHz.

increases and T_1 decreases significantly between x = 0.5 and x = 0.70 with a maximum and minimum, respectively, at x = 0.75. Since $D \sim \tau_c^{-1}$ and $\gamma_I M_{2d}^{1/2}\tau_c \ll 1$ for these samples, the BPP model gives

$$\tau_c^{-1} = \gamma_H^2 M_{2D} T_2 \qquad (17)$$

and hydrogen diffusion rate at room temperature is seen to reach an apparent maximum at $VH_{0.75}$. The decrease in T_1 is also consistent with this behavior.

To further evaluate diffusion behavior in $\dot{V}H_x$, the temperature dependences of the proton T_1 for the nine compositions in Figure 16 were obtained between ~100K and ~500K. Representative results are given in Figures 17 and 18 where the major features are:

1. A sharp break in the T_1 curves near ~450K due to the transition to the α-phase.
2. Evidence for phase-transitions near or below 200K in several of the samples.
3. Composition-dependent minima in the T_1 relaxation times indicate that diffusion is the dominant spin relaxation mechanism between ~220K and ~450K.

Since the hydrogen mobility in α-phase VH_x is extremely rapid[1,2], diffusion does not contribute to T_1 in these phases. Furthermore, the phase transitions near 200K complicates the analyses. However, a tentative evaluation of diffusion in β-phase VH_x has been performed using the basic BPP model. The T_1 data was first corrected for the conduction electron interactions before deducing τ_c with the expression

$$T_{1d}^{-1} = 1.491 T_{1Min} \left[\frac{y_H/3}{1+(y_H-y_v)^2} + \frac{y_H}{1+y_H^2} + \frac{2y_H}{1+(y_H+y_v)^2} \right]^{-1} \qquad (18)$$

where $y_H = (\omega_H \tau_c)^{-1}$, $y_v = (\omega_v \tau_c)^{-1}$, and T_{1Min} is the T_{1e} corrected minimum T_1 values. The temperature dependences of some of the τ_c values are given in Figure 19. In contrast to the Arrhenius behavior found for PdH_x (Figure 6) and TiH_x (Figure 8), the τ_c values for VH_x give different E_a above and below T_{1min}. These energies are shown in Figure 20. A significant decrease of E_a is observed as x increases with a possible minimum near $VH_{0.70}$ for the E_a below the T_{1min}. Again, this is in sharp contrast to the composition independent E_a values for PdH_x[13,14] and TiH_x[9,15,16]. The T_{1d} minimum for the V-H dipolar interactions of Equation 18 occurs when

$$\omega_H \tau_c = 0.922. \qquad (19)$$

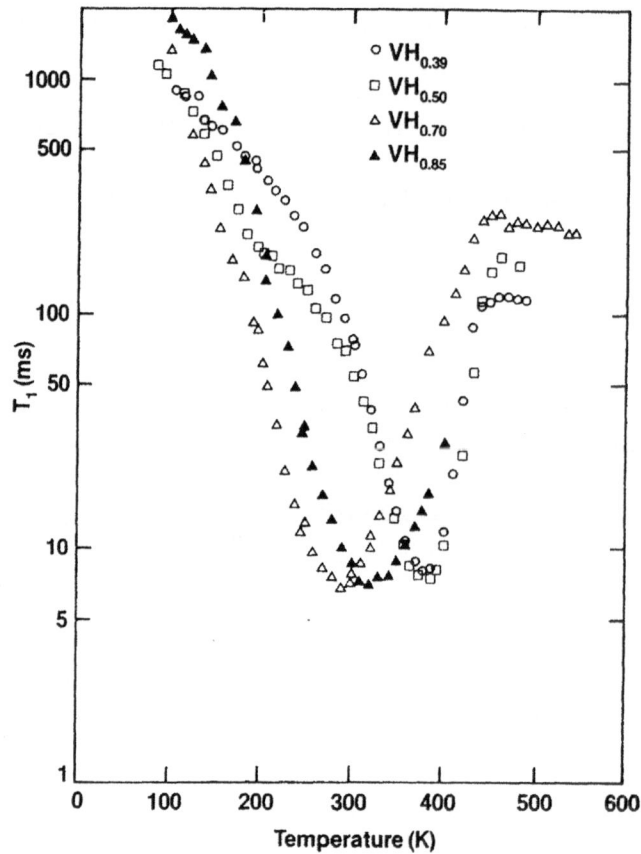

Fig. 17.   Proton T_1 in VH_x as measured at 34.5 MHz.

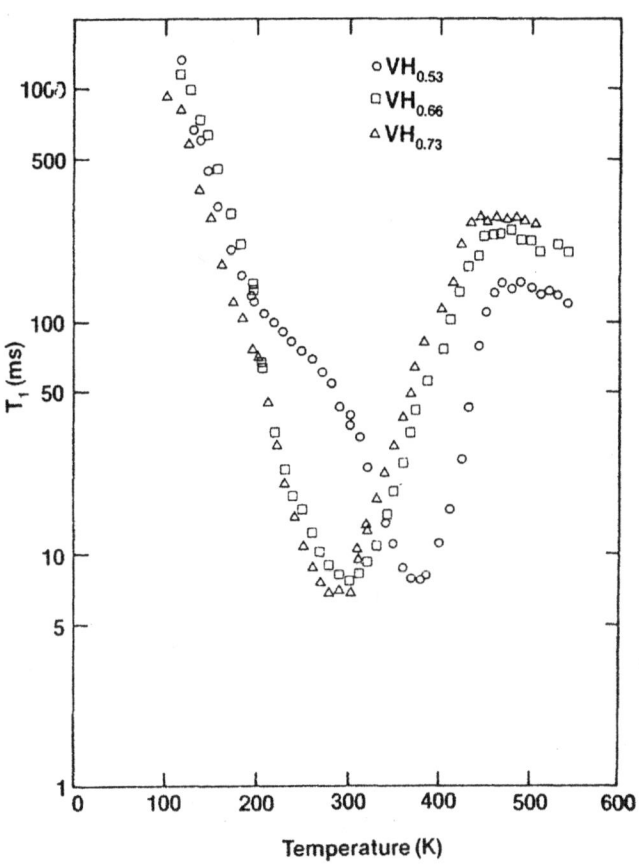

Fig. 18.   Additional proton T_1 in VH_x as measured at 34.5 MHz.

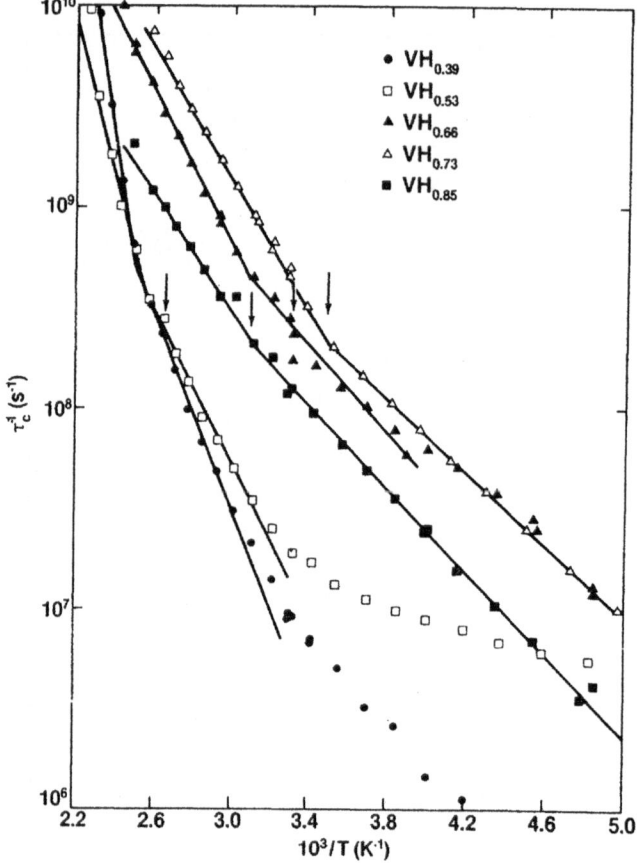

Fig. 19. Temperature dependence of τ_c in several VH_x samples as calculated with BPP model using proton T_1 data conduct-ion electron interactions.

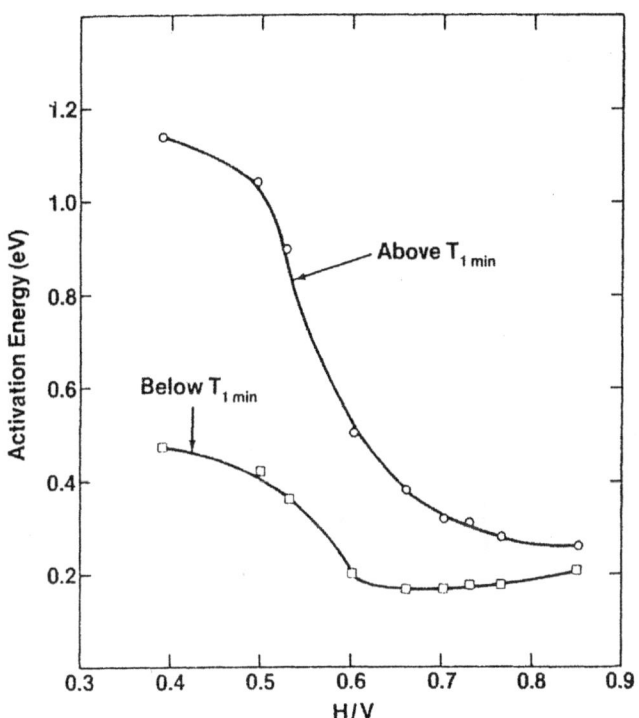

Fig. 20. Activation energies for hydrogen diffusion in β-VH$_x$.

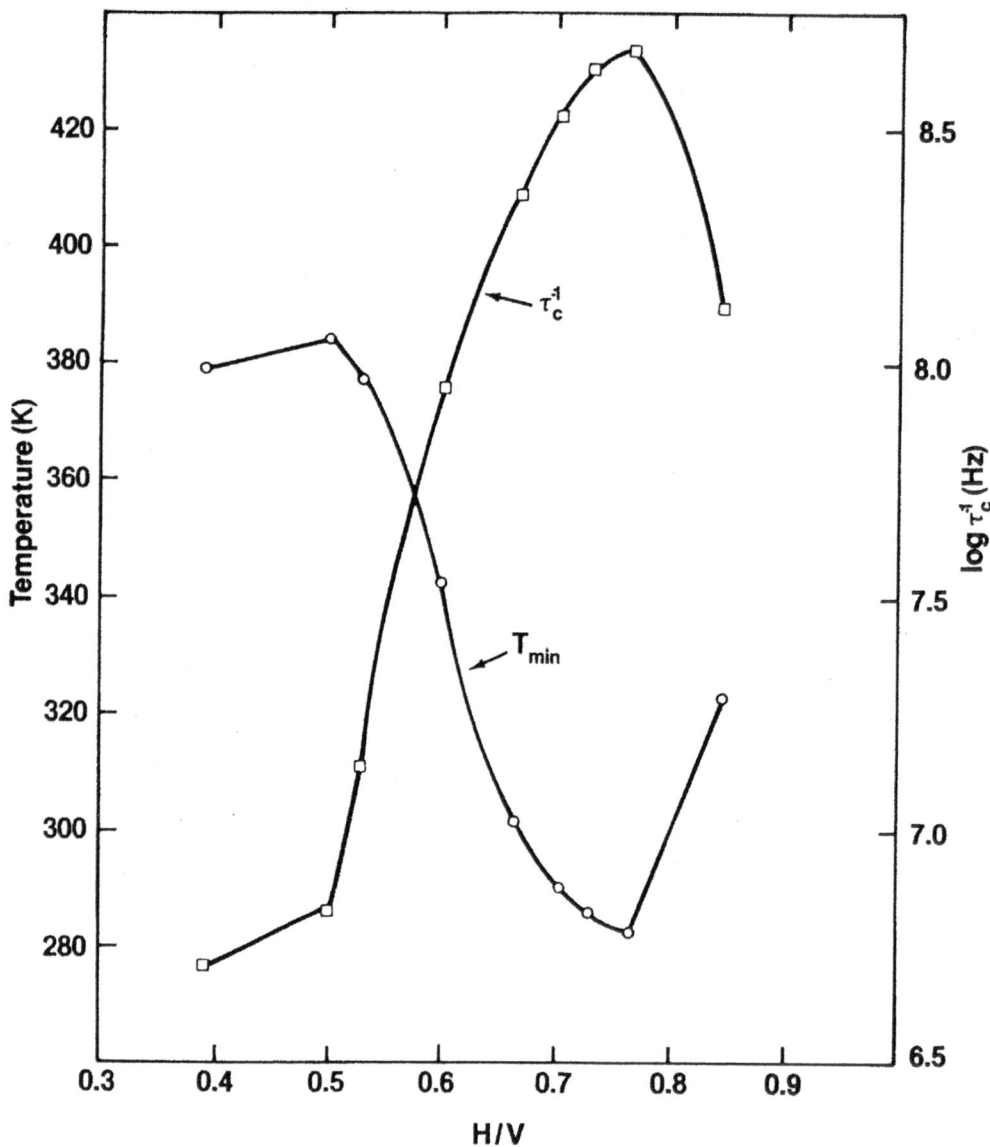

Fig. 21. Dependences of temperatures for T_{1d} minima and room temperature (300K) τ_c^{-1} values on hydrogen content in β-VH$_x$.

Hence, for any VH_x composition the T_1 minimum at a constant frequency defines a unique τ_c value. Figure 21 plots the temperature (T_{Min}) of T_{1d} minimum as well as log (τ_c^{-1}) at 300K versus hydrogen content for the nine VH_x samples. Since $D \sim \tau_c^{-1}$, both results verify the observations from Figure 16 that hydrogen mobility increases with x before reaching a maximum at $VH_{0.75}$ and decreases at higher stoichiometries.

Although much effort has been spent on developing a quantitative model for the diffusion mechanisms in β-VH_x, several factors contribute to make this a formidable task that remains to be completed. From Figures 13 and 15, it is apparent that for an H-atom to go from one O_{z1} site (i.e., the dominant equilibrium position when x < 0.7) to another it must also pass through several metastable tetrahedral sites as well as, possibly, at least one O_{z2} site. This process would be an extension of the one given by Figure 1b. Furthermore, the low-temperature (i.e., $\gtrsim$230K) diffusion behavior is complicated by the partially characterized phase transitions[21] indicated in Figure 14. Because the $\beta_1 \rightarrow \beta_2$ transition is second-order, disordering of H-atoms from O_{z1} to O_{z2} sites will occur well below T_c and this almost certainly results in a strong temperature dependent modification to the diffusion process. Consequently, it is not so surprising that the vacancy diffusion model which was quite successful for PdH_x and γ-TiH_x is inadequate for β-VH_x.

A possible model for hydrogen diffusion in β-VH_x between ~230K and ~450K is presented. When x < 0.5, only the O_{z1} sites are occupied (neglecting the $\beta_1 \rightarrow \beta_2$ transition effects for the moment) and mobility is restricted. However, as x increases above 0.5 to 0.7, the O_{z2} sites become partially occupied and provide an easier diffusion pathway. This view is supported by the recent theoretical analysis of Fedders[28] who identified some consequences of diffusion jumps involving partial occupancy of an intermediate inequivalent site on nuclear spin relaxation correlation functions. For T_1 relaxation by dipolar coupling with immobile metal spins which closely corresponds to β-VH_x, Fedders obtained[28]

$$T_1^{-1} \sim \exp \left[-(E_p+E_c)/k_BT\right] = \exp \left[-E_a/k_BT\right] \qquad (20)$$

at temperatures just below the T_1 minimum. Here, E_p is the energy to hop from the preferred O_{z1} sites and E_c is the energy to populate the metastable sites (i.e., O_{z2} sites where tetrahedral sites are neglected for simplicity) whose occupancy is proportional to $\exp (-2E_c/k_BT)$. Consequently, the net effect of increasing x up to ~0.7 is to reduce E_c until it vanishes (which is the point at which H-atoms are randomly distributed on both O_{z1} and O_{z2} sites and $\beta_1 \rightarrow \beta_2$). From Equation 20, E_a will subsequently decrease in this stoichiometry range providing E_p remains constant (or, possibly even decreases itself) which is observed in Figure 20.

For x > 0.7, there is no way to distinguish between the O_{z1} and O_{z2} sites and E_a remains very nearly constant and is now similar to the vacancy mechanism. Furthermore, increasing x beyond 0.75 fills up the O_z vacancies with the subsequent decrease in the diffusion rate for x > 0.75. The disordering from the $\beta_1 \rightarrow \beta_2$ transition may be responsible for the anomalously large E_a deduced from the τ_c data above the T_1 minima temperatures for samples with x < 0.7. Since the O_{z2} are being populated[23] by a temperature dependent disordering process, the diffusion process involving O_{z2} sites will be strongly enhanced. However, further studies that can accurately determine the relative population in the O_{z1} and O_{z2} (as well as, possibly, tetrahedral) sites as functions of temperature and stoichiometry are needed. More extensive knowledge of the phase compositions, structures, and diffusion behavior below 200K would also be valuable.

ISOTOPE EFFECTS FOR DIFFUSION IN THE METAL HYDRIDE PHASES

The large mass differences of the hydrogen isotopes will be readily apparent in the diffusion behavior. Isotope substitution on diffusion in the low concentration metal hydrides has had extensive experimental studies[1,2]. Much of this interest arises from efforts to identify and subsequently explain various non-classical diffusion mechanisms which often manifest themselves with large isotope dependent E_a and D_o values. Theoretical models have been developed[22] to rationalize many of the experimental observations in the infinite dilution limit; however, many questions remain to be resolved. For an ideal classical jump process (such as in Figure 1a) isotope substitution will not alter the observed E_a for the two isotopes A (lighter) and B (heavier) while the preexponential factor obeys

$$\frac{D_{oA}}{D_{oB}} \sim \frac{\tau_{oA}^{-1}}{\tau_{oB}^{-1}} \sim \sqrt{\frac{M_B}{M_A}} \tag{21}$$

where M_A and M_B are the atomic masses. Thus, in the classical diffusion limit the lighter species A always diffuses more rapidly by a factor $\sqrt{M_B/M_A}$ at all temperatures. However, various quantum diffusion processes have been identified to give[22] $E_{aA} < E_{aB}$ or $E_{aA} > E_{aB}$ as well as preexponential factors $\sim (M_B/M_A)^n$ with $n \neq 1/2$ but positive, negative, or zero. The character of the specific isotope effects are very sensitive to both host metal crystal structure and symmetry of occupied interstitial site.

In contrast to the many experimental studies[1,2] of diffusion isotope effects in the dilute hydrogen contents in metals, very few studies have been reported[1,2] in the high concentration hydride phases with essentially no formulation of detailed theoretical models as well. Table 3 summarizes the few NMR studies that permit comparison of diffusion isotope effects in the metal

TABLE 3. ISOTOPE EFFECTS FOR HYDROGEN DIFFUSION IN HIGH CONCENTRATION METAL HYDRIDE PHASES FROM NMR MEASUREMENTS OF T_1 VALUES.

System	Metal Lattice Structure	Dominant H-Isotope Sites	E_a (eV)	τ_o^{-1} (Hz)	Temperature Range (K)	Reference
γ-TiH$_{1.55}$	f.c.c.	Tet	0.507	1.8×10^{13}	300-733	Korn & Zamir[9]
γ-TiT$_{1.50}$	f.c.c.	Tet	0.50	2.1×10^{13}	463-743	Weaver[29]
α'-PdH$_{0.76}$	f.c.c.	Oct	0.22 (2)	5×10^{12}	200-333	Schoep et al.[13]
α'-PdD$_{0.70}$	f.c.c.	Oct	0.206 (10)	3×10^{12}	170-370	Bogdan et al.[30]
β-NbH$_{0.78}$	b.c.o.	Tet	0.22	1×10^{12}	225-300	Bohn[31]
β-NbD$_{0.78}$	b.c.o.	Tet	0.31	5×10^{12}	225-300	Bohn[31]
β-VH$_{0.50}$	b.c.t.	Oct	0.41 (2)	5×10^{13}	320-390	Bowman et al.[32]
β-VT$_{0.50}$	b.c.t.	Oct	0.32(3)	5×10^{12}	290-345	Bowman et al.[32]

hydride phases. Both γ-TiH(T)$_x$ and β-PdH(D)$_x$ show nearly classical isotope effects with E_a being nearly isotope independent and the τ_o^{-1} ratios are close to the ratio in Equation 21. However, both of the hydrides formed from the bcc host metal show very large deviations with classical predictions. Furthermore, the E_a and τ_o ratios are in opposite directions for β-NbH(D)$_x$ and β-VH(T)$_x$. The β-NbH(D)$_x$ isotope effects are similar to those found[1,2] for the bcc α'-phases which is reasonable since the hydrogen isotopes occupy tetrahedral sites in both situations and nearly the same quantum diffusion mechanisms may apply.

The diffusion behavior in β-VH$_{0.50}$ and β-VT$_{0.50}$ is most unusual where $E_{aT}/E_{aH} = 0.78$ and in temperatures below the T_1 minima the BPP-model calculations yield absolute τ_c^{-1} that are ~5-10 times larger for VT$_{0.50}$ relative to VH$_{0.50}$. Since $D \sim \tau_c^{-1}$, larger τ_c^{-1} and smaller E_a imply T-atoms in β-VT$_{0.50}$ diffuse much more rapidly than H-atoms in β-VH$_{0.50}$. This is an inverse isotope effect where heavier isotopes generally[1,2] diffuse more slowly. Since β-VH$_{0.50}$ and β-VT$_{0.50}$ are isomorphic, the basic diffusion jump processes must be similar. However, there is strong evidence from NMR studies[32] that T-atoms in β-VT$_{0.50}$ are inherently more disordered with occupancy of some sites in addition to the predominant O_{z1} down to ~150K. A rather complex process involving atomic exchange of T-atoms between these inequivalent sites has been proposed[32] to contribute to the β-VT$_{0.50}$ diffusion in a manner completely analogous to the one described in the previous section for VH$_x$ (x > 0.5). In β-VT$_{0.50}$ the partial occpancy of intermediate sites involved in diffusion would again give a reduced E_c value for Equation 20 resulting in a smaller apparent E_a and larger τ_c^{-1} compared to β-VH$_{0.50}$. Further experiments and analyses are required before a quantitative model of these isotope effects can be established.

CONCLUDING REMARKS

The present paper has attempted an introductory description of diffusion behavior in metal hydride phases with an emphasis on the relationship between structure and the atomic jump processes. This field of research is currently quite active and many hydride phases formed from metal and intermetallic compounds are being investigated by the experimental techniques described in this paper. Besides characterizing hydrogen mobility in hydrides with potential technological applications, much effort is directed towards improved understanding of the atomic diffusion processes and the specific roles of variations in structure and partial substitution of metal atoms.

ACKNOWLEDGEMENTS

The author appreciates the collaboration of A. Attalla, G. Bambakidis, B. D. Craft, A. Maeland, M. W. Pershing, and W. E.

Tadlock in obtaining some of the results presented in this paper. This work has been supported by the Chemical Sciences Branch, Office of Basic Energy Sciences, U. S. Department of Energy.

REFERENCES

1. J. Völkl and G. Alefeld in Diffusion in Solids, Recent Developments, Ed. A. S. Nowick and J. J. Burton (Academic, New York, 1975), p. 231.
2. J. Völkl and G. Alefeld in Hydrogen in Metals I - Basic Properties, Ed. by G. Alefeld and J. Völkl (Springer-Verlag, Berlin, 1978), p. 321.
3. R. M. Cotts, Ber. Bunsenges. Physik. Chem. 76, 760 (1972).
4. R. M. Cotts, in Hydrogen in Metals I - Basic Properties, Ed. by G. Alefeld and J. Völkl (Springer-Verlag, Berlin, 1978), p. 227.
5. K. Sköld, in Hydrogen in Metals I - Basic Properties, Ed. by G. Alefeld and J. Völkl (Springer-Verlag, Berlin, 1978), p. 267.
6. A. Abragam, Principles of Nuclear Magnetism (Oxford, London, 1961).
7. H. Y. Carr and E. M. Purcell, Phys. Rev. 94, 630 (1954); S. Meiboom and D. Gill, Rev. Sci. Instru. 29, 688 (1958).
8. D. Zamir and R. M. Cotts, Phys. Rev. 134, A666 (1964); D. Zamir and R. M. Cotts, Proc. XIII Colloque Ampere (North Holland, Amsterdam (1964)), p. 276.
9. C. Korn and D. Zamir, J. Phys. Chem. Solids 31, 489 (1970).
10. E. O. Stejskal and J. E. Tanner, J. Chem. Phys. 42, 288 (1965).
11. W. D. Williams, E. F. W. Seymour, and R. M. Cotts, J. Magn. Reson. 31, 271 (1978).
12. R. F. Karlicek, Jr. and I. J. Lowe, J. Magn. Reson. 37, 75 (1980).
13. G. K. Schoep, N. J. Poulis, and R. R. Arons, Physica 75, 297 (1974).
14. P. P. Davis, E. F. W. Seymour, D. Zamir, W. D. Williams, and R. M. Cotts, J. Less-Comm. Met. 49, 159 (1976).
15. A. Schmolz and F. Noack, Ber. Bunsenges, Phys. Chem. 78, 339 (1974).
16. L. D. Bustard, R. M. Cotts, and E. F. W. Seymour, Zeitz. Phys. Chem. N. F. 115, 247 (1979); L. D. Bustard, R. M. Cotts, and E. F. W. Seymour, submitted to Phys. Rev.
17. C. L. Bisson and W. D. Wilson, in Effects Hydrogen Behavior Proc. Int. Conf. 1975, A. W. Thompson and I. M. Berstein (Eds.) AIME (1976), p. 416.

18. C. F. Melius and T. H. Upton, Bull. Am. Phys. Soc. 23, 234 (1978).
19. R. C. Bowman, Jr., A. Attalla, and A. J. Maeland, Solid State Comm. 27, 501 (1978).

20. A. Santoro, A. Maeland, and J. J. Rush, Acta Cryst. $\underline{B34}$, 3059
     (1978).

21. T. Schober and H. Wenzl, in Hydrogen in Metals II - Application
     Orientation Properties, Ed. by G. Alefeld and J. Völkl
     (Springer-Verlag, 1978), p. 11.

22. K. W. Kehr, in Hydrogen in Metals I - Basic Properties, Ed.
     by G. Alefeld and J. Völkl (Springer-Verlag, Berlin, 1978),
     p. 197.

23. G. Bambakidis, M. W. Pershing, and R. C. Bowman, Jr., Scrip.
     Met. $\underline{13}$, 441 (1979).

24. R. C. Bowman, Jr. and W. E. Tadlock (to be published).

25. H. Asano and M. Hirabayashi in 2nd Int. Cong. Hydrogen in
     Metal, Paris, 1977, 106.

26. R. C. Bowman, Jr., A. Attalla, and W. E. Tadlock, Int. J.
     Hydrogen Energy $\underline{1}$, 421 (1977).

27. Y. Fukai and S. Kazama, Acta Met. $\underline{25}$, 59 (1977).

28. P. A. Fedders, Phys. Rev. $\underline{B18}$, 1055 (1978).

29. H. T. Weaver, Phys. Lett $\underline{35A}$, 417 (1971); H. T. Weaver and
     J. P. Van Dyke, Phys. Rev. $\underline{B6}$, 694 (1972).

30. M. Bogdan, V. Simplaceanu, and D. Lupu, in Proceedings of
     XXth Congress AMPERE, Ed. by E. Kundla, E. Lippman, and T.
     Saluvere (Springer-Verlag, Berlin, 1979), p. 131.

31. H. G. Bohn, Jül-Ber. Jül-853-FF (Report), (1972).

32. R. C. Bowman, Jr., A. Attalla, and B. D. Craft (to be
     published).

AN INTRODUCTION TO HYDROGEN IN ALLOYS[*]

D. G. Westlake

Materials Science Division
Argonne National Laboratory
Argonne, IL   60439

ABSTRACT

Substitutional alloys, both those that form hydrides and those that do not, are discussed, but with more emphasis on the former than the latter. This overview includes the following closely related subjects: 1) the significant effects of substitutional solutes on the pressure-composition-temperature (PCT) equilibria of metal-hydrogen systems, 2) the changes in thermodynamic properties resulting from differences in atom size and from modifications of electronic structure, 3) attractive and repulsive interactions between H and solute atoms and the effects of such interactions on the pressure dependent solubility for H, 4) "H trapping" in alloys of Group V metals and its effect on the terminal solubility for H (TSH), 5) some other mechanisms invoked to explain the enhancement (due to alloying) of the (TSH) in Group V metals, and 6) "H-impurity complexes" in alloys of the metals Ni, Co, and Fe. Some results showing that an enhanced TSH may ameliorate the resistance of a metal to hydrogen embrittlement are presented. Recent studies of resistivity and of elastic constants are summarized as just two examples of the important effects that interstitial hydrogen can have on the properties of metals and their alloys. Finally, a potential hydrogen storage material, ZrNi, is considered. From empirical rules, predictions are made regarding both the particular sites that can be occupied by hydrogen in this intermetallic compound and the resultant stoichiometries.

---

[*] Work supported by the U.S. Department of Energy.

INTRODUCTION

As the title of this contribution suggests, it is meant to introduce the reader to a few aspects of the behavior and effects of hydrogen in alloys. Because of space considerations, it could not be a comprehensive review; some properties and some theories and experimental techniques have not even been mentioned. The list of references is by no means complete either. Those that are given here, however, refer to many others, and the interested reader may use the present list as a nucleus for subsequent growth.

PCT EQUILIBRIA

For the study of metal-hydrogen (M-H) systems, one of the most useful diagrams is that showing pressure-composition-temperature (PCT) equilibria. Figure 1 is a schematic representation of such a diagram[1] for $Zr-ZrH_2$. At low H concentrations, the isotherms are shown as straight-lines in accord with Sieverts' Law (Equation 1).

$$P^{1/2} = K^{-1}n, \qquad\qquad (1)$$

where $n \equiv H/M$ (hydrogen to metal ratio). K is the equilibrium

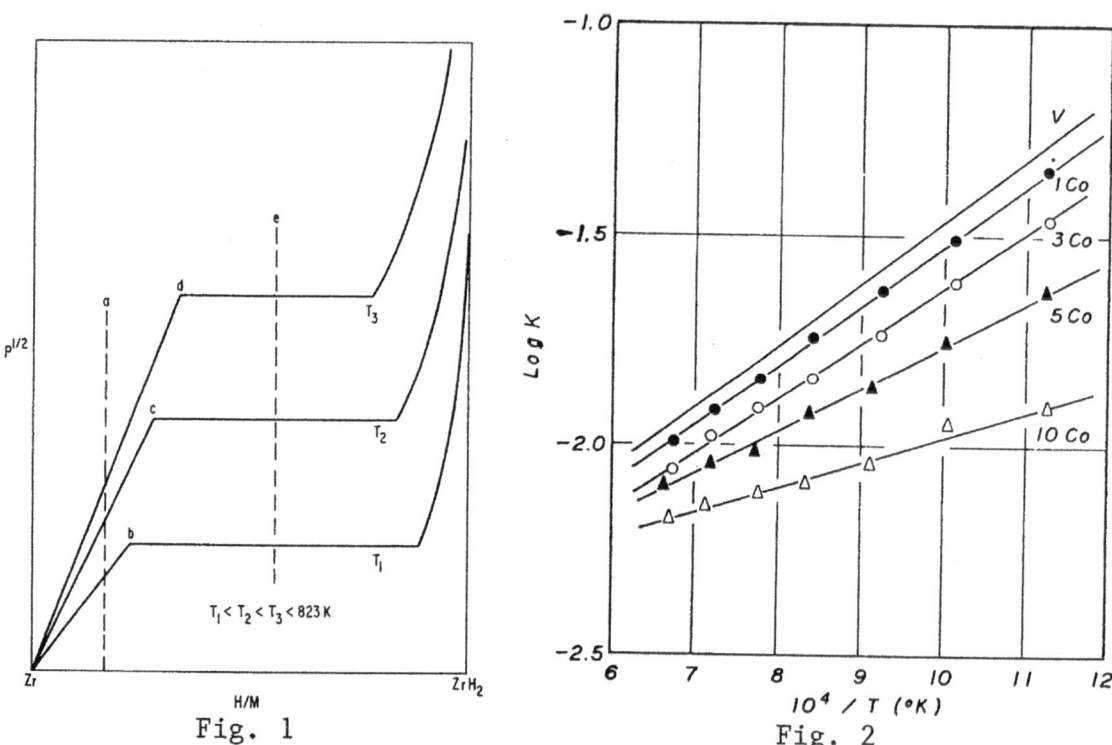

Fig. 1

Fig. 2

Fig. 1. Schematic pressure-composition isotherms for the $Zr-ZrH_2$ system.

Fig. 2. Temperature dependence of hydrogen solubility in various V-Co alloys.[2]

constant for the reaction 1/2 H_2 (gas) $\rightarrow$ H (in solution in metal) and is given by

$$K = n_s \exp (-\overline{\Delta H}/RT) \exp (\overline{\Delta S}/R) .$$

Here, n_s is the maximum value of n derived from the crystal structure of the solvent, $\overline{\Delta H}$ is the relative partial molar enthalpy of solution of hydrogen in the metal, and $\overline{\Delta S}$ is the relative partial molar entropy of solution. For a given isochore (a, for example), one can plot ℓn K vs 1/T to obtain $\overline{\Delta H}$ from the slope and $\overline{\Delta S}$ from the intercept.

The points b,c, and d in Fig. 1 represent the terminal solubilities for hydrogen (TSH) at three different temperatures. A plot of ℓn (TSH) vs 1/T yields ΔH_s, the heat of solution of the hydride phase in the solid solution. The compositions of the two phases are given by the respective extremities of the plateau. For an isochore such as e, one can obtain ΔH_f (the heat of formation of the hydride phase from the saturated solid solution) from the slope of a plot of ℓn $P^{1/2}$ vs 1/T.

The value of K (in Equation 1) changes with the addition of alloying elements. Examples are shown in Figs. 2[2] and 3[3] for V-Co and Mo-Ti alloys, respectively. For these experiments P = 1 atm., so the numerical value of K is the same as the solubility, n. In Fig. 2, if we use the isotherm at 1000°K, we obtain the open triangular symbols shown in the top half of Fig. 4. With the addition of Co to V and the concomitant increase in e/a ratio, the value of K decreases. Similar behavior is observed with the addition of Fe and Cr to V. Addition of Ti, a group IV element, however, increases the value of K. The way these changes affect $\overline{\Delta H}$ is shown at the bottom of Fig. 4. Obviously, $\overline{\Delta H}$ becomes more negative with decreasing values of e/a and less negative with the addition of elements that increase the value of e/a. Figure 5 shows that the variation of $\overline{\Delta S}$ with e/a is relatively small, but certainly significant. For both $\overline{\Delta H}$ and $\overline{\Delta S}$, these are the same qualitative trends reported by Lynch et al.[4] for V-Cr alloys.

Burch and Mason[5] have calculated the contributions of the enthalpies of lattice expansion and of electronic effects to the relative partial molar enthalpy of hydrogen in body-centered cubic metals. They found that the enthalpy due to lattice expansion falls in the range 50-100 kJ per g-atom of H, while the electronic contribution is roughly -1500 kJ(g-atom)$^{-1}$. If we take the V-Cr-H system as an example, we would conclude that the observed changes in $\overline{\Delta H}$ are effected primarily by modification of electronic structure and not by the 5% difference in atom size between V and Cr. A similar explanation would seem to be fitting for the observed[3,6-8] changes in $\overline{\Delta H}$ for the Nb-Mo-H system.

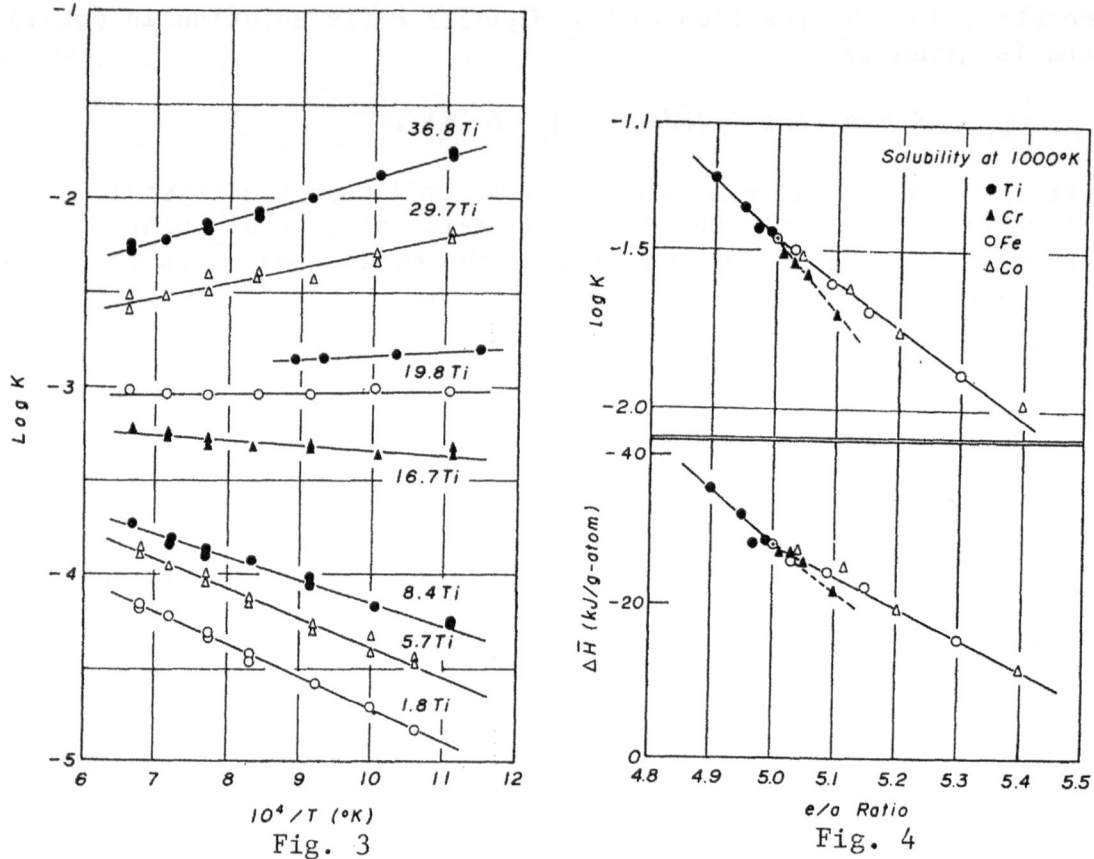

Fig. 3. Temperature dependence of hydrogen solubility in various Mo-Ti alloys.[3]

Fig. 4. Top: Dependence of the hydrogen solubility at 1000°K on the electron to atom ratio for various alloys of vanadium.[2] Bottom: Dependence of the relative partial molar enthalpy for solution of H in vanadium alloys on the electron to atom ratio.[2]

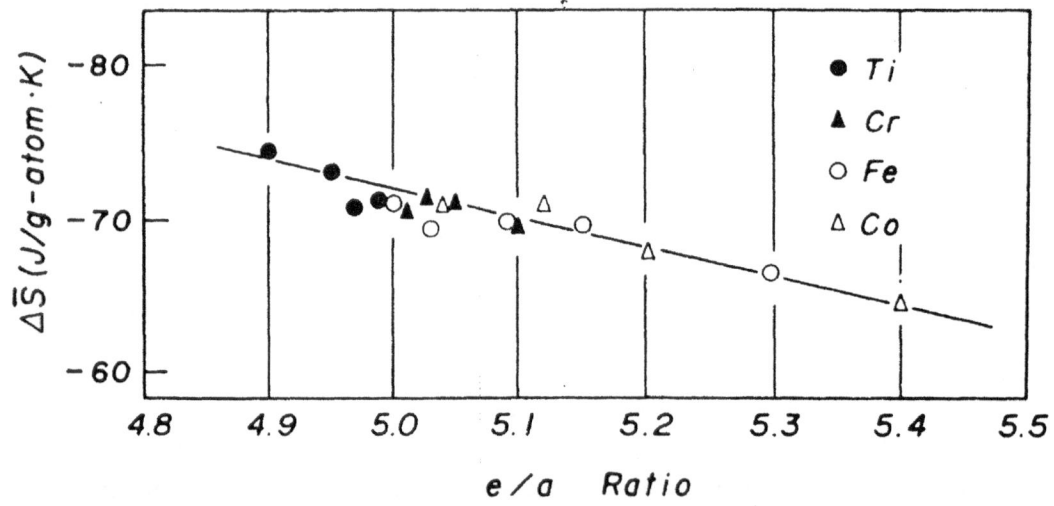

Fig. 5. Dependence of the relative partial molar entropy for solution of H in vanadium alloys on the electron to atom ratio.[2]

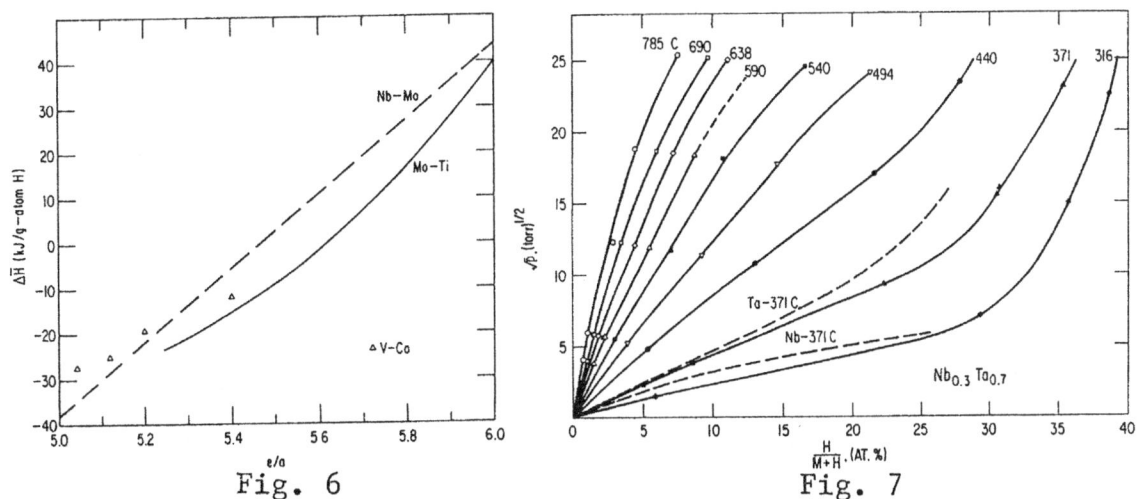

Fig. 6. Variation of the relative partial molar enthalpy for
        solution of H with electron to atom ratio in Nb-Mo[8],
        Mo-Ti[3] and V-Co[2] alloys.

Fig. 7. Pressure-composition isotherms for the $Nb_{0.3}Ta_{0.7}$-H
        system.[9]

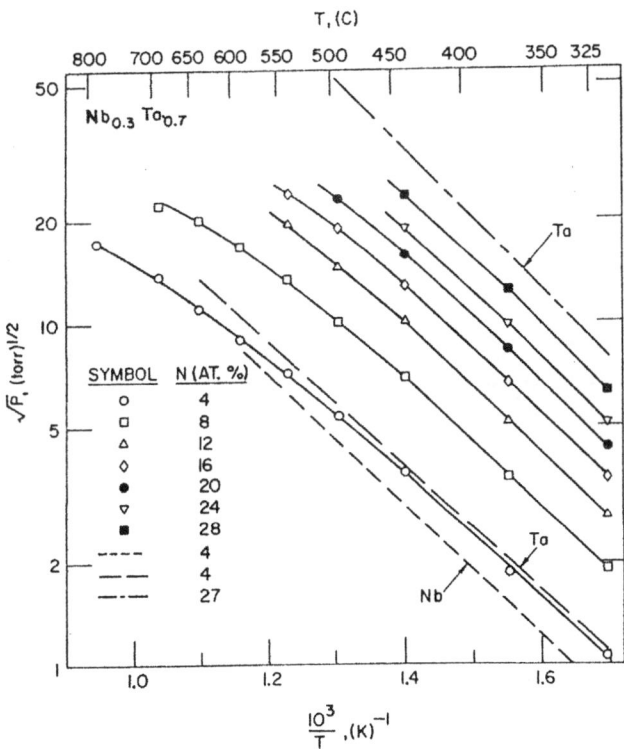

Fig. 8. van't Hoff plots for the $Nb_{0.3}Ta_{0.7}$-H system[9]. One
        isochore for Nb and two for Ta are shown for
        comparison.[10a]

The case of Mo-Ti alloys shown in Fig. 3 is an especially interesting one. At low concentrations of Ti, plots of ℓn K vs $1/T$ have negative slope meaning, of course, that $\overline{\Delta H}$ is endothermic. At 19.8 at.% Ti ($e/a = 5.60$), the slope approximates zero, and it becomes positive for higher Ti concentrations. The variation of $\overline{\Delta H}$ with e/a is shown in Fig. 6[3]. The four triangles (from Fig. 4) are for the V-Co system[2] and the dashed line is for the Nb-Mo system[8]. Obviously, for given e/a ratios, the values of $\overline{\Delta H}$ for the three systems are similar. Eguchi and Morozumi[3] have shown that $\overline{\Delta S}$ remains constant in Mo-Ti alloys for $5.3 < e/a < 6$.

PCT results for the $Nb_{0.3}Ta_{0.7}$-H system[9] are shown in Fig. 7. For comparison, we show that the isotherm at 371°C falls between those[10a] for the pure metals Ta and Nb. This collection of isotherms yields the van't Hoff plots shown in Fig. 8, from which we obtain the values of $\overline{\Delta H}$ and $\overline{\Delta S}$ shown in Figs. 9 and 10, respectively. Obviously, both $\overline{\Delta H}$ and $\overline{\Delta S}$ remain virtually unchanged in this alloy system for which the apparent e/a ratio remains constant at 5.

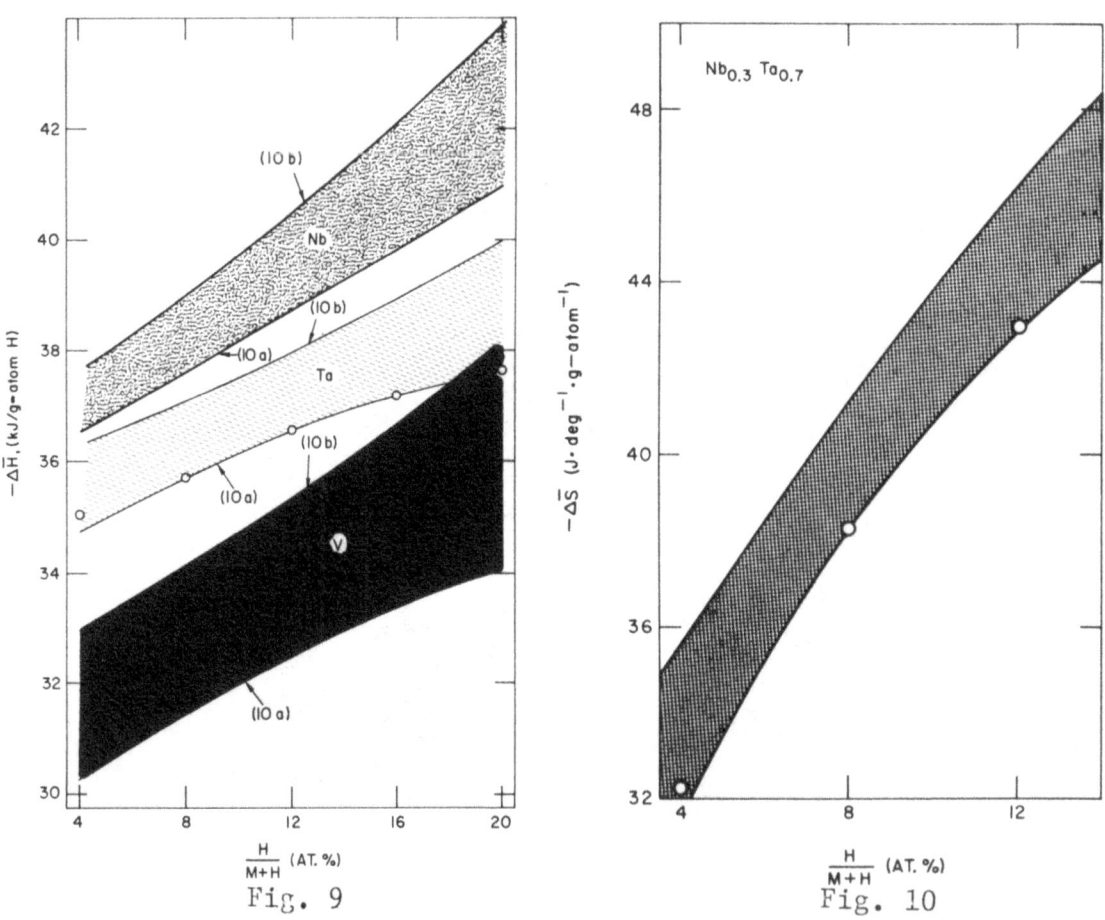

Fig. 9   Fig. 10

Figs. 9 & 10. Relative partial molar enthalpies and entropies for various hydrogen concentrations in $Nb_{0.3}Ta_{0.7}$ (o) compared with published ranges for Nb, Ta, and V.[9]

Similar studies of metals like Fe and Ni and of their alloys are made more difficult by the fact that the pressure dependent solubilities for H are very low. For example, in a study of the Fe-Ni-H-system[11], the solubilities (H/M) at 1 atmosphere H_2 pressure were less than 5×10^{-4}. As a result, the thermodynamic data obtained in such studies show considerable experimental scatter.

TERMINAL SOLUBILITY FOR HYDROGEN

In recent years, the terminal solubility for hydrogen (TSH) in V, Nb, and Ta has been studied by several techniques: resistometry[12], differential thermal analysis[13], optical metallography[14], internal friction[15], and others. For a given temperature T, the TSH represents the H concentration of the α phase of a metal when it is in equilibrium with the hydride phase. In other words, a plot of T vs TSH is the solvus of the metal-metal hydride phase diagram. At $273^{\circ}K$, the TSH is 100 H/(M+H) $\simeq$ 1.5 at.% for V, ~ 2.5 at.% for Nb and ~9 at.% for Ta.

Only recently were any attempts made to determine the influence of alloying on the TSH[9,16-20]. Two factors that could

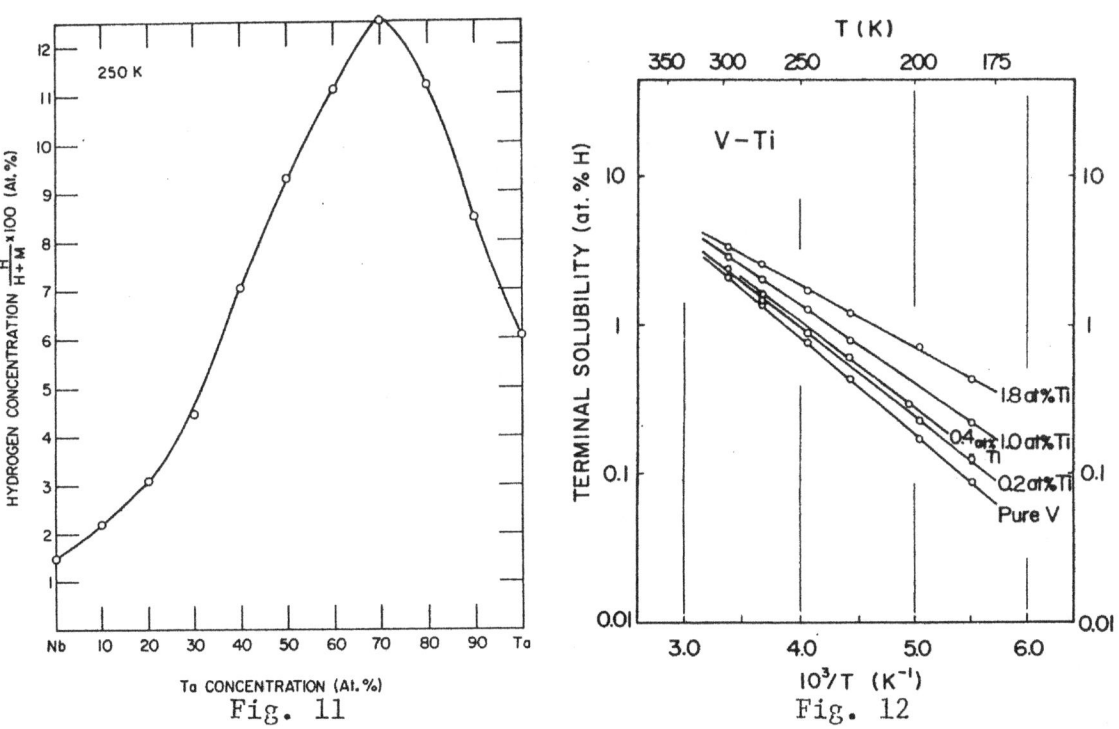

Fig. 11

Fig. 12

Fig. 11. Terminal solubilities for H in α-phase Nb-Ta alloys at $250^{\circ}K$.[9]

Fig. 12. Temperature dependence of the terminal solubilities for H in various V-Ti alloys.[19]

be important are electronic structure and atomic volume. As an
initial step in differentiating between these, Westlake and
Miller[9] undertook the investigation of Nb-Ta alloys. These two
metals were selected because their Goldschmidt atomic diameters
differ by only 0.06% and they are isovalent. Both the lattice
parameter[9] and the electron density-of-states at the Fermi
surface[21] are linear functions of composition in the Nb-Ta
system. Because of the simplicity of this system, it was
anticipated that the TSH would increase monotonically with
additions of Ta to Nb, thereby providing a base line for
subsequent measurements on other alloys in which e/a and atomic
volume would be allowed to vary, one at a time. In fact, however,
the TSH exhibits a maximum at all temperatures for the alloy
$Nb_{0.3}Ta_{0.7}$. The results at $250^{\circ}K$ are shown in Fig. 11.

In the preceding section, it was pointed out that
both $\overline{\Delta H}$ and $\overline{\Delta S}$ for the solution of H in $Nb_{0.3}Ta_{0.7}$ are nearly the
same as those for the pure metals. Thus, the observed enhancement
of TSH does not seem to be attributable to changes in the
thermodynamic properties of the solid solutions. Variations in
the elastic properties of the alloys and in the mechanical
constraint to hydride formation could conceivably affect the TSH,
but their roles have been minimized[9].

Before further discussion of the mechanisms for enhancement
of the TSH, let us consider some of the other systems that have
been investigated. Titanium additions have important effects on
both V^{19} and Nb^{20}, as shown in Figs. 12 and 13, respectively. The
linear relationship in Fig. 12 allows the calculation of ΔH_s, the
heat of solution of the hydride phase in the saturated solid
solution. For V, $\Delta H_s = 12.3$ kJ/g-atom H, in agreement with Ref.
12, and the addition of Ti to V causes ΔH_s to decrease.

It is from solvi such as those shown in Fig. 13, that we
obtain isothermal values of TSH for an alloy system. Our results
at $250^{\circ}K$ for Nb-V alloys are presented in Fig. 14 and compared
with the results for Nb-Ta alloys (shown also in Fig. 11). While
the enhancement of TSH in Nb-Ta alloys is very significant, it
appears small relative to that in the Nb-V system. The TSH in
$Nb_{0.5}V_{0.5}$ at $250^{\circ}K$ is unknown but apparently very high; this alloy
was charged to 30 at.% H and slowly cooled all the way to $77^{\circ}K$
without the initiation of hydride precipitation. A few
measurements have been made on Nb-Mo alloys, also[18,20]. Figure 15
compares the TSH's in the four different Nb alloy systems at
$230^{\circ}K$. For the enhancement of TSH, at all concentrations, Ta is
the least effective solute in Nb. At low concentrations, V is the
most effective, but Mo becomes most effective at a solute
concentration of 10 at.%.

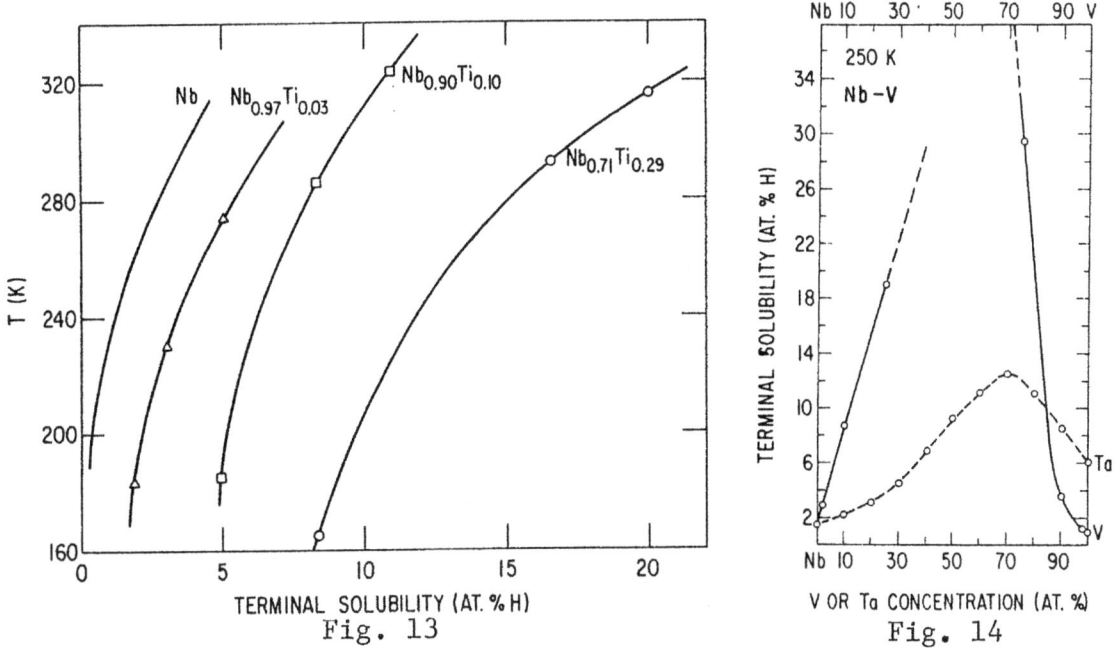

Fig. 13

Fig. 14

Fig. 13. Partial phase diagrams (hydrogen solvi) for Nb-Ti-H alloys.[20]

Fig. 14. Terminal solubilities for hydrogen at 250°K in Nb-V and Nb-Ta alloys.[20]

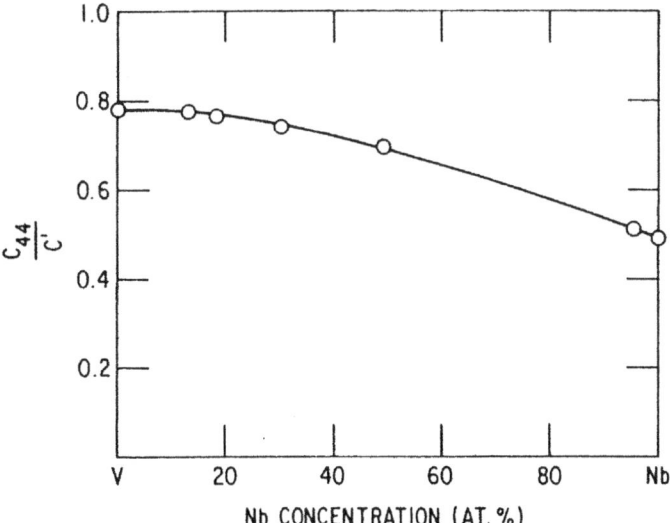

Fig. 15. Effect of metal solutes in Nb on the terminal solubility for hydrogen at 230°K.[20]

Several explanations for the observed enhancement of TSH are conceivable, and it seems likely that there may be more than one mechanism involved. Matsumoto et al.[16] suggested an attractive interaction between one solute atom and one H atom arising from the elastic strain around an undersized solute atom. One can rationalize H "trapping" at oversized solutes, also[9]. For V-Ti alloys with less than 4% Ti, Tanaka and Kimura[19] reported that H is bound to single Ti atoms with an energy of 0.15 eV. After studying the effects of several solutes on H diffusivity in V, however, they have concluded that trapping does not correlate well with atom-size and that the binding may be of electronic origin. Moreover, they have shown that the apparent binding energy between Ti and H in a V solvent increases for Ti concentrations greater than 3.9 at.%, and this correlates well with the onset of Ti segregation that they observe in this alloy system. Further evidence that "one-on-one" H-solute interactions cannot be the only mechanism involved in the enhancement of TSH is obvious in Figs. 13 and 14. At solute concentrations greatly in excess of 10 at.%, the TSH continues to increase rapidly with solute additions, despite the fact that, at these concentrations, even in a random solid solution, very few solute atoms would not have at least one other solute atom for a nearest neighbor.

The literature contains other reports of clustering and short-range order in bcc solid solutions. Rudman[22] used x-ray diffuse scattering to show that, in Nb, Ti solute atoms cluster moderately ($\alpha_1 \simeq 0.10 \pm 0.05$, where α_1 is the short range order parameter for the first coordination sphere), a phenomenon he attributed almost wholly to the valence difference between Nb and Ti, because he considered the atom-size difference so small as to be nearly negligible. He quoted from other sources that clustering occurs in Mo-W alloys, also, and short range order (SRO) occurs in Ti-Mo alloys. With the limited information available to him, he postulated, for bcc alloys of these transition metals, "a general tendency toward clustering with the superposition of a size-difference driving force toward ordering". Later studies, in which the same experimental technique was used, seem to support the postulate. In $Nb_{0.50}Ta_{0.50}$, with the atom size difference less than 0.06%, clustering has been reported[23], and in $Nb_{0.50}V_{0.50}$ and $Ta_{0.50}V_{0.50}$, for which the atom-size difference is nearly 8%, SRO has been reported[24]. Investigations of the electronic structures of Nb alloys have recently been reported[25,26] and it could be fruitful to try to correlate these with studies of local order and with the enhancement of TSH. Abell[27], in a theoretical treatment, has invoked a d-band resonance state localized on the nearest-neighbor metal atoms of an interstitial H, to explain distortion of the cubic lattice, hydrogen diffusion and excess partial entropy. He suggests that, within the framework of his model, the enhancement of TSH due to alloying is expected.

As one tries to rationalize the observations, an important fact to bear in mind is that the TSH is a point on the solvus of the metal-metal hydride phase diagram. It is, therefore, a function of the properties of the hydride as well as the properties of the metal alloy. One may not be able to fully comprehend the mechanism of enhanced TSH's without investigating the properties of the relevant ternary hydrides. Stalinski and Nowak[28] have determined crystal structures and lattice parameters for the hydrides of Nb-Ti alloys, but little else is known about any of the properties of other alloy hydrides discussed in this section.

HYDROGEN TRAPPING

There is ample evidence that H can be "trapped" by other interstitial impurities in Nb metal. Chen and Birnbaum[29] have discussed some of the published reports and also presented their results from internal friction and stress-relaxation measurements. One O atom, occupying an octahedral interstice, traps one H atom, and the observed relaxations are due to stress-induced hopping of the H around the O. They were not able to ascertain the sites occupied by the H or the jump paths. For a different case, that of an O-H pair in V metal, Ozawa et al[30] used channeling of ^{3}He$^+$ ions to show that the H and the O occupy nearest-neighbor octahedral interstices.

As mentioned in the preceding section, the enhancement of TSH in both Nb-V and Nb-Mo alloys has been attributed to H trapping by the substitutional solutes[16-18]. The binding energies for V-H and Mo-H pairs have been estimated to be 0.2 eV[18] and 0.07 eV[16], respectively. These values, however, are now being contested. Pick and Welch[31] have developed a theoretical model for the Nb-V-H system from which they predict the pressure-composition isotherm for 423°K shown in Fig. 16. For $Nb_{0.94}V_{0.06}$, their calculated isotherm (423°K) deviates in a positive sense, from the calculated isotherm for pure Nb[10a] when the V-H binding energy is taken to be 0.2 eV. At this temperature, their experimental plot does not exhibit a positive deviation from Sievert's law, as did their calculated plot, but this does not allow one to conclude that there is no trapping at lower temperatures. They have concluded, however, that, if trapping occurs, the binding energy of the V-H complex must be considerably less than the assumed 0.2 eV. From other measurements of PCT, they have estimated that the critical temperature for the $\alpha + \alpha'$ phase field is reduced from 443°K to slightly less than 418°K by the addition of 6 at.% V, and this allows them to calculate a binding energy of ~ 0.07 eV. This is in reasonably good agreement with the value 0.09 eV obtained from NMR measurements by Matsumoto[17] and with 0.095 eV obtained from quasielastic neutron scattering measurements[32]. If we use their values of the pressure for H/M = 0.05 and T = 418, 433, and

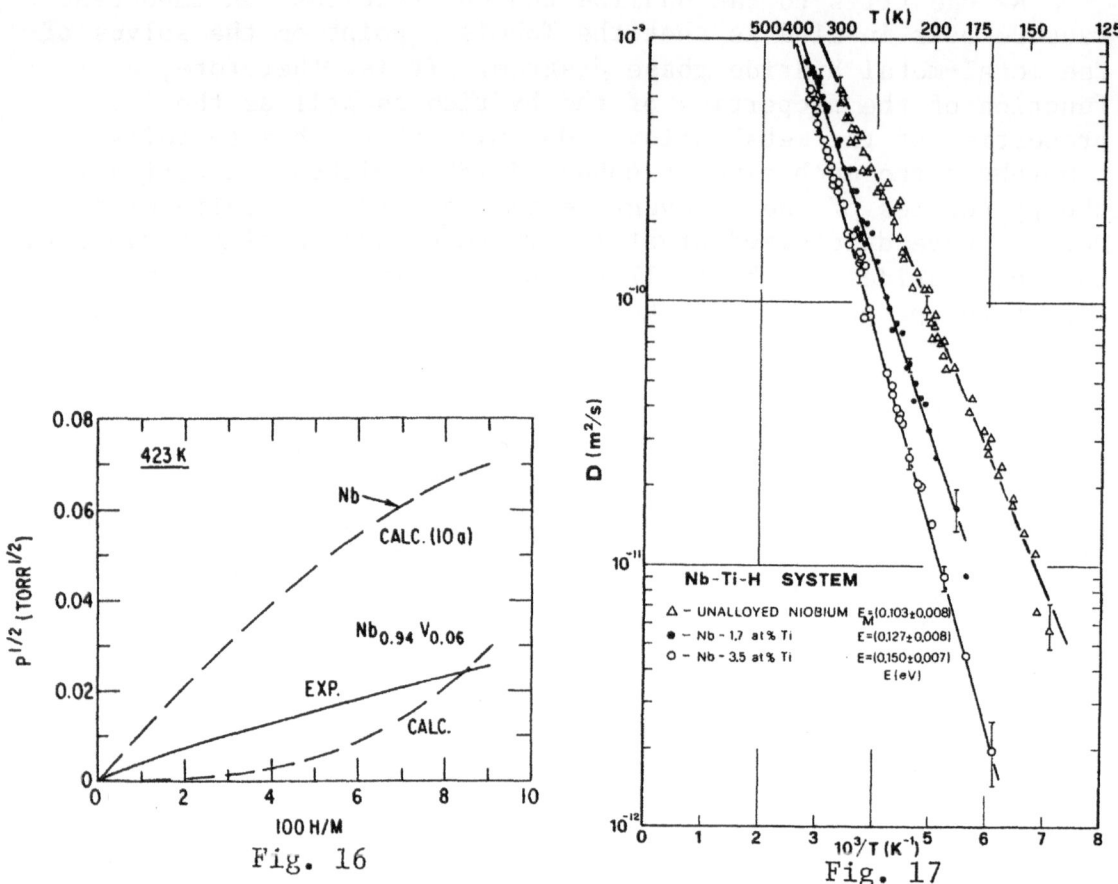

Fig. 16. Experimental pressure-composition diagram for the Nbv$_{0.94}$V$_{0.06}$-H system at 423°K[31] compared with calculated plots for the Nb-H[10a] and Nb$_{0.94}$V$_{0.06}$-H[31] systems.

Fig. 17. Temperature dependence of hydrogen diffusivity in Nb-Ti alloys.[33]

473° K, we can estimate that the value of $\overline{\Delta H}$ is roughly −44 kJ/g-atom H compared with −36.7 kJ/g-atom H for pure Nb[10a]. The increase in enthalpy is consistent with trapping.

As was mentioned earlier, Inoue et al.[6] reported that $\overline{\Delta H}$ for the solution of H in Nb becomes less negative as e/a is increased by the addition of Mo. They conclude from this that the interaction between Mo and H must be repulsive, a contradiction to the trapping model[16,18] for Nb-Mo-H alloys.

The occurrence of H trapping in Nb-Ti alloys has been inferred from studies of diffusion. Measurement of the Gorsky effect in Nb-Ti-H alloys has produced the plots of H diffusivity shown in Fig. 17.[33] The apparent activation energy for H diffusion increases with increasing Ti concentration. The word apparent is used because, in this case, a given value does not represent the activation energy for just one type of jump.

Instead, each H atom makes at least four distinguishable types of jump[34]: from one host lattice site to another, from a host lattice site to a Ti trap site, from one trap site to another, and from a trap site to a host lattice site. From their determination of D as a function of Ti concentration, Cannelli and Cantelli[33] have estimated a Ti-H binding energy $\Delta E_B \geqslant 0.06$ eV. In their studies of internal friction[35], for a given concentration of H, the addition of 2 at.% Ti to Nb caused the hydride precipitation temperature to be markedly lower. In Nb with 4.2 at.% H, they observed hydride precipitation at about 300°K, but, with a similar H concentration in $Nb_{0.95}Ti_{0.05}$, they could not detect the formation of any hydride down to 50°K. It was concluded that the partial removal of H from solid solution in a trapped state was the cause of the effective increase in terminal solubility.

Krönmuller et al.[36] have measured the magnetic after-effect in Ni, Co, and Fe alloyed with transition metals and charged with H. This proves to be a highly sensitive technique for detection of Snoek-type relaxations arising from the interaction between "metal-hydrogen (M-H) complexes" and domain walls. They found that, in bcc binary alloys, M-H complexes were formed if the substitutional solute atom has a smaller number of d-electrons than the host metal. Thus, in an Fe host, Ti and Zr solutes formed M-H complexes, but Ni, Co, and Pd solutes did not. In all fourteen of the Ni-based fcc alloys that they studied, M-H complexes were formed, regardless of the number of d-electrons in the solute atom.

In the body centered lattice, the reorientation process of M-H complexes has been modeled[36] as shown in Fig. 18. In a tetragonal configuration, sites 1, 2, and 3 are nearest neighbor octahedral interstices for the solute (Ti) atom, whereas sites 4-9 are next nearest neighbors. Two jump frequencies, indicated on the energy diagram, are involved as the H occupies the two types of site, alternately. For a monoclinic configuration (Fig. 18b) also, there would be two jump freqencies, one for the tetrahedral-tetrahedral (t-t) jumps in a single (100) plane and another for t-t jumps out of that plane. Magnetic after-effect measurements cannot distinguish between the two configurations.

In a face-centered lattice (Fig. 19), a M-H complex in a tetragonal configuration would reorient by the motion of H^+ from one nearest neighbor octahedral site to another (1,2,3,or 4)[37]. The jump path, however, would probably be through a tetrahedral site, so two jump frequencies would be involved in this lattice, also.

Hohler and Krönmuller[37] have shown that Q, the activation energy for the reorientation of M-H complexes in Ni, is not correlated with the number of outer (d + s + p) electrons, but is

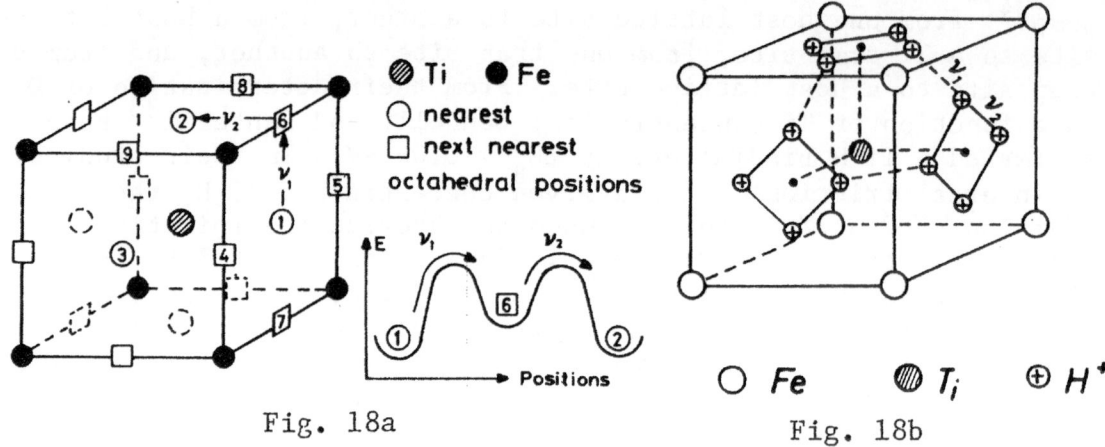

Fig. 18a   Fig. 18b

Fig. 18.   M-H complexes in b.c.c. alloys.[36]   (a) H in octahedral
          interstices.   Jumps of the type $1 \to 6$ and $6 \to 2$ occur
          with frequencies ν_1 and ν_2, respectively.   (b) H in
          tetrahedral interstices.   Diagram indicates the 12
          different orientations of the M-H axis.

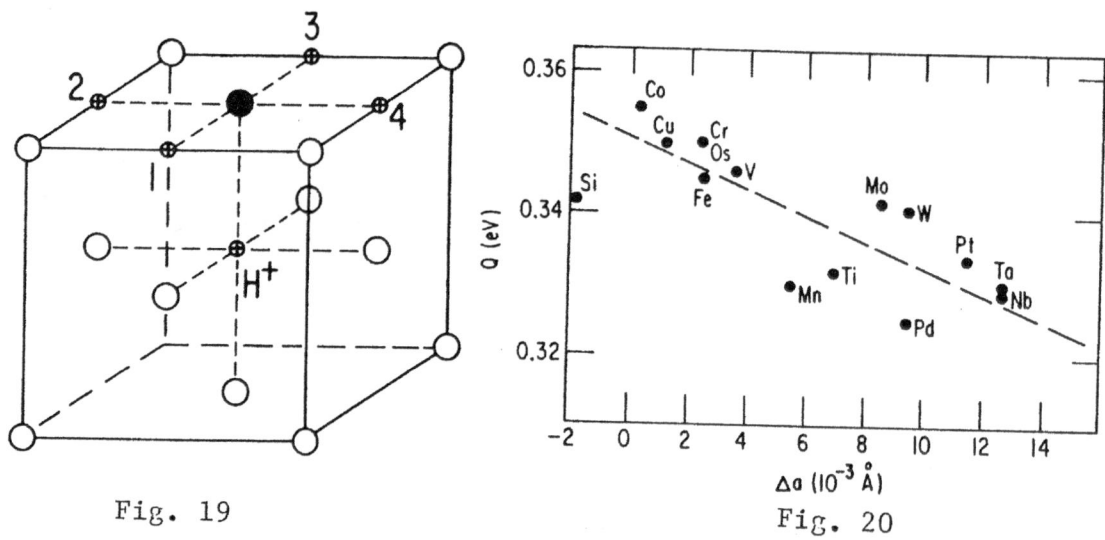

Fig. 19   Fig. 20

Fig. 19.   M-H complexes in f.c.c. alloys.[37]   Reorientation of the
          M-H axis would occur by diffusion of H^+ from the
          occupied octahedral site to one of the four numbered,
          unoccupied sites.
Fig. 20.   Dependence of the activation energy Q for the
          reorientation of M-H complexes in Ni on Δa, the lattice
          expansion caused by a 2% concentration of the indicated
          solute.[37]

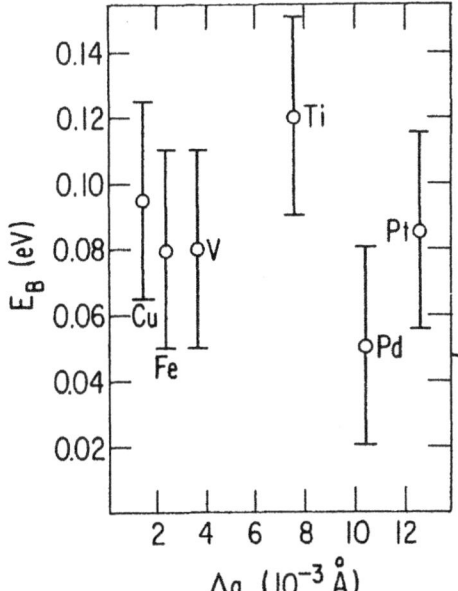

Fig. 21. A plot of the binding energy E_B for M–H complexes in Ni
vs Δa, the lattice expansion caused by a 2%
concentration of the indicated solute. No correlation
is indicated.[37]

a function of Δa, the lattice expansion caused by a 2 at.%
concentration of the solute M (Fig. 20). From this they conclude
that elastic effects should make a contribution to the binding
energy of M–H complexes. There must be electronic contributions
as well, however, because E_B, the binding energy, does not
correlate well with Δa (Fig. 21). They suggest that the
electronic and elastic contributions to the binding energy may be
of the same order of magnitude. One can see, also, that the range
of E_B for all solutes is from 0.05 to 0.12 eV. For D (deuterium),
the values of E_B and Q were always, respectively, about 0.02 eV
and 0.045 eV higher than for H.

ELASTIC CONSTANTS

     In an attempt to correlate elastic properties with the
maximum in TSH for Nb-Ta alloys (Fig. 11), the elastic moduli of
Nb, Ta, and seven alloys were measured by Fisher et al.[38]
(Fig. 22). The bulk modulus, K, is linear with concentration and
both of the principle shear moduli, C' and C_{44}, deviate only
slightly from linear behavior. Figure 23 shows that both Nb and
Ta are highly anisotropic, elastically; the ratio C_{44}/C' is much
greater than 1.0 for Ta and much less than 1.0 for Nb. It was
suggested,[9] by an extrapolation of the earlier results of
Armstrong and Mordike[39] (solid circles in Fig. 23), that the
maximum in TSH at a Ta concentration of 70 at.% might be

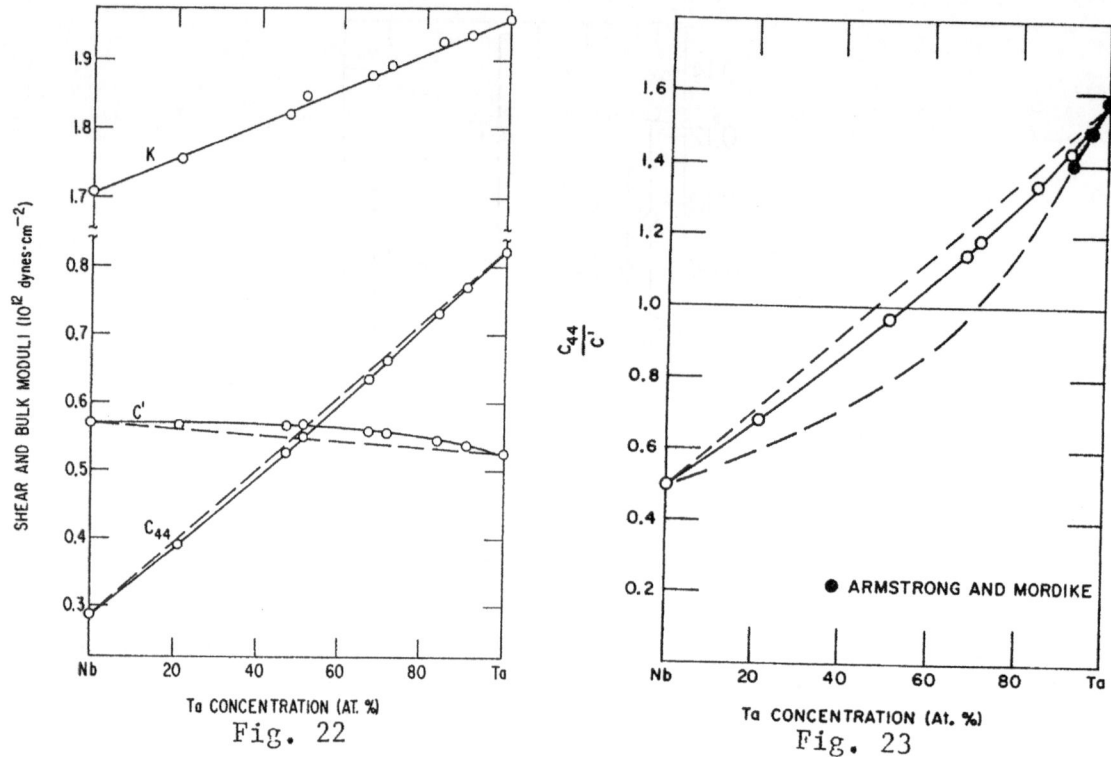

Fig. 22.   Variation of the shear and bulk moduli in Nb-Ta single
            crystals with Ta concentration (298°K). Results for C'
            and C_{44} deviate only slightly from linear plots.[38]
Fig. 23.   Deviation from elastic isotropy ($C_{44}/C' = 1$) in Nb-Ta
            alloys at 298°K.[38]

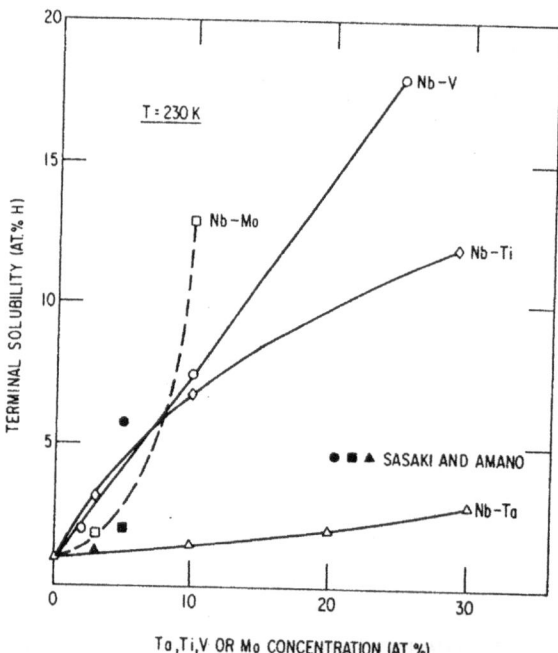

Fig. 24.   Variation of the ratio C_{44}/C' with Nb concentration in
            V-Nb alloys at 298°K.[38]

correlated with isotropic elastic behavior, i.e., $C_{44}/C' = 1$.
Instead, Fisher et al.[38] have shown that elastic isotropy occurs
in the alloy containing about 54 at.% Ta. They have found, also,
that introduction of H shifts isotropy to even lower Ta
concentrations. In a similar study[38] of Nb-V single crystals, it
was found that C_{44}/C' increased monotonically, but not linearly,
from 0.492 for Nb to 0.778 for V (Fig. 24). There seems,
therefore, to be no correlation between elastic properties and the
enhancement of TSH shown in Figs. 11 and 14.

It has been known[40,41] for several years that, in V, Nb, and
Ta metals, H reduces C', increases C_{44} and effects little change
in K. Only recently have the effects of H been studied in
alloys. For the Nb-Ta system, Figs. 25 and 26 show, respectively,
the relative changes in C' and C_{44} normalized to 1 at.% H. For
one particular alloy, $Nb_{33}Ta_{67}$, the changes effected in the moduli
at 298°K by increasing concentrations of H or D are shown in
Fig. 27. The analogous results for one alloy in the Nb-V system
are reproduced in Fig. 28. Some possible explanations for the
isotope effect have been developed[38], but these will not be
presented here.

Alberts et al.[42] have discussed the existing controversy over
whether the effect of H on C' can be interpreted as a Snoek
relaxation resulting from a tetragonal distortion around the
interstitial atom. The change in C', according to the Snoek

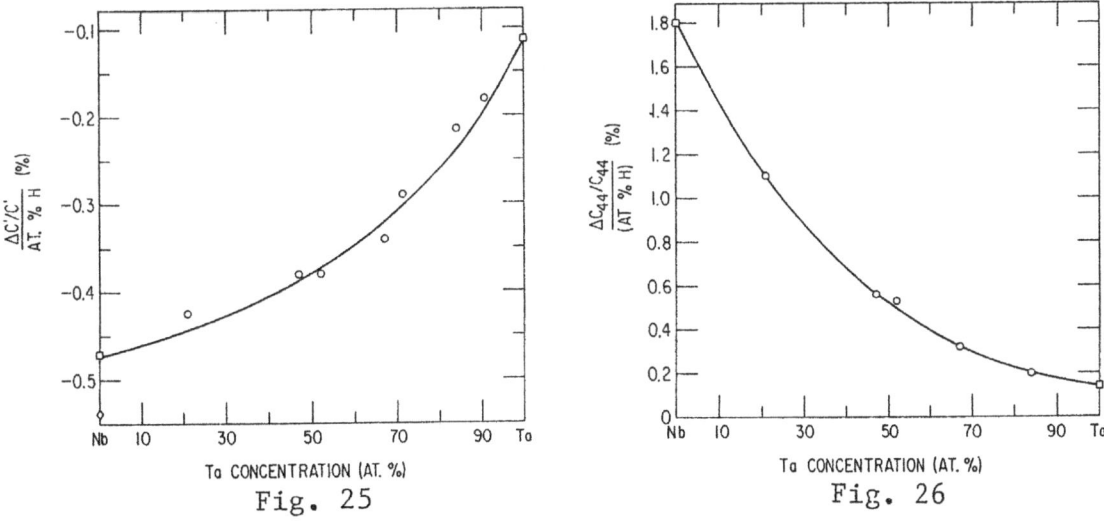

Fig. 25   Fig. 26

Fig. 25. Variation of the relative change in C' (normalized to
1 at.% H) with Ta concentration in Nb-Ta alloys at
298°K.[38]

Fig. 26. Variation of the relative change in C_{44} (normalized to
1 at.% H) with Ta concentration in Nb-Ta alloys at
298°K.[38]

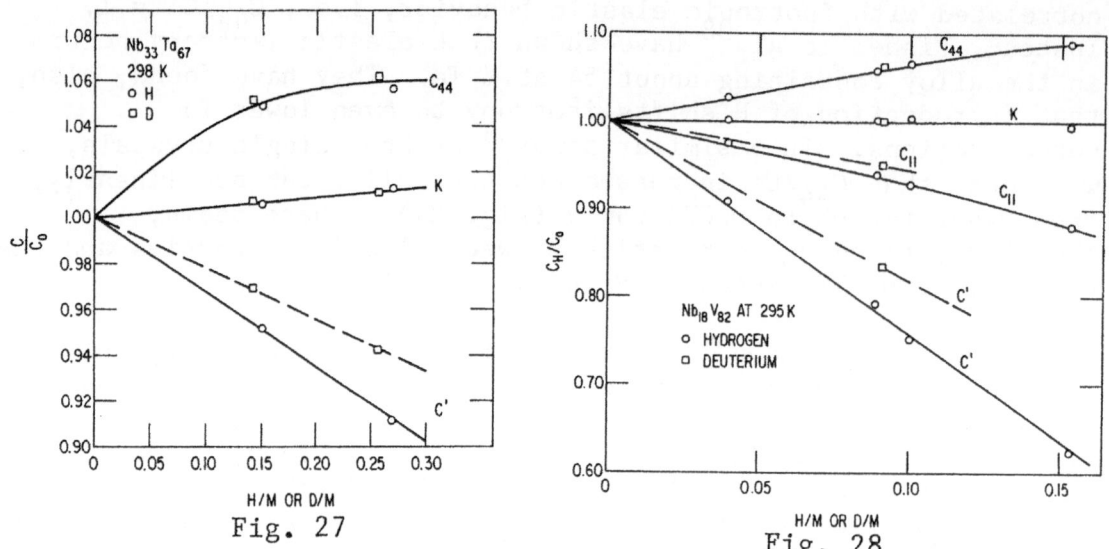

Fig. 27.   Plots of normalized elastic constants versus hydrogen or
           deuterium concentrations in $Nb_{0.33}Ta_{0.67}$.
Fig. 28.   Plots of normalized elastic constants versus hydrogen or
           deuterium concentrations in $Nb_{0.18}V_{0.82}$.

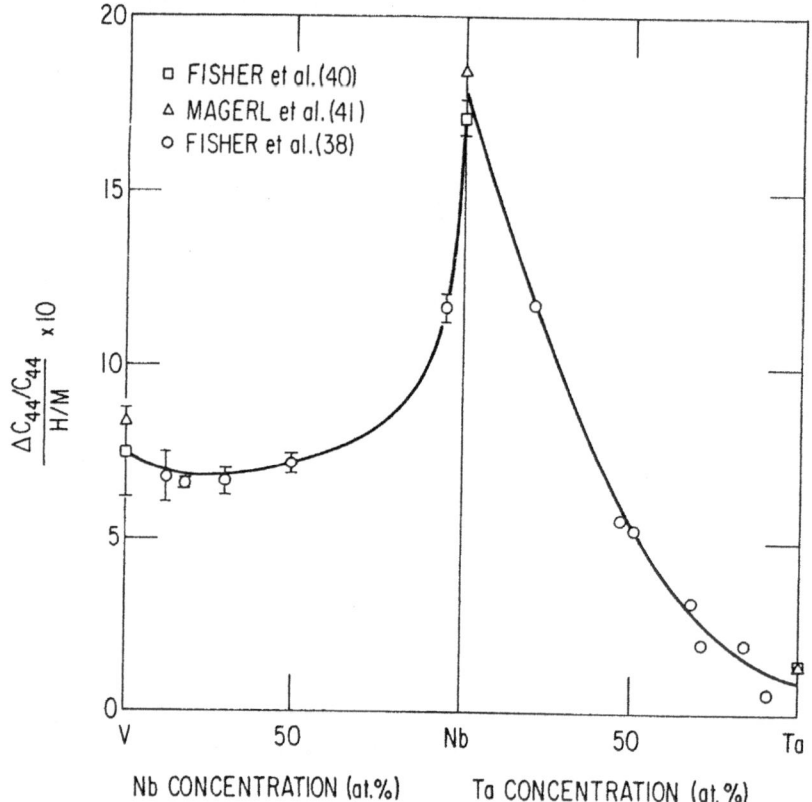

Fig. 29.   Variation of the relative change in C_{44} (normalized to
           unit hydrogen concentration) with metal solute
           concentration in V–Nb and Nb–Ta alloys at 298°K.

model, is given by:

$$\Delta C'/C' = - [2\ N(V_0 C')/3\ kT](\lambda_1 - \lambda_2)^2,$$

where N is the mole fraction of interstitial atoms, V_0 is the atomic volume of the bcc lattice, λ_1 is the strain along one cube axis and λ_2 is the strain along the two orthogonal axes. Any change in C_{44} cannot be attributed to a Snoek mechanism.

Alberts et al.[42] argue that, under increased hydrostatic pressure, the tetragonal distortion parameter ($\lambda_1 - \lambda_2$) should become larger. Accordingly, $\Delta C'$ should also be larger. They measured the elastic constants of both pure and hydrogenated crystals of V and $Nb_{53}Ta_{47}$ and found that the value of $\Delta C'$ for unit concentration of H became larger by 1.4% for V and by 2.0% for $Nb_{53}Ta_{47}$ when the pressure was increased from 0 to 3 kbar. They have considered other contributions but have concluded that the largest must arise from the effect of pressure on ($\lambda_1 - \lambda_2$). In both materials, the value of ΔC_{44} was unaffected by pressure. These observations are consistent with the Snoek model.

The increase in C_{44} due to solute H has been attributed to electronic effects[38]. Figure 29 shows the variation in the relative change in C_{44} per unit H concentration for the V-Nb and Nb-Ta systems. The maximum occurs for Nb and, thus, coincides with the maximum in the density of electron states at the Fermi surface.

RESISTIVITY DUE TO HYDROGEN

Many of the values of TSH in alloys have been determined resistometrically[9,12,16,18,20]. By careful examination of the contribution made by solute H to the resistivity, ρ, of transition metal alloys[43], one can gain considerable insight regarding the types of interaction involved.

Matthiessen's rule states that, in a given temperature range, $d\rho/dT$ for a metal remains unchanged by the addition of solute atoms. Westlake and Miller[43] have shown that, for H solute in Nb-V, Nb-Ti and Nb-Mo alloys, Matthiessen's rule is not obeyed. While there are many possible causes for failure to conform to this rule, their results lend support to some electronic and trapping effects.

The behaviors of Nb-Mo-H and Nb-Ti-H alloys are shown in Figs. 30a and 30b. The addition of Mo to Nb causes $d\rho/dT$ to decrease and addition of H to the Nb-Mo alloy causes a continuation of the same decreasing trend. Addition of Ti to Nb decreases $d\rho/dT$, also, but the effect of H in these Nb-Ti alloys is an enhancement of $d\rho/dT$. Within the approximation of a rigid

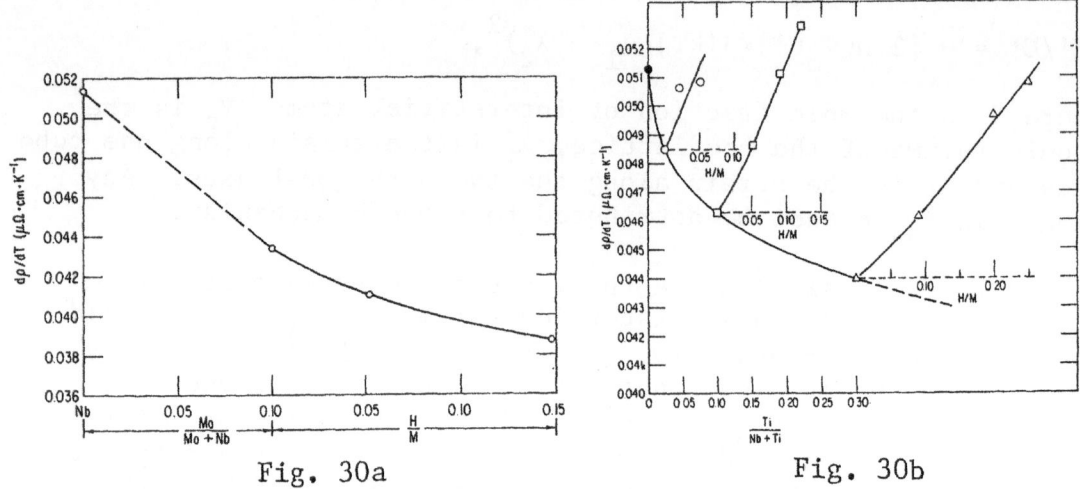

Fig. 30a   Fig. 30b

Fig. 30.   (a) Temperature derivative of the resistivity in Nb-Mo-H
           alloys for 240 ≤ T ≤ 300°K.   (b) Temperature derivative
           of the resistivity in Nb-Ti-H alloys for 295 ≤ T ≤
           340°K.

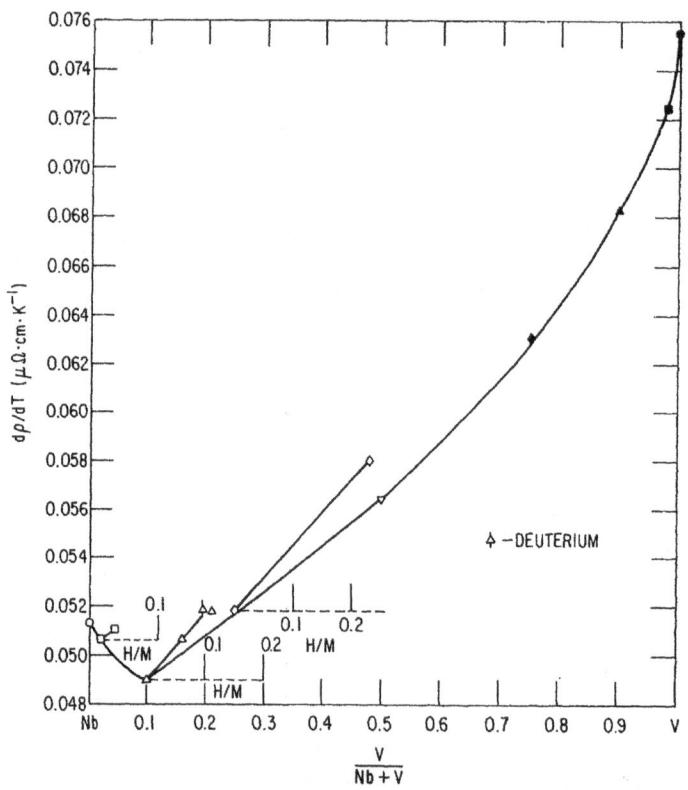

Fig. 31.   Temperature derivative of the resistivity in Nb-V-H
           alloys for 295 ≤ T ≤ 350°K.[43]

band model, these observations seem to suggest that the values
of $d\rho/dT$ are related to the density of electron states at the
Fermi level. The density is at or near a maximum for Nb and would
be reduced by the addition of Ti, Mo or H, but for Nb-Ti alloys,
it would be increased by solute H.

Hydrogen trapping may also make a significant contribution to
some of the observed deviations from Matthiessen's rule.
According to Pfeiffer and Wipf[44], the fraction of H atoms that are
weakly trapped by nitrogen atoms in Nb increases with decreasing
temperature, and, in the trapped state, $\Delta\rho_H$ (the contribution of H
to the resistivity) is reduced by 40%. Thus, in Nb-N-H
alloys, $\Delta\rho_H$ decreases with decreasing temperature. Such behavior
makes $d\rho/dT$ greater for a hydrogenated alloy than for a
hydrogen-free alloy.

Figure 31 shows that $d\rho/dT$ decreases slightly as V is added
to Nb, but then increases for V concentrations greater than
10 at.%. Solute H causes $d\rho/dT$ to increase, just as it did in
Nb-Ti alloys. In Ta metal and in Nb-Ta alloys, the change
in $d\rho/dT$ due to additions of H were small, and maybe
insignificant. These observations seem to suggest that trapping
may occur in Nb-V and Nb-Ti alloys, but not in Nb-Mo alloys, and
maybe not in Nb-Ta alloys. Attractive interactions might be
expected to arise from the size difference ($\sim$8%) between Nb and V
and from, the effective valence difference between Nb and Ti. On
the other hand, Nb and Ta are isovalent and have virtually the
same atom size. The possible repulsive interaction between Mo and
H in Nb was discussed earlier. Further evidence against a
trapping model for the Nb-Mo-H system arises from the following
comparison: for H/M = 0.05, the values of $\Delta\rho_H$ at room temperature
in Nb, $Nb_{0.9}V_{0.1}$, $Nb_{0.9}Ti_{0.1}$ and $Nb_{0.9}Mo_{0.1}$ are, respectively,
3.10, 2.45, 2.45, and 3.95 $\mu\Omega$ cm. Thus, the presence of either V
or Ti solute atoms reduces $\Delta\rho_H$, an effect that is consistent with
a trapping mechanism, but the presence of Mo solute atoms
anomalously increases $\Delta\rho_H$.

HYDROGEN EMBRITTLEMENT

The topic of hydrogen embrittlement may be somewhat outside
the scope of this NATO Advanced Study Institute on Metal
Hydrides. On the other hand, the problem of reduced ductility in
hydrogenated transition metals and their alloys is so closely tied
to the nucleation and growth of hydride particles, that it seems
to be deserving of at least a short discussion here. The
phenomenon of stress-induced hydride precipitation[45-47], which is
essential to the generalized model we proposed for hydrogen
embrittlement[48], has now been demonstrated experimentally[49,50].
Because of this phenomenon, the brittle hydride phase is nucleated
at one or more structural inhomogeneities during mechanical

testing. In all cases, therefore, the temperature for the onset
of H embrittlement during cooling is a few degrees higher than the
temperature for initiation of hydride precipitation in an
unstressed specimen. We know now that we can lower this latter
temperature by alloying, and this could very well be accompanied
by increased resistance to H embrittlement.

In Fig. 15, we can see the relative importance of Ta, V, Ti,
and Mo for enhancement of the TSH in Nb. Sasaki and Amano[18] have
measured the reduction in area (R.A.) as a function of temperature
in Nb, $Nb_{0.95}V_{0.05}$, $Nb_{0.95}Mo_{0.05}$, $Nb_{0.97}Ta_{0.03}$, and in each of
these materials with 1 at.% H. The effect of Ta was minimal, as
might be expected from the small change in TSH due to 3 at.% Ta
(Fig. 11). The addition of V, however, lowered the embrittlement
temperature by at least $50^{\circ}K$, as shown in Fig. 32a. The small
arrows indicate the temperatures for initiation of hydride
precipitation in the unstressed materials. The recovery of the
ductility below 140 K can be attributed to the fact that trapping
reduces the effective diffusivity of H. In Fig. 32b, the arrows
indicate the considerable lowering of the temperature for the
initiation of hydride precipitation when 5% Mo was added to Nb,
but the temperature for the onset of hydrogen embrittlement
remained essentially, and anomalously, unchanged.

Tanaka and Kimura[51] have measured the total strain to
fracture in hydrogenated V-Ti alloys. In Fig. 33, their results
for V-3.9 at.% Ti (with and without 0.55 at.% H) are compared with
the analogous results for unalloyed V. Obviously, the enhancement
of TSH[19] due to the addition of Ti to V is accompanied by
appreciably increased resistance to H embrittlement.

For the Nb and V alloys discussed above, the reaction with H_2
is exothermic and the hydride phase formed is relatively stable.
It is now becoming realized, however, that hydride formation may
play a role in the hydrogen embrittlement of some metals and
alloys, under certain conditions, even if the hydride is quite
unstable. One example is that of Ni-base superalloys, which are
candidates as structural materials in the hostile chemical
environments of sour-gas wells. Baranowski[52] has shown that it
requires a H_2 pressure in excess of 6 kbar to bring about the
formation of NiH at $298^{\circ}K$. Yet, in a recent study of Hastelloy
Alloy C-276 (~55% Ni)[53], it was found that a fcc hydride was
formed on the surface during electrolytic charging. Obviously
then, these are conditions of very high H fugacity. The hydride
formed on this alloy had the same structure as NiH, but its
lattice parameter was somewhat different. Like NiH, it was an
unstable hydride in that it decomposed completely in 20 h at room
temperature. Decomposition left the surface covered with a
network of cracks. Because local cathodic conditions and high H
fugacities can exist at the tips of microcracks and surface flaws

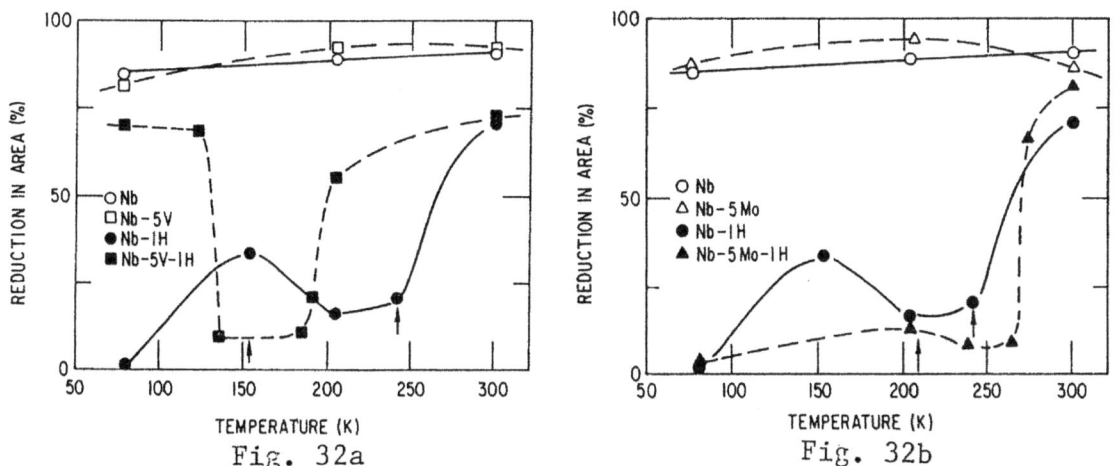

Fig. 32a   Fig. 32b

Fig. 32. The effects of hydrogen additions on the ductility of
          niobium below room temperature, with and without
          solute V(a) or Mo(b).[18]

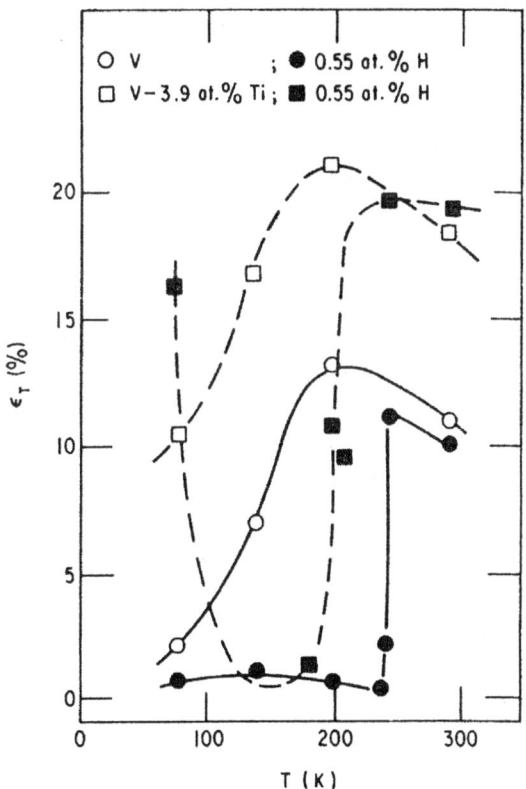

Fig. 33. The effects of hydrogen on the total strain to fracture
          of vanadium below room temperature, with and without
          solute titanium.

even in anodically polarized specimens[54], it seems that the role
of hydride formation in the hydrogen embrittlement of such alloys
warrants further study.

HYDRIDES OF INTERMETALLIC COMPOUNDS

All of the materials being seriously considered as media for
the chemical storage of hydrogen are alloys and, therefore, fall
within the purview of the title for this contribution. On the
other hand, many of the properties of these particular alloys have
been discussed as separate topics by other researchers at this
NATO Advanced Study Institute. On the subject of the hydrides of
intermetallic compounds, therefore, we shall limit our remarks to
our recently completed consideration of the stoichiometries and
the sites occupied by H in the hydrides of ZrNi[55].

Several possible criteria have been considered by different
researchers to be useful in the prediction of H site occupation
and the resulting hydride stoichiometry. Lundin et al.[56], for
example, reported a correlation between the radii of tetrahedral
holes and the stability of both hexagonal AB_5 and cubic AB
intermetallic compounds. Another suggestion of the importance
of hole size was made by Thompson et al.[57] for a solution of
deuterium in the α phase of FeTi. Only octahedral sites
coordinated by two Fe atoms and 4 Ti atoms are occupied by D. In
a hard-sphere model, this colinear Fe site can accommodate an atom
of radius $r = 0.31$ Å, whereas the analogous Ti site would limit r
to 0.09Å.

A phenomenological model was proposed by Jacob et al.[58] for
pseudobinary Laves phase compounds to explain the capacity for
absorption of hydrogen on the basis of nearest-neighbor-atom
species. The same criterion is essential to another model[59,60]
used to predict the sites to be occupied by H in AB_2 Laves phases
(both hexagonal and cubic) from the values of $\Delta H'$ assigned to each
site. $\Delta H'$ is the sum of the enthalpies of formation of elementary
(imaginary, binary) hydrides formed by the A and B atoms
surrounding a given site, and these enthalpies are calculated from
the theory of Miedema (references listed)[59].

From theoretical considerations, Switendick[61] has concluded
that, in a stable hydride, H-H interatomic distances can be no
shorter than about 2.1 Å . In accord with this, Didisheim et
al.[62,63] have suggested that, in hexagonal and cubic Laves phases,
H's do not occupy nearest neighbor tetrahedral (t) sites
(1.3 Å apart), but do occupy next nearest neighbor t sites, which
are about 2.0 Å apart. Shoemaker and Shoemaker[64] considered two
particular AB_2 Friauf-Laves phases and were led to postulate the
operation of the following exclusion rule: "two tetrahedra having
a triangular face in common may not both accommodate hydrogen

atoms at their centers". In these phases, the centers of such tetrahedra are less than 1.6 Å apart, while for tetrahedra not sharing a face the distance is at least 2.2 Å.

What we have chosen to study[55] is the ZrNi-H system with its orthorhombic metal sublattice (space group Cmcm). Figure 34 shows this rather complex structure and a sampling of the possible sites for H. Using only the criteria of interstitial-hole size and H-H interatomic distance, we have sorted through these sites and combinations of sites to make predictions of occupancy and the resultant stoichiometries. As an example, let us consider the case of $ZrNiH_3$.

From the estimated, expanded lattice parameters of $ZrNiH_3$, we have calculated the H-H distances and hole radii listed in Table 1. One can see that, for ★ and ● sites, the hole radius, is just that which seems to be minimum for typical stable hydrides (0.37-0.39 Å)[54]. The radii of the other three sites, however are even larger. One can eliminate the ● sites from further consideration, because the H-H distance for nearest neighbor ● sites is only 1.07 Å.

A combination of ★ and ◆ would give the correct number of sites and the respective hole radii, 0.39 and 0.72 Å, are acceptable. The ◆ - ● and the ◆ - ★ distances are more than adequate, but the ★ - ★ distance is somewhat less than Switendick's[61] minimum value of 2.1 Å, so let us consider other possibilities.

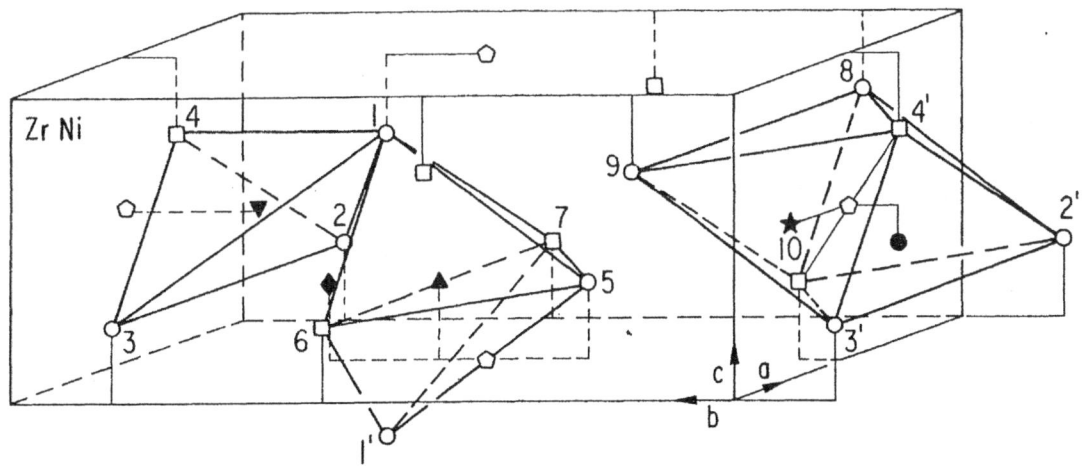

Fig. 34. Structure of ZrNi with examples of the interstitial sites possibly available for occupation by hydrogen shown as solid symbols. The symbols O and □ represent zirconium and nickel, respectively. As an aid in perspective, four face centers are shown as ⬠ .

Table 1. Hole sizes and H–H nearest-neighbor distances for H in designated interstitial sites in $ZrNiH_3$.

Site	No. Sites Per Zr Atom	Hole Radius in Å	H–H distance in Å				
			◆	▲	▼	★	●
◆	1	0.72	2.87	1.61	1.43	2.67	2.23
▲	1	~0.55	1.61	2.58	2.46	1.52	2.39
▼	2	~0.50	1.43	2.46	2.08	2.21	1.30
★	2	0.39	2.67	1.52	2.21	1.97	1.38
●	2	0.38	2.23	2.39	1.30	1.38	1.07

There is only one ◆ site per Zr atom, so, if it is occupied in $ZrNiH_3$, then some other sites must be occupied, as well. We note, however, that the distances ◆ – ▲ and ◆ – ▼ are only 1.61 and 1.43 Å, respectively. We submit, therefore, that occupation of ◆ would preclude occupation of either ▲ or ▼ sites.

Examination of Table 1 reveals that a combination of ▲ and ▼ yields just the required number of sites, and both hole sizes and H–H distances are well above the minima for stable hydrides. The resulting prediction of full occupancy of ▲ and ▼ sites in $ZrNiH_3$ is in complete agreement with experimental observations (references listed).[55] Using analogous rationales for several stoichiometries, we have made the following predictions:

1. A monohydride, ZrNiH, should exist with H occupation of either ◆ or ▲ sites.

2. $ZrNiH_2$ probably does not occur.

3. Little, if any, H solubility is to be expected in $ZrNiH_3$ because of the short H–H distances (▲ – ◆, ▲ – ★, ▼ – ◆, ▼ – ●).

4. The compound $ZrNiH_{4-x}$ may exist at high H_2 pressures with occupancy of ▼ and ★ sites.

CONCLUSIONS

In general, when Group V metals (V, Nb, Ta) are alloyed with other metals, for each alloy the logarithm of the hydrogen solubility is linear with reciprocal temperature. The resulting relative partial molar values of both enthalpy and entropy become less negative as e/a (electron to atom ratio) increases. For a given alloy system, the isothermal hydrogen solubilities decrease monotonically with increasing values of e/a.

The terminal solubilities for hydrogen in Group V metals are enhanced markedly by the addition of solute atoms from Groups IV, V or VI. No single explanation for the enhancement is universally accepted, and it may very well be that more than one mechanism is operative.

One of the phenomena possibly influencing the terminal solubility for H in alloys is "hydrogen trapping" or the formation of "M-H complexes" between H and the metal solute atom. There is considerable evidence that such complexes do exist in alloys whether the relative partial molar enthalpy is negative, as it is for Nb, or positive, as it is for Ni. In Ni, apparently, the elastic and the electronic contributions to the binding energy of M-H complexes are of the same order of magnitude.

In alloys of Group V metals, the principal shear moduli vary almost linearly with solute concentration, and there are no anomalies in elastic properties to which one can attribute the observed enhancements in terminal solubility for hydrogen. The addition of H to alloys, as well as to pure metals of Group V, causes important decreases in C' and increases in C_{44}. Whether Snoek relaxation is responsible for the effect of H on C' is still contested, but the change in C_{44} apparently arises from electronic effects.

The contribution of H to the resistivity of Nb alloys does not obey Matthiessen's rule. Such nonconformity is common and has many possible causes, but the behavior of both Nb-V and Nb-Ti alloys is consistent with the formation of M-H complexes.

Group V metals containing H are known to be embrittled by the precipitation of a hydride phase. Some alloying elements with demonstrated capacity for impeding hydride formation in such a metal have been shown to enhance its resistance to embrittlement. On the other hand, the addition of Mo to Nb anomalously increases the terminal solubility for H without any improvement in ductility. Unstable hydrides can be produced even in some metals and alloys not usually thought of as hydride formers when conditions of very high H fugacity exist. Because in many practical situations high H fugacities do exist at crack tips or surface flaws, the role of hydride formation in the hydrogen embrittlement of such materials should be reassessed.

A number of models and criteria have been used in the past to explain the observed stoichiometries and H-site occupations in hydrides of various intermetallic compounds. Presently, an examination of the structure of $ZrNiH_3$ has shown that one can make predictions agreeing with experimental observations using only empirical rules regarding the sizes of occupied interstices in stable hydrides and a theoretically determined minimum H-H distance. Predictions made here regarding other stoichiometries for hydrides of ZrNi require experimental verification.

REFERENCES

1.   D. G. Westlake, Enthalpy data for the zirconium-hydrogen system, J. Nuc Mat. 7:346 (1962).
2.   T. Eguchi and S. Morozumi, Influence of alloying elements on the solubility of hydrogen in vanadium, J. Jap. Inst. Met. 38:1025 (1974).
3.   T. Eguchi and S. Morozumi, Solubility of hydrogen in molybdenum and its alloys, J. Jap. Inst. Met. 38:1019 (1974).
4.   J. F. Lynch, J. J. Reilly and F. Millot, The absorption of hydrogen by binary vanadium-chromium alloys, J. Phys. Chem. Solids 39:883 (1978).
5.   R. Burch and N. B. Mason, The relative importance of geometric and electronic contributions to the thermodynamic properties of body-centered cubic metal hydrides, J. Less-Common Met. 63:57(1979).
6.   Akihiko Inoue, Masahiro Katsura and Tadao Sano, The solubility of hydrogen in Nb-Mo alloy, J. Less-Common Met. 55:9 (1977).
7.   D. W. Jones, N. Pessall and A. D. McQuillan, Correlation between magnetic susceptibility and hydrogen solubility in alloys of early transition elements, Phil. Mag. 6:455 (1961).
8.   H. Katsuta and Rex B. McLellan, Thermodynamics of molybdenum-niobium-hydrogen ternary solid solutions, J. Phys. Chem. Solids 40:845 (1979).
9.   D. G. Westlake and J. F. Miller, Terminal solubility of hydrogen in Nb-Ta alloys and characterization of the solid solutions, J. Less-Common Met. 65:139 (1979)
10a. E. Veleckis and R. K. Edwards, Thermodynamic properties in the systems vanadium-hydrogen, niobium-hydrogen, and tantalum-hydrogen, J. Phys. Chem. 6:83 (1969).
10b. O. J. Kleppa, P. Dantzer and M. E. Melnichak, High-temperature thermodynamics of the solid solutions of hydrogen in bcc vanadium, niobium and tantalum, J. Chem. Phys. 61:4048 (1974).

11. S. W. Stafford and Rex B. McLellan, The thermodynamic properties of the Fe-Ni-H ternary system, Acta Met. 24:553 (1976).

12. D. G. Westlake, A resistometric study of phase equilibria at low temperatures in the vanadium-hydrogen system, Trans. TMS-AIME 239:1341 (1967).

13. T. Schober and A. Carl, A differential thermal analysis study of the vanadium-hydrogen system, Phys. Stat. Sol. (a) 43:443 (1977).

14. D. H. Sherman, C. V. Owen and T. E. Scott, The effect of hydrogen on the structure and properties of vanadium, Trans. TMS-AIME 242:1775 (1968).

15. G. Cannelli and R. Cantelli, A study of the effects of deuteride precipitation in tantalum by freqency and internal friction measurements, Appl. Phys. 3:325 (1974).

16. T. Matsumoto, Y. Sasaki, and M. Hihara, Interaction between interstitial hydrogen and substitutional solute atoms in solid solutions of niobium-base ternary alloys, J. Phys. Chem. Solids 36:215 (1975).

17. Takehiko Matsumoto, NMR study of interaction between interstitial hydrogen and substitutional vanadium atoms in niobium metal, J. Phys. Soc. Japan 42:1583 (1977).

18. Y. Sasaki and M. Amano, Hydrogen solubility and embrittlement in Nb-V, Nb-Mo and Nb-Ta alloys, in: Proceedings of the 2nd International Congress on Hydrogen in Metals", Paris, Vol. 1, Pergamon Press, New York (1978).

19. S. Tanaka and H. Kimura, Solubility and diffusivity of hydrogen in vanadium and its alloys around room temperature, Trans. Japan Inst. Met 20:647 (1979).

20. J. F. Miller and D. G. Westlake, Enhanced terminal solubilities for hydrogen in niobium alloyed with vanadium, titanium and molybdenum, Trans. Japan Inst. Met., Supplement, 21: 153 (1980).

21. J. M. Corsan and A. J. Cook, Specific heat and superconductivity of binary alloys containing V, Nb, and Ta, Phys. Stat. Sol. 40:657 (1970).

22. P. S. Rudman, X-ray diffuse-scattering study of the Nb-Ti bcc solution, Acta Met. 12:1381 (1964).

23. Ya. S. Umanskii and V. I. Fadeeva, Peculiarities of the atomic structure of a Ta-Nb solid solution, Sov. Phys.- Cryst. 11:193 (1966).

24. Farid A. Khavadzha, V. M. Silonov and A. A. Katsnel'son, Short-range order in the systems Nb-V, Ta-V, and Nb-Ta, Sov. Phys. J. 20:5 (1977).

25. E. Colavita, A. Franciosi, R. Rosei, F. Sacchetti, E. S. Giuliano, R. Ruggeri, and D. W. Lynch, Electronic structure of Nb-Mo alloys, Phys. Rev. B22:4864 (1979).

26. E. S. Black, D. W. Lynch and C. G. Olson, Optical properties
    (0.1-25 eV) of Nb-Mo and other Nb-based alloys, Phys.
    Rev. B16:2337 (1977).

27. G. C. Abell, Quasimolecular Jahn-Teller resonance states in
    the bcc metallic hydrides of vanadium, niobium, and
    tantalum, Phys. Rev. B, 20:4773 (1979).

28. Bohdan Stalinski and Bogdan Nowak, On the structure of
    titanium-niobium-hydrogen alloys, Bull. de l'Acad. Pol.
    des Sci., Ser. des sci. chim. 25:451 (1977).

29. C. G. Chen and H. K. Birnbaum, Low-temperature H-O and H-N
    relaxations, Phys. Stat. Sol (a) 36:687 (1976).

30. K. Ozawa, S. Yamaguchi, Y. Fujino, O. Yoshinari, M. Koiwa and
    M. Hirabayashi, Channeling studies on the trapping of
    deuterium in vanadium by oxygen interstitials, Nucl.
    Instr. and Meth. 149:405 (1978).

31. M. A. Pick and D. O. Welch, Hydrogen absorption in the
    niobium-vanadium system, Z. für Phys. Chem. 114:37
    (1979).

32. D. Richter, Hydrogen diffusion and trapping in bcc and fcc
    metals, Report No. BNL-26132 (1979).

33. G. Cannelli and R. Cantelli, Hydrogen diffusion in niobium-
    titanium alloys, in: "Proceedings of the 2nd
    International Congress on Hydrogen in Metals", Paris,
    Vol. 1, Pergamon Press, New York 1978).

34. K. E. Blazek, The effect of a substitutional solute element
    on the diffusivity of an interstitial solute element in a
    dilute ternary alloy, Trans. Japan Inst. Met. 19:253
    (1978).

35. G. Cannelli and R. Cantelli, Anelasticity in niobium-titanium
    alloys, in: "Internal Friction and Ultrasonic Attenuation
    in Solids", R. R. Hasiguti and Nobuo Mikoshiba, eds.,
    University of Tokyo Press, Tokyo (1977).

36. H. Krönmuller, B. Hohler, H. Schreyer and K. Vetter,
    Investigation of hydrogen-impurity complexes in
    transition metals, Phil. Mag. B37:569 (1978).

37. B. Hohler and H. Krönmuller, Investigation of hydrogen-
    impurity complexes in transition metals, Z.für Phys.
    Chem. 114:93 (1979).

38. E. S. Fisher, J. F. Miller, D. G. Westlake, and
    H. L. Alberts, to be published.

39. D. A. Armstrong and B. L. Mordike, The influence of alloying
    and temperature on the elastic constants of tantalum,
    J. Less-Common Met. 22:265 (1970).

40. E. S. Fisher, D. G. Westlake and S. T. Ockers, Effects of
    hydrogen and oxygen on the elastic moduli of vanadium,
    niobium, and tantalum single crystals, Phys. Stat. Sol.
    (a) 28:591 (1975), and E. S. Fisher, Effects of hydrogen
    and UHV annealing on the elastic moduli of tantalum,
    Scripta Met. 11:685 (1977).

41. A. Magerl, B. Berre and G. Alefeld, Changes of the elastic constants of V, Nb, and Ta by hydrogen and deuterium, Phys. Stat. Sol. (a) 36:161 (1976).

42. H. L. Alberts, E. S. Fisher, K. W. Katahara and M. H. Manghnani, The effect of hydrostatic pressure on the elastic constants of pure and hydrogenated single crystals of V and $Nb_{53}Ta_{47}$, J. Phys. F: Met. Phys. 9:L209 (1979).

43. D. G. Westlake and J. F. Miller, Resistivity due to hydrogen in transition metal alloys, J. Phys. F: Met. Phys. 10:859 (1980).

44. G. Pfeiffer and H. Wipf, The trapping of hydrogen in niobium by nitrogen interstitials, J. Phys. F: Met. Phys. 6:167 (1976).

45. T. W. Wood and R. D. Daniels, The influence of hydrogen on the tensile properties of columbium, Trans. TMS-AIME, 233:898 (1965).

46. W. T. Chandler and R. J. Walter, Hydrogen effects in refractory metals, in: "Proc. AIME Symposium on Refractory Metal Alloys", I. Machlin, R.T. Begley and E. D. Weisert, eds., Plenum, New York (1968).

47. T. G. Oakwood and R. D. Daniels, The ductile-brittle-ductile transition in columbium-hydrogen alloys, Trans. TMS-AIME, 242:1327 (1968).

48. D. G. Westlake, A generalized model for hydrogen embrittlement, Trans. ASM 62:1000 (1969).

49. S. Gahr, M. L. Grossbeck and H. K. Birnbaum, Hydrogen embrittlement of Nb. I-Macroscopic behavior at low temperatures, Acta Met. 25:125 (1977).

50. M. L. Grossbeck and H. K. Birnbaum, Low temperature hydrogen embrittlement of Nb. II-Microscoic observations, Acta Met 25:135 (1977).

51. S. Tanaka and H. Kimura, Hydrogen embrittlement of vanadium-titanium alloys, Trans. Japan Inst. Met., Supplement, 21:513 (1980).

52. B. Baranowski, Metal-hydrogen systems in the high pressure range, Z. für Phys. Chem 114:59 (1979).

53. Ellina Lunarska-Borowiecka and Nicholas F. Fiore, Hydride formation in a Ni-base superalloy, to be published.

54. H. W. Pickering and R. P. Frankenthal, On the mechanism of localized corrosion of iron and stainless steel, II. Morphological studies, J. Electrochem Soc. 119:1304 (1972).

55. D. G. Westlake, Stoichiometries and interstitial site occupation in the hydrides of ZrNi and other isostructural intermetallic compounds, J. Less-Common Met. 75:177 (1980).

56. C. E. Lundin, F. E. Lynch and C. B. Magee, A correlation
     between the interstitial hole sizes in intermetallic
     compounds and the thermodynamic properties of the
     hydrides formed from those compounds, J. Less-Common Met.
     56:19 (1977).

57. P. Thompson, F. Reidinger, J. J. Reilly, L. M. Corliss and
     J. M. Hastings, Neutron diffraction study of α-iron
     titanium deuteride, J. Phys. F: Metal Phys. 10:L57
     (1980).

58. I. Jacob, D. Shaltiel, D. Davidov, and I. Miloslavski,
     A phenomenological model for the hydrogen absorption
     capacity in pseudobinary Laves phase compounds, Solid
     State Comm. 23:669 (1977).

59. I. Jacob and D. Shaltiel, Hydrogen sorption properties of
     some AB_2 Laves phase compounds, J. Less-Common. Met
     65:117 (1979).

60. J. Shinar, I. Jacob, D. Davidov, and D. Shaltiel, Hydrogen
     sorption properties in binary and pseudobinary
     intermetallic compounds, in: "Proc. Int. Symp. on
     Hydrides for Energy Storage, Geilo, Norway, 1977,
     A. F. Andresen and Arnulf Maeland, eds., Pergamon Press,
     New York (1977); I. Jacob, J. M. Block, D. Shaltiel and
     D. Davidov, On the occupation of interstitial sites by
     hydrogen atoms in intermetallic hydrides: A quantitative
     model, Solid State Comm. 35:155 (1980).

61. A. C. Switendick, Theoretical studies of hydrogen in metals:
     Current status and further prospects, Report No. SAND
     78-0250 (1978).

62. J.-J. Didisheim, K. Yvon, D. Shaltiel, and P. Fischer, The
     distribution of the deuterium atoms in the deuterated
     hexagonal Laves-phase $ZrMn_2D_3$, Solid State Comm. 31:47
     (1979).

63. J.-J. Didisheim, K. Yvon, D. Shaltiel, P. Fischer, P. Bujard
     and E. Walker, The distribution of the deuterium atoms in
     the deuterated cubic Laves-phase $ZrV_2D_{4.5}$, Solid State
     Comm. 32:1087 (1979).

64. David P. Shoemaker and Clara Brink Shoemaker, Concerning
     atomic sites and capacities for hydrogen absorption in
     the AB_2 Friauf-Laves phases, J. Less-Common Met. 68:43
     (1979).

HYDROGEN ABSORPTION IN METALLIC GLASSES

Arnulf J. Maeland

Corporate Research and Development
Allied Corporation
Morristown, New Jersey 07960

ABSTRACT

Metallic glasses have become an important class of new materials in recent years. By employing cooling rates in excess of 10^6 °C/sec, amorphous structures have been produced in many alloy systems over a wide range of compositions which, in some cases, encompasses intermetallic compound compositions. We have studied hydrogen absorption in such metallic glass systems and, where possible, compared the hydrogen absorption characteristics to those of the crystalline counterparts. The results are discussed with reference to the relative importance of electronic structure and crystal structure in hydrogen absorption.

INTRODUCTION

Although glassy or amorphous metallic alloys were prepared as early as 1930 by electrolytic deposition techniques,[1] it is only in recent years that they have emerged as an important class of new materials. The breakthrough which made this possible came in 1959 when Duwez, Willens, and Klement developed a technique for obtaining ultrafast cooling rates.[2,3] Their contribution to the development of metallic glasses earned them the 1980 American Physical Society International Prize for new materials. Subsequent investigations coupled with improvements in rapid cooling techniques have produced a large number of metallic glasses whose physical and chemical properties have been and continue to be a subject of considerable interest.[4]

The absorption of hydrogen in metallic glasses, is of
interest for a number of reasons. 1) The electronic structure
and the crystal structure, i.e., the type and size of the interstitial sites in the lattice, are both important considerations
in trying to understand hydrogen absorption in crystalline metals
and alloys. The study of hydrogen absorption in metallic glasses
offers a unique opportunity to eliminate crystallographic considerations,[5] because it is possible in certain systems to prepare
glassy and crystalline materials of the same composition and
compare their hydrogen absorption characteristics. 2) Hydrogen
can be used as a probe to study the local environment in the
metallic glass.[6] 3) Metallic glasses with alloy compositions
which do not have single phase crystalline counterparts can in
many cases be prepared and studied. 4) The investigation of the
effect of hydrogen on the mechanical properties of metallic
glasses may ultimately lead to a better understanding of hydrogen
embrittlement.[7] The emphasis here is on 1) and 2). However,
before proceeding, a short summary of some of the important
features of metallic glasses will be given.

METALLIC GLASSES

The familiar silicate glasses are amorphous materials in the
sense that long range order in their structure is lacking. The
bonding is covalent, with the valence electrons localized in the
bond between individual atoms, giving the bond a strong, directional character which explains most of their common properties.
They are brittle, generally good insulators, often transparent to
visible light, high strength materials, and they show excellent
corrosion resistance. Metals and alloys are, by contrast,
normally crystalline solids, i.e., long range order characterizes
their structure. They are typically malleable, ductile, opaque,
and good conductors of heat and electricity. These properties
are due to metallic bonding which is nondirectional and
collective; atoms donate a fraction of their electrons to form a
sea of electrons which gives both overall cohesion and ease of
electron movement. Metallic glasses have some of the properties
normally associated with metals and alloys, but at the same time
have certain glass-like properties. They are ductile, opaque,
and have relatively high electrical conductivity, but their
structure is amorphous and they exhibit good corrosion
resistance.

Techniques for Preparing Metallic Glasses

Normal glasses are prepared from the melt at low cooling
rates. It is in fact difficult in many cases to avoid glass
formation on cooling. The rate of crystallization is very slow
because the network of covalent directional bonds in the liquid

must be broken in order for the atoms to rearrange themselves
into a crystal lattice. This requires time and a substantial
energy input (activation energy) and in the normal cooling
process the configuration of the atoms in the liquid becomes
frozen in the glass. In a metal, however, the transition from
liquid to crystalline solid requires a much smaller activation
energy because of the nondirectional nature of the bonding. In
the usual cooling process the atoms readily arrange themselves in
the orderly array of the crystal lattice. Duwez, Willens, and
Klement[2,3] demonstrated that by using ultrafast cooling rates,
the amorphous state of some metals and alloys could be retained
on quenching from the molten to the solid state. In their
quenching technique molten metal was accelerated to high speed by
high pressure helium gas and allowed to impinge on a rotating
copper surface held at room temperature or below. Cooling rates
in excess of 10^6 degrees/second can be reached by this technique
which is the basis of the melt spinning process of today. Thin
ribbons, wire or sheet of metallic glasses are produced by
this continuous process. Other methods for preparing metallic
glasses include vapor deposition on a cold surface, sputtering,
electrodeposition, and irradiating with high energy particles to
disrupt the lattice.

Chemical or Compositional Classification of Metallic Glasses

Chemical classification of alloy compositions which readily
form metallic glasses is, at present, incomplete. Polk[8] has
outlined four major categories of alloy systems in which metallic
glasses have been reported. Polk's classification is summarized
in Table I together with representative examples, typical
compositions, and method of preparation. Several metallic glasses
are now commercially available. Allied Corporation, for example,
produces METGLAS® alloys 1816MB ($Fe_{0.40}Ni_{0.38}Mo_{0.04}B_{0.18}$),
2605SC ($Fe_{0.81}B_{0.135}Si_{0.035}C_{0.02}$) and 2605CO
($Fe_{0.67}Co_{0.18}B_{0.14}Si_{0.01}$).

Table I. Major Categories of Metallic Glasses Based on
Chemical Classifications of Constituents [8]

Alloy System	Examples	Typical Composition (at. %)	Method of Preparation
1. M-X	Pd-Si, Au-Si Fe-P, Fe-P-C	15-25X	LQ,ED
2. M'-M	Zr-Cu (Ti-Cu) Ti-Ni	30-65Cu 30-40Ni	LQ,SP LQ

Alloy System	Examples	Typical Composition (at. %)	Method of Preparation
3. Miscellaneous systems			
a) A-B	Mg-Zn	25-35Zn	LQ
b) M'-A	(Ti-Zr)-Be	20-60Be	LQ
c) Actinide-M	U-V, U-Cr	20-40M	LQ
4. Metallic glasses not formed by LQ			
M-rare earth metal	Co-Gd	~20Gd	SP

M = Late transition Metal (Mn, Fe, Co, Ni groups). Also noble
      metal.
M' = Early transition Metal (Sc, Ti, V groups).
X = Metalloid B, C, Si, Ge, P.
A = Li, Mg groups; B = Cu, Zn, Al groups.
LQ = Liquid quench; ED = electrodeposition; SP = sputtering.

Structure and Structure Models

The absence of long range order in metallic glasses is
experimentally established by x-ray, neutron, and electron
diffraction methods. It is also observed that their densities
are only 1-2 percent lower than those measured in crystalline
materials with similar compositions. By contrast, fused silica
has a density which is 20 percent less than quartz.[4c] This
implies that the spatial packing of atoms in the metallic glasses
is not radically different from the packing in crystals. The
near-neighbor pair interaction must be very similar and certain
kinds of short-range order are preserved in the metallic glass.[9]

Three major models have been proposed to describe the
metallic glass structure, the dense random packing model, the
microcrystalline model, and the random packing of atomic
clusters.[4,9] The dense random packing model was originally
proposed by Bernal[10] to account for the structure of liquids and
describes an assembly of tetrahedral units, distorted in various
ways, and arranged in an irregular and continuous fashion. The
microcrystalline model contains the idea that the structure is
that of tiny crystallites, randomly oriented and embedded in a
matrix which is completely amorphous. Finally, in the random
packing of atomic clusters model groups of atoms (clusters) are
packed together in arrays that are regular, but not crystalline.
The configurations in these clusters may be similar to those

found in crystalline materials or they may contain atomic
arrangements which are symmetry forbidden in periodic systems.[9,11]

Physical and Chemical Properties

It has already been mentioned that the density of a metallic
glass is only a little less than that of the crystalline material
of the same composition. The heat capacity and the compressibility
are also only slightly different, while the electrical resistivity
of the glass may be as much as three times higher than the
crystalline counterpart. The temperature coefficient of
resistivity, however, is very small and in some cases zero or
negative at room temperature.

Many metallic glasses have unusual magnetic properties.
They are quite soft magnetically which means that they show a
large response of the magnetization to a small applied field.
This is due to the fact that the glassy metals have no long range
order and therefore respond isotropically. Their isotropic nature,
combined with the non-directional nature of the bonding, also
account for their ductility and high strength.

Metallic glasses are chemically homogeneous on scales
greater than a few atomic diameters. Unlike crystalline materials
they have no grain boundaries which can act as chemically active
sites, and should therefore show improved resistance to corrosion.
This appears indeed to be the case. Especially high corrosion
resistance has been observed in metallic glasses formed of
transition metal-metalloid alloys containing chromium.[12]

Applications of Metallic Glasses

Metallic glasses are currently being used as brazing alloys,
high-strength fibers (composites), magnetic shielding materials,
and as harmonic generation targets in anti-theft detectors.[13]
The most promising application by far appears to be their use as
core material in power transformers. The magnetization in these
devices reverses direction twice during each cycle of the
alternating current (120 times a second), and power losses occur
with each reversal. The accumulated power loss can be substantial.
Since the metallic glasses are isotropic, the magnetization can
be rotated at a much smaller cost in energy than in crystalline
materials which are anisotropic. Other applications utilizing
the magnetic "softness" and high resistivity of metallic glasses
include their use in magnetic tape recorders and magnetic disk
memories.[4a]

HYDROGEN ABSORPTION IN METALLIC GLASSES

The Ti-Cu system was chosen for our initial study.[5,18] The choice was dictated by two major considerations. First, an examination of Table I shows that glasses selected from categories 2-4 might be expected to absorb large quantities of hydrogen if our knowledge of hydrogen absorption in crystalline materials can be extrapolated to metallic glasses. These glasses contain at least one major component which is a hydride former. The Ti-Cu glass system is found in category 2. Second, and most importantly, the phase diagram for the Ti-Cu system[14] shows that several intermetallic phases fall in the glass-forming composition range. It is, therefore, possible in this system to prepare glasses and crystalline materials of the same composition and to compare their hydrogen absorption characteristics.

Experimental

The metallic glass samples were vacuum-cast ribbons approximately 2.1 mm wide and 0.05 mm thick. Prior to hydriding, the ribbons were surface cleaned by abrasion with emery paper and ultrasonically cleaned in acetone followed by ether. A 1-3 g sample was then placed in a vacuum system which was subsequently evacuated to 1.3×10^{-4} Pa. Pure hydrogen, generated by the decomposition of TiH_2 was introduced into the samples at room temperature to avoid crystallization. The amount of hydrogen absorbed was calculated, using the ideal gas law, from the weight of the sample, the known volume of the system and the change in hydrogen pressure. The compositions were also checked by thermal analysis and the two methods agreed within experimental error (H/metal ratios agreed to within ±0.05).

The intermetallic compounds, TiCu and Ti_2Cu shown in Figures 1 and 2, were prepared by arc melting the appropriate amounts of the constituent metals under an argon atmosphere. The

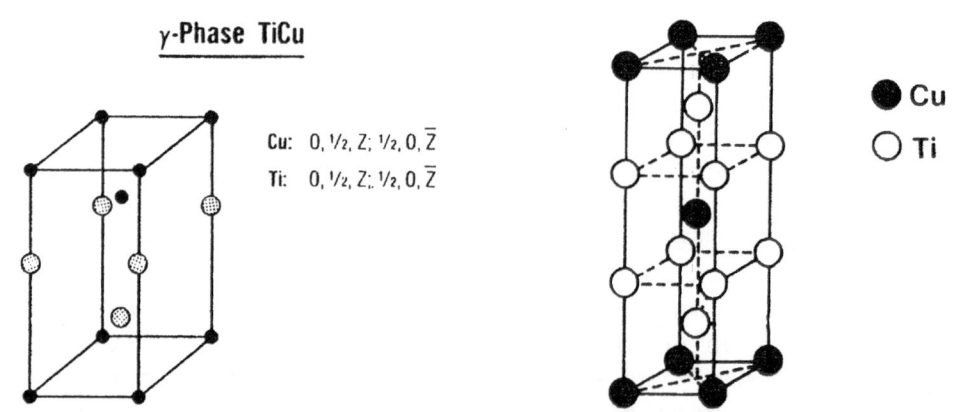

γ-Phase TiCu

Cu: 0, ½, Z; ½, 0, $\bar{Z}$
Ti: 0, ½, Z; ½, 0, $\bar{Z}$

● Cu
○ Ti

Figure 1.   Structure of γ-phase   Figure 2.   Structure of Ti_2Cu,
TiCu, B11 type.   C11b type.

purities of the starting materials were 99.99% for titanium and 99.9% for copper. The resulting buttons were remelted at least four times to ensure homogeneity. Lattice parameters, determined from x-ray powder patterns, were in good agreement with published values.[15-17] Prior to exposure to hydrogen, the samples were broken up into small chunks and degassed by heating to 500-600°C in vacuum (1.3 x 10^{-4} Pa). After cooling to room temperature, hydrogen was admitted at approximately atmospheric pressure. Absorption started slowly at room temperature but proceeded rapidly when the samples were heated to 150-200°C. When absorption was complete, the samples were slowly cooled to room temperature and were removed for x-ray examination. The compositions were calculated as before from the initial and final pressures, the known volume of the system and the weight of the samples. The samples were handled and kept in an argon dry box.

Results and Discussion

The hydrogen absorption capacities of four glass compositions in the Ti-Cu system are shown in Table II; the data refer to room temperature and atmospheric pressure of hydrogen. The corresponding data on the two intermetallic compounds, TiCu and Ti_2Cu are also included.

Table II. Hydrogen Absorption Capacities at Room Temperature and Atmospheric Pressure of Hydrogen

Glass composition	H/Metal	Corresponding intermetallic compound	Ref.	H/Metal
$Ti_{0.35}Cu_{0.65}$	0.37	Non-existent	14	---
TiCu	0.68	TiCu	14	0.47
$Ti_{0.60}Cu_{0.40}$	0.96	Non-existent	14	---
$Ti_{0.65}Cu_{0.35}$	1.15	Ti_2Cu	14	0.92

X-ray diffraction patterns of the metallic glasses, taken before and after hydrogen absorption, verified that no crystallization had taken place. Small shifts in the broad maxima in the patterns to lower angles were observed, however, indicating that hydrogen absorption had caused volume expansion of the glass. In the TiCu glass, for example, the broad amorphous peak at 2θ = 41.3° (Cu, Kα radiation) shifted to 40.2° in $TiCuH_{1.33}$ (see Table IV), corresponding to an increase in volume of about 8.5%. The x-ray data on crystalline TiCu and TiCuH are shown in Table III. The volume increase is the same, but it must be recognized that the glass contains a third more hydrogen so that the volume

increase per hydrogen atom is less in the glass than in the
crystalline counterpart. This may be due to the fact that the
glass has a slightly more open structure and the introduction of
hydrogen requires less of an expansion.

Table III. X-ray Data on TiCu and TiCuH

	Lattice Parameter (nm)	Volume (nm)^3	% Volume Increase
TiCu	a = 0.311 c = 0.589	0.0570	---
TiCuH	a = 0.302 c = 0.678	0.0618	8.5

The hydrided metallic titanium-copper glasses are quite
stable with respect to crystallization at room temperature. The
TiCuH$_{1.33}$ glass, for example, was still amorphous after a storage
period of two years in the dry box at room temperature (Table IV).
However, thermal analysis data indicated that decomposition to
form crystalline titanium hydride took place at relatively low
temperatures.[18] Figure 3 shows the crystallization of the
uncharged TiCu glass; the crystallization takes place in two

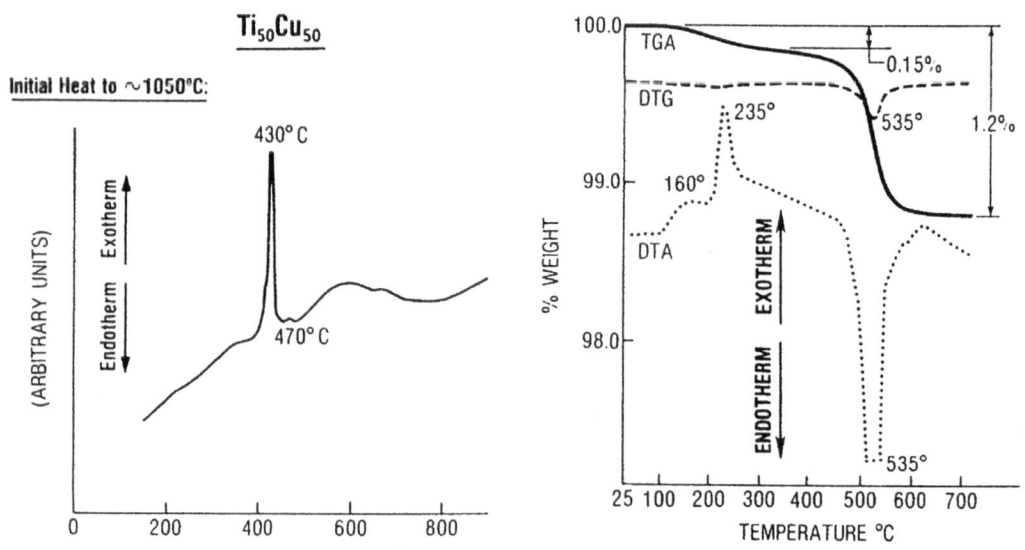

Figure 3. The differential thermal analysis curve for an uncharged
(no hydrogen) TiCu metallic glass sample. The 24.3 mg sample was
initially heated to approximately 1050°C at a rate of 25°C min^{-1}
in purified helium in an alumina crucible.[18]

Figure 4. Differential thermal analysis (...), differential
thermogravimetric analysis (---), and thermogravimetric (——)
data for the metallic glass hydride TiCuH$_{1.33}$. The 24 mg sample
was heated at a rate of 25°C in purified helium in an alumina
crucible.[18]

stages as indicated by the two exothermic peaks at 430 and 470°C
in the DTA curve. Figure 4 shows thermal analysis data on the
same glass after equilibrating in hydrogen at room temperature
and atmospheric pressure of hydrogen. The two peaks at 430 and
470°C are no longer present. Instead we see two exothermic
peaks at 160° and 235°C and a very strongly endothermic reaction
at 535°C. The two exothermic peaks are associated with the
decomposition of the metallic glass hydride and the formation of
crystalline titanium hydride while the strongly endothermic
reaction at 535°C, accompanied by the maximum weight loss
indicated in the differential thermogravimetric analysis curve,
is due to the reaction of TiH_{2-x} with copper to form TiCu and
hydrogen.

A similar low temperature decomposition reaction takes place
in the $Ti_{0.65}Cu_{0.35}H_{1.1}$ glass as indicated by the exothermic peak
at 150°C shown in Figure 5.

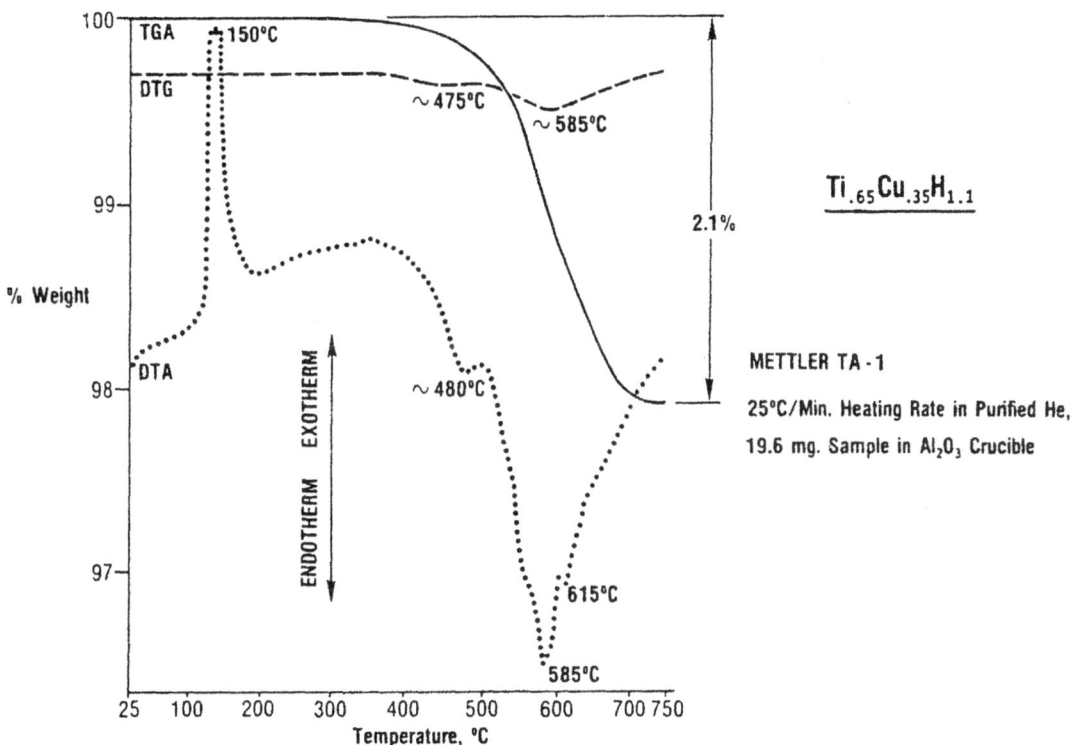

Figure 5.   Thermal analysis data on $Ti_{0.65}Cu_{0.35}H_{1.1}$ glass.[5]

The fact that the strongly endothermic reaction at 535°C is
associated with the reaction of crystalline TiH_{2-x} with elementary
copper is confirmed by the thermal analysis data shown in Figure 6.
The data were obtained under identical conditions of heating rate
and helium flow using an equimolar mechanical mixture of $TiH_{1.98}$
and copper powder. The thermal effects in this temperature

TiH₂ + Cu, 355-81 (9-2910)

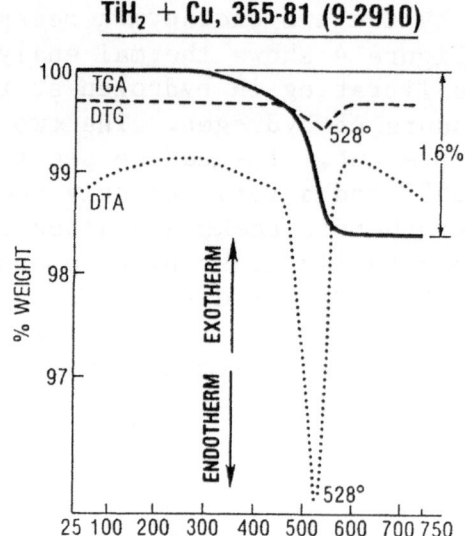

Figure 6. Differential thermal analysis (...), differential thermogravimetric analysis (---) and thermogravimetric analysis (——) data for a mixture of $TiH_{1.98}$ and copper metal. The 16.6 mg sample was heated at a rate of 25°C min⁻¹ in purified helium in an alumina crucible.[18]

region are essentially the same, suggesting similar reactions in the two samples (Figures 4 and 6).

X-ray diffraction studies on thermally treated metallic glasses support the decomposition mode suggested by the thermal analysis data.[18] Samples were sealed in evacuated glass ampules and heated to a constant temperature for various periods of time. The ampules were opened in the dry box and x-ray samples were prepared in nujol to protect them from the atmosphere. The results of these measurements are summarized in Table IV. No change in the x-ray pattern was observed when the hydrogen free metallic glass TiCu was heated to 245°C for 24 hours. The hydrided metallic glass $TiCuH_{1.33}$, however, decomposed completely into titanium hydride and elementary copper when subjected to the same thermal treatment. Partial decomposition of the hydride occurred on heating to 134°C for 63 hours while no change was observed on heating to 90°C for 18 hours. Reheating the same sample to 170°C for 16 hours caused partial decomposition.

Further support for the suggested decomposition mode comes from recent nuclear magnetic resonance studies.[19] Near 147–157°C an anomaly in the proton relaxation times of glassy $TiCuH_x$ was observed; the spin-spin relaxation time, T_{2m}, determined by the Carr-Purcell-Meiboom-Gill pulse sequence technique[20] decreased irreversibly by a factor of about ten.[19] Both slowly diffusing

Table IV. X-ray Diffraction Data on Thermally Treated
Metallic Glasses TiCu and $TiCuH_{1.33}$

Sample	Heat Treatment	Time	X-ray Pattern
CuTi	None (as cast)	--	Broad amorphous peak, $2\theta = 41.3°$
CuTi	245°C	24 hrs.	No change
$CuTiH_{1.33}$	None (as prepared)	--	Broad amorphous peak, $2\theta = 40.2°$
$CuTiH_{1.33}$	None (stored at room temperature	2 years	No change
$CuTiH_{1.33}$	90°C	18 hrs.	No change
$CuTiH_{1.33}$	Reheated to 170°C	16 hrs.	Medium peak at $2\theta = 35.0°$ (TiH_2); medium broad peak at $2\theta = 40.8°$; (TiH_2 and metallic glass hydride); weak sharp peak at $2\theta = 43.2°$ (copper)
$CuTiH_{1.33}$	134°C	63 hrs.	Broad amorphous peak at $2\theta = 40.5°$; very weak peak at $2\theta = 35.1°$ (TiH_2); very weak peak at $2\theta = 43.2°$ (copper) and very weak peak at $2\theta = 50.4°$ (copper)
$CuTiH_{1.33}$	245°C	18 hrs.	Strong peak at $2\theta = 35.1°$ (TiH_2); medium peak at $2\theta = 43.2°$ (copper); very weak peak at $2\theta = 50.4°$ (copper)

protons, associated with TiH_{2-x}, and rapidly moving protons
ascribed to the metallic glass hydride were observed,[19] again
suggesting incomplete or partial decomposition.

Decomposition reactions also occur at relatively low
temperatures, approximately 200°C, in the intermetallic compounds
TiCuH[21] and $Ti_2CuH_{2.80}$.[22] The low values of the enthalpy of
formation for the intermetallic compounds[22] and the structure of
the hydrides[21] are major factors in the relative ease of decomposition. The enthalpy of formation of TiCu is estimated to be

−25 kJ/g atom and for Ti_2Cu only −23 kJ/g atom using the method given in reference 23. The structure of TiCuH[24] is shown in Figure 7. Hydrogen is located in tetrahedral sites, surrounded by four titanium atoms; the titanium-hydrogen distances are very close to those in TiH_2[25] where hydrogen is also in tetrahedral sites. The Ti_2Cu structure contains four tetrahedral sites surrounded completely by titanium atoms as shown in Figure 8. These are the preferred sites for hydrogen occupation; when completely filled the composition would be Ti_2CuH_2. The $Ti_2CuH_{2.80}$ structure has not yet been determined, but an inelastic neutron-scattering study[26] has indicated that hydrogen atoms reside in both tetrahedral and octahedral sites. Since relatively little thermal energy is necessary to cause precipitation of TiH_2 in these structures, the observed decomposition temperature is low (~200°C).

The observation of rapidly moving protons in the metallic glass hydride[19] referred to above, requires a further comment. Our general observation has been that hydrogen diffusion in metallic glasses is slower than in the corresponding crystalline compounds. In the case of TiCuH this is not so. The TiCuH structure, Figure 7, can be described as alternating layers of TiH_2 and Cu atoms. This corresponds to a two-dimensional analog of metal hydrides with the calcium fluoride type structure.[27] This structural pecularity makes hydrogen diffusion in crystalline TiCuH quite difficult. Compared to $TiH_{1.96}$, for example, the diffusion coefficient in TiCuH is 9.9×10^{-16} cm^2/sec at 300°K vs. 2.1×10^{-13} in $TiH_{1.96}$ and the activation energy is 0.79 eV vs. 0.50 eV.[27] The amorphous nature of the metallic glass removes the structural restriction to diffusion and hydrogen diffusion in the metallic glass is considerably larger in this particular case.

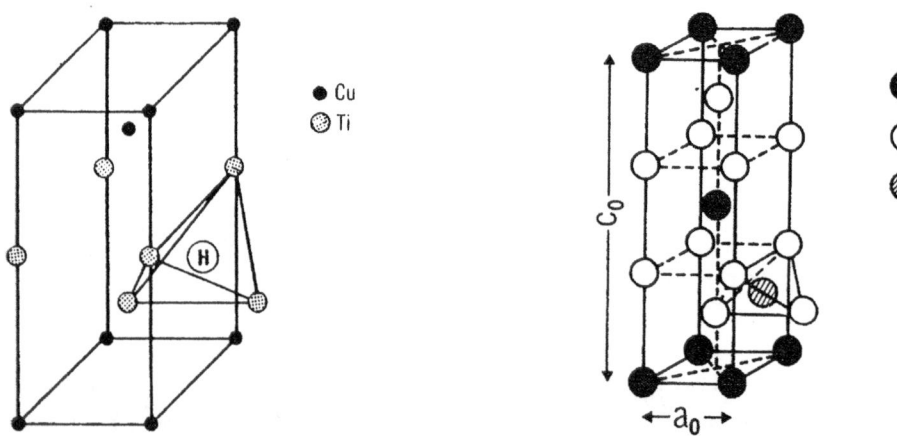

Figure 7. The TiCuH structure   Figure 8.   The Ti_2CuH_2 structure

The metallic Ti-Cu glass alloys had larger hydrogen absorption capacities than their crystalline counterparts. According to Table II, for example, glassy TiCu absorbed approximately one third more hydrogen than crystalline TiCu under similar conditions of temperature and hydrogen pressure. We believe this points out the relative importance of electronic structure vs. crystal structure in hydrogen absorption. To a first approximation the electronic structure of a metallic glass is quite similar to that of the corresponding intermetallic compound.[28] If we assume that the maximum hydrogen absorption capacity is determined by the electronic structure, we might expect similar hydrogen absorption capacities in the amorphous and crystalline states. However, maximum hydrogen absorption in the crystalline state may not always readily occur because of the crystal structure, i.e., the type and size of the interstitial sites in the lattice. In the case of TiCuH, for example, the hydrogen atoms are located at the centers of slightly distorted tetrahedra of titanium atoms; crystallographically the limiting composition is TiCuH. Absorption of hydrogen beyond TiCuH would involve placing hydrogen atoms in higher energy sites. In the metallic glass structure, however, recent inelastic neutron scattering results[6] indicate that, although on the average the hydrogen atoms are in tetrahedral-type sites, there is a wide distribution of local environments for hydrogen atoms in the metallic glass. The experimental results are shown and compared to the results on crystalline TiCuH in Figure 9. The broad band of hydrogen vibrations peaks at an energy of about 145 meV which is nearly the same energy as in the crystalline counterpart. The width of the band, however, is much larger, about four times that observed in polcrystalline TiCuH, suggesting considerable fluctuations in local symmetry around the average site. Such fluctuation in local symmetry may increase the number of sites available for hydrogen occupation (with respect to the crystalline state) by lowering the energy. It may, therefore, be possible to find a higher hydrogen absorption capacity in certain metallic glass structures than in their crystalline counterparts.

It is interesting to compare TiCu hydride to palladium hydride. The hydrogen atoms in palladium hydride occupy the octahedral sites in the face centered cubic palladium lattice. If all the sites were filled, the composition would be PdH and the structure that of sodium chloride. It is difficult, however, to fill the octahedral sites much beyond the composition $PdH_{0.6}$. To do so requires high pressures of hydrogen and/or low temperatures. The reason for this behavior is that there are only 0.6 empty electronic states below the Fermi level in palladium hydride; these electronic states consist of 0.36 holes in the d band plus 0.24 additional s-like states due to the presence of hydrogen in the lattice.[29] Electrons from hydrogen atoms fill these states

at the composition $PdH_{0.6}$. To add additional hydrogen requires
putting electrons into higher energy levels. The maximum hydrogen
content in palladium hydride is, therefore, in contrast to TiCuH
limited by the electronic structure rather than the crystal
structure.

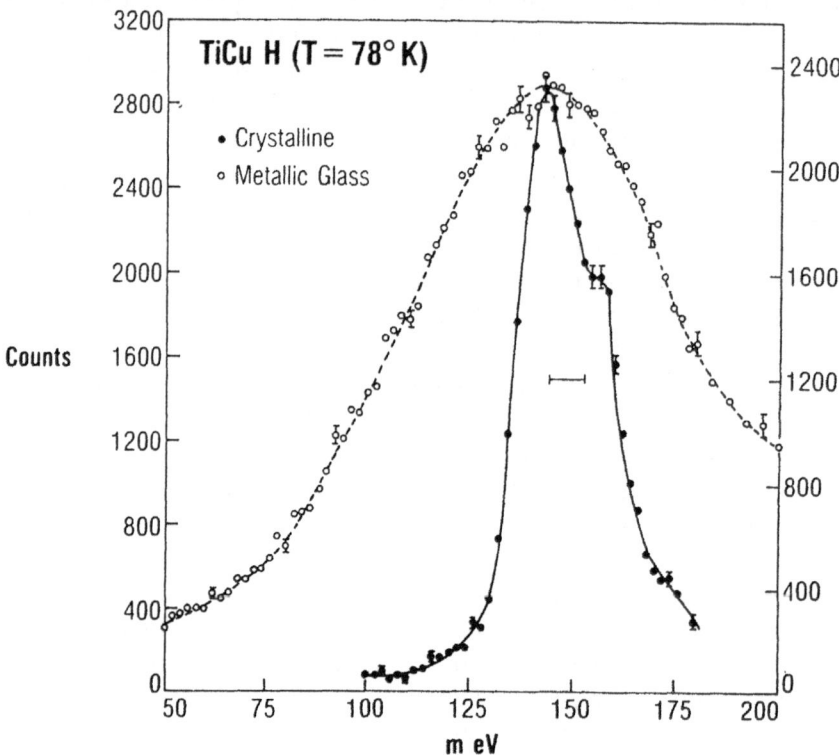

Figure 9. Inelastic neutron scattering spectra of crystalline
$TiCuH_{0.93}$ and amorphous $TiCuH_{1.33}$.

SUMMARY

1. Some metallic glasses can absorb large amounts of hydrogen
    without crystallization.

2. Hydrogen absorption leads to volume expansion of the glass.

3. The diffusion of hydrogen in metallic glasses is generally
    slower than in the corresponding crystalline compounds. The
    TiCu glass hydride is an exception.

4. Hydrogen absorption capacity may be larger in certain
    metallic glasses than in their crystalline counterparts.

5. Hydrogen can be used effectively as a probe of the metallic
    glass structure.

REFERENCES

1. R. Brill, Z. Kristallogr., 75, 217 (1930).
2. P. Duwez, R. H. Willens, and W. Klement, J. Appl. Phys., 31, 1136 (1960).
3. W. Klement, R. H. Willens, and P. Duwez, Nature, 187, 869 (1960).
4. a) P. Chandhari, B. C. Giessen, and D. Turnbull, Sci. Amer. 242, 98 (1980).
    b) S. Takayama, J. Mat. Sci., 11, 164 (1976).
    c) J. J. Gilman, Physics Today, 28, 46 (1975).
5. A. J. Maeland, in A. F. Andresen and A. J. Maeland, eds., Hydrides for Energy Storage, Proc. Int. Symp., Geilo, 1977, Pergamon, Oxford, 1978, pp. 447-462.
6. J. J. Rush, J. M. Rowe, and A. J. Maeland, J. Phys. F: Metal Physics, 10, L283 (1980).
7. R. K. Viswanadham, J. A. S. Green, and W. G. Montague, Seripta Met., 10, 229 (1976).
8. D. E. Polk and B. C. Giessen, "Overview of Principles and Applications," J. J. Gilman and H. J. Leamy, eds., Metallic Glasses, ASM Materials Science Seminar Series, ASM, Metals Park, Ohio, 1977.
9. D. S. Boudreaux, "Structure of Metallic Glass Alloys," Chapter Two in R. Hasegawa, ed., The Magnetic Chemical and Structure Properties of Glassy Metallic Alloys, CRC Press, Inc., Boca Raton, Florida, U.S.A., 1981.
10. J. D. Bernal, Nature, 185, 68 (1960).
11. J. J. Gilman, Phil. Mag. B, 37, 577 (1978).
12. T. Masumoto and K. Hashimoto, Ann. Rev. Mater. Sci., 8, 215 (1978).
13. D. Nathasingh and C. H. Smith, B2, Proceedings of Powercon 7, March 25-27, 1980, San Diego, California.
14. M. Hansen, Constitution of Binary Alloys, McGraw Hill, New York, 1958.
15. M. H. Mueller and H. W. Knott, Trans. Metall. Soc. AIME, 227, 674 (1963).
16. N. Karlson, J. Inst. Met., 79, 391 (1951).
17. J. M. Vitek, Z. Metallkd., 67, 559 (1976).
18. A. J. Maeland, L. E. Tanner, and G. G. Libowitz, J. Less-Common Met., 74, 279 (1980).
19. R. C. Bowman, Jr. and A. J. Maeland, paper presented at American Physical Society Meeting, Chicago, Illinois, March 19-23, 1979, to be published.
20. H. Y. Carr and E. M. Purcell, Phys. Rev., 94, 630 (1954); S. Meiboom and D. Gill, Rev. Sci. Instrum., 29, 688 (1958).
21. A. J. Maeland, Preparation and Properties of TiCuH, Adv. Chem. Ser., 167, 302-311 (1978).

22. A. J. Maeland and G. G. Libowitz, J. Less-Common Met., $\underline{74}$, 295 (1980).

23. A. R. Miedema, J. Less-Common Met., $\underline{46}$, 67 (1976).

24. A. Santoro, A. J. Maeland, and J. J. Rush, Acta Cryst. Sect. B, $\underline{34}$, 3059 (1978).

25. S. S. Sidhu, L. Heaton, and D. D. Zauberis, Acta Cryst., $\underline{9}$, 612 (1959).

26. J. J. Rush and A. J. Maeland, unpublished results, 1979.

27. R. C. Bowman, Jr., A. Attalla, and A. J. Maeland, Solid State Commun., $\underline{27}$, 501 (1978).

28. a) A. Amamou, Solid State Commun., $\underline{33}$, 1029 (1980);
     b) P. Steiner, M. Schmidt, and S. Hufner, Solid State Commun., $\underline{35}$, 493 (1980).

29. A. C. Switendick, Ber. Bunsenges. Phys. Chem., $\underline{76}$, 535 (1972).

HYDRIDE FORMATION AT HIGH HYDROGEN PRESSURE

Bogdan Baranowski

Institute of Physical Chemistry
Polish Academy of Sciences,
01-224 Warsaw, ul.Kasprzaka 44/52

ABSTRACT

The transition from low concentrated solutions of hydrogen in metals to hydride phases is treated from the point of view of the thermodynamic stability condition. The necessity for an increase of the hydrogen concentration in the metallic matrix is exposed. The available methods for creation of active hydrogen are reviewed. The simplicity of various kinetic procedures is outlined. The preference of the high pressure equilibrium method is clearly demonstrated. Recently used high pressure devices are shortly described whereby the limitations and difficulties are discussed. Properties possible to be directly investigated "in situ" conditions are enumerated. Examples of hydrides requiring high pressure of gaseous hydrogen for formation under equilibrium conditions are given. Transition metals like nickel, chromium, cobalt, manganese, molybdenium and rhodium as well as their alloys are mentioned. Some secondary properties possible to be measured are described. Conclusions and further perspectives of the high pressure technique are discussed.

1. THERMODYNAMIC CONDITIONS FOR A HYDRIDE FORMATION

The appearance of a metallic hydride is usually preceded by the existence of a dilute solution of statistically distributed hydrogen particles in the metallic matrix. These solutions are mostly termed as α-phase. The simple ideal distribution law between the gaseous hydrogen and the metallic phase is hereby fulfilled - termed by historical reasons as Sievert's law. As long as this law holds, no phase transition can be expected because inside ideal solutions the condition for thermodynamic stability

can never be violated. Let us explain this statement in a more
detailed way. From the point of view of non-equilibrum thermodynamics [1] the molar flow of the mobile hydrogen particles inside
the practically rigid metallic matrix can be described by

$$J_H = -L_{HH} \text{ grad } \bar{\mu}_H \tag{1}$$

where J_H denotes the mentioned flow (for time and surface unit),
L_{HH} is the phenomenological coefficient, grad $\bar{\mu}_H$ is identical with
the gradient of the molar electrochemical potential of hydrogen.
Due to the positive character of the entropy production, the inequality holds

$$L_{HH} > 0. \tag{2}$$

Neglecting the electrostatic part of the electrochemical potential,
equation [1] can be written for isobaric and isothermal conditions

$$J_H = -\frac{L_{HH} RT}{x_H} \left(1 + \frac{\partial \ln f_H}{\partial \ln x_H} \right) \text{ grad } x_H \tag{3}$$

where R and T are the gas constant and the absolute temperature, f_H
and x_H denote the activity coefficient and the molar ratio of hydrogen. Let us introduce the notation

$$D_H = \frac{L_{HH} RT}{x_H} \tag{4}$$

and call it the Einstein diffusion coefficient of the hydrogen particles in the metal considered. Thus equation [3] reads

$$J_H = -D_H \left(1 + \frac{\partial \ln f_H}{\partial \ln x_H} \right) \text{ grad } x_H. \tag{5}$$

As D_H is always positive (see [4]), the inversion of the flow J_H
can be caused by the term $1 + \frac{\partial \ln f_H}{\partial \ln x_H}$ only. In ideal solutions,
the term equals 1, thus is always positive. Thus the value of the
term $\frac{\partial \ln f_H}{\partial \ln x_H}$ is crucial for our discussion. What is the physical
mean of the flow inversion [5]? The negative value of the term
$1 + \frac{\partial \ln f_H}{\partial \ln x_H}$ has the consequence that fluctuations of hydrogen con-

centration don't relax to uniform distribution, but contrary to
this normal expectation, the hydrogen particles exhibit the
temdency to cluster formation. This tendency is equivalent with
the formation of a miscibility gap, or in other words, with the
creation of a new phase which is characterized by a higher concentration of hydrogen than the previous low concentration α-phase.
It follows clearly from the above discussion that systems fulfilling
Sievert's law cannot violate the above condition of thermodynamic
stability. First of all we have to deviate from the ideal behavior
in order to expect a phase separation. Well, it is a common feature
that dilute solutions are mostly ideal. Thus the unavoidable condition for shifting the system to the non-ideal behavior is the
increase of hydrogen concentration in the bulk metal. This again
requires, from simple thermodynamic reasons, the increase of the
hydrogen activity in the supplying phase.

The above reasoning, explaining the thermodynamical requirements for a phase instability, did not take into account the
contribution of the elastic energy of the lattice to the problem
considered. As the solution of hydrogen in a metal is usually
connected with volume changes, this should be included, through
the elastic energy, in the free energy of the system. Such contribution may be crucial for the nucleation kinetics.

2. FORMATION OF HIGHLY ACTIVE HYDROGEN

Which possibilities exist for the formation of thermodynamically active hydrogen? To facilitate the understanding of the
problems involved, let us write down two reactions which occur both
in the bulk gaseous phase as well as on the surface of the interacting metal:
   a) Dissociation of molecular hydrogen to atoms:

$$H_2 \rightleftharpoons 2H \qquad\qquad\qquad [6]$$

   b) Ionization of hydrogen atoms:

$$H \rightleftharpoons H^+ + e. \qquad\qquad\qquad [7]$$

In both cases the equilibrium is shifted, at normal temperature
and pressure conditions, radically to the left-hand side of
equations [6] and [7]. But as hydrogen molecules cannot be
accumulated by the metallic lattice, even due to simple geometrical
reasons, one of the above reactions has to occur at the metal surface before the bulk absorption process can take place. And here
we meet the question of why electrical discharge in low pressure
gaseous hydrogen or cathodic deposition of hydrogen may be so
efficient in respect to hydride formation. As the concentration
of protons or hydrogen atoms is extremely low in normal gaseous

hydrogen in equilibrium conditions, any increase of concentration of both species, what takes place in gaseous discharge or
during the electrode reaction, is equivalent with an increase
of the effective thermodynamic activity of hydrogen. The only
problem remains that the recombination kinetics of protons to
atoms and atoms to diatomic molecules is usually very fast. Therefore the competitive processes of diffusion of the active particles
into the bulk phase and the kinetics of the lattice reconstruction,
characteristic for the formation of a hydride phase, may not be
sufficient under the investigated non-equilibrium conditions. In
other words, the problem of reaching the region of instability of
the metallic lattice remains a competition between different relaxation times, governing the element steps of creation and recombination of the active particles and the steps of the phase transition.
As the relaxation processes mentioned are very sensitive to the
local structure and catalytic steps, we are faced with a rich variety of effects and poor reproducibility. But let us remark here
that for pure preparative purposes, all kinetic, non-equilibrium
methods of formation of active hydrogen are very effective. Due
to the simplicity and often a high efficiency, a wide field of
applications is available. This is especially true in respect to
the cathodic discharge of hydrogen and its simultaneous absorption
by the metallic cathode. Some of the above aspects were discussed
previously [2].

But being interested in well defined and reproducible conditions for active hydrogen formation, we are left with one method
only, namely, the application of the high pressure technique.
What are the advantages of this method?

From the P, V, T data of pure hydrogen or pure deuterium in
the gaseous state, the crucial property for our purposes, namely
the activity of these gaseous components, can be extracted. It is
common to present it in terms of fugacity, that is fictive
pressure which has to replace the formula for the chemical potential or molar free energy of the component considered. Thus in
all thermodynamic equations, where in the ideal case the hydrostatic pressure of the gas was applied, the fugacity is coming in,
playing the corresponding role. In other words, we have to multiply
the hydrostatic pressure by a coefficient, called the fugacity
coefficient, in order to get numerically the activity term,
replacing the hydrostatic pressure of ideal gases. This fugacity
coefficient is in condensed pure gases usually larger than one.
It means that real gases are more thermodynamically active than
the corresponding values for the ideal case. In hydrogen and deuterium the fugacity coefficient can easily exceed the order of
magnitude 10^2 or even 10^4 in the suitable pressure range [2].
Thus a very active hydrogen is available if a suitable high pressure
of the gas can be realized. This active hydrogen phase can be

maintained in contact with the metal investigated for an arbitrary
time interval. In such conditions, even slow processes can be
investigated under objective and reproducible conditions and the
results achieved in different laboratories can be easily checked
and compared. Any kinetic method for formation of active hydrogen
does not exhibit such a possibility as it is mostly hopeless to
repeat all the required details for these purposes. Changing in a
continuous way the pressure of gaseous hydrogen, we are able to
work with a well defined thermodynamic activity of this component.

3. HIGH PRESSURE DEVICES USED

The principal difficulty in treating high pressure gaseous
hydrogen is to avoid any contact with steel elements. As these
elements are mostly included in such devices, one has to look for
other alloys for these purposes.

The development of high pressure devices, suitable for gaseous
hydrogen in the pressure range up to 30 kbar at room temperature
took us a long time. The details of the different steps can be
followed in a sequence of papers [3-8]. Let us present on Fig. 1
the last development which is used in our recent measurements.

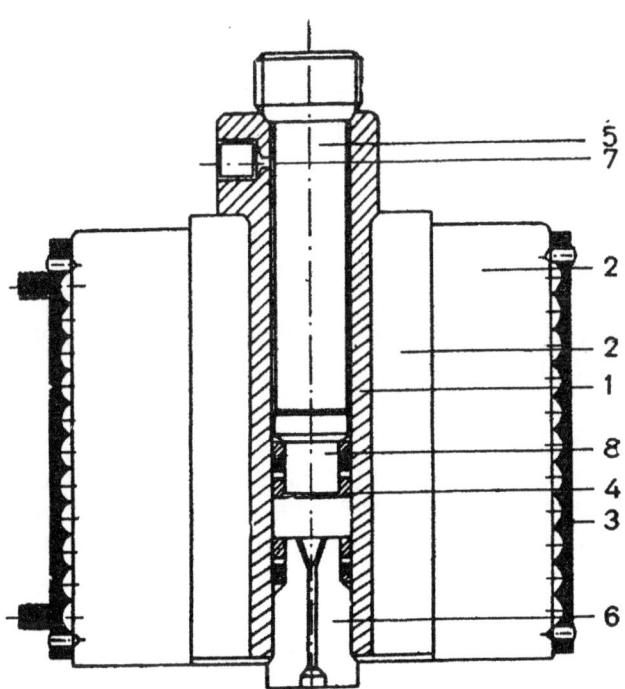

Fig. 1. Scheme of the high pressure devide for gaseous hydrogen.

It consists of an internal beryllium bronze cylindrical vessel [1]
which is externally supported by two steel rings [2]. The total
vessel may be thermostated by an external heating ring or circula-
ting fluid [3]. The hydrogen working volume [4] is placed between
the mobile piston [5] and the closing stopper [6]. The inlet capillary [7] is located in the upper part of the beryllium vessel
[1]. The force transmitting piston [5], composed of steel, is
supplied by a beryllium bronze ending [8]. Both the stopper [6]
and the piston [8] are sealed by Bridgman-type metallic sealings.
The lower pressure range is sealed by soft o-rings. The samples
to be investigated are placed on the stopper inside the hydrogen
working volume. At the starting moment the end of the mobile piston [8] is placed above the inlet capillary, thus allowing the
entrance of the hydrogen gas from an outside multiplier into the
working volume. The initial pressure of hydrogen or deuterium lies
between 10^2-10^3 bars, depending on the final pressure required.
After the initial supply of the gas, the inlet capillary is cut
off by an appropriate movement of the piston which is connected
with a hydraulic press. All further increase of the pressure is
carried out by a suitable movement of the piston [5]. After a
careful polishing of the internal vessel [1], a manifold movements
of the piston in both directions are possible, without losing the
necessary sealing. The upper limit of the pressure possible to
be achieved is determined by the mechanical properties of the
beryllium bronze vessel. Mostly the range of 15 kbar is possible
to be reached if not exceeding temperatures of about 150°C.
Pressures up to 30 kbar can be reached in a hydrogen working volume
if surrounded by an organic liquid inside a steel vessel [2].

Recently, hydrogen pressures up to 70 kbar were reported but
unfortunately, no details concerning the devices used were published [9].

A more detailed description of the high pressure devices used
in our laboratory can be found in a recent review article [10].

4. PROPERTIES INVESTIGATED "IN SITU" CONDITIONS AS INDICATORS OF
    HYDRIDE FORMATION

The working volume in high pressure devices is, for obvious
reasons, rather limited. In our cases, gaseous hydrogen volumes
of about 10 to values below 1 cm^3 are available. Any pressure
increase is equivalent with a reduction of the working volume,
corresponding to the compression of gaseous hydrogen. Therefore,
the measuring device must be prepared for minimal working volume
conditions. Further on so far, we did not use any optical windows
in our measurement. Thus all information required could be transmitted from the working volume by electrical methods, or the displacement of the piston was registered in a careful way. The

limitations of the working volume mean that a radical minia-
turization of the common devices, used at normal pressure
measurements, is unavoidable. As the realization of this
condition is sometimes equivalent with less accurate results, we
may finally meet the situation, especially at extreme high
pressure, that only rough numbers are coming out, being sufficient
for some preliminary pruposes but not acceptable for a careful and
systematic investigation. We face here one of the limitations for
the front profile of the high pressure research.

The simplest property possible to be measured "in situ"
conditions is the electrical resistance. As our samples exhibited
mostly metallic conductivity, the classical four pole technique
could be used here. This enabled us to measure the changes of the
electrical resistance with a very high resolution, quite comparable
with that achieved at normal pressure conditions. As small metallic
strips can be used, simultaneously even more than 10 different
samples can be investigated, applying a common current circuit and
different potential wires. In order to measure as large resistances
as possible or in other words, as large potential differences as
possible, high currents and small thickness of the samples had to be
taken into account. The first condition is limited mainly by the
Joule heating, which can create temperature differences and conju-
gated thermopowers, which may introduce a considerable systematic
error. In our measurements we mostly did not exceed 100 mA. The
small thickness of the samples has the advantage to reduce the
kinetics of hydrogen absorption and phase transformations. In some
cases, like nickel hydride [11], the limited penetration depth of
the hydride phase makes the application of small thicknesses unavoidable. In our laboratory we usually work in the range of $[1-10] \times 10^{-4}$
cm of sample thickness. Rolled down metallic foils fulfill this
requirement. The formation of a hydride phase manifests itself by
a discontinuity of the electrical resistance. This has to be
expected as a discontinuous change of the lattice parameter takes
place, or sometimes even a lattice reconstruction. On the other
hand, the hydride formation is connected with a discontinuous
increase of the number density of hydrogen particles inside the
metallic matrix that changes both the density of scattering centers
for the electrons - a considerable increase of the residual resistance is the result - and the band structure of the electrons. The
first aspect increases the resistance, the second may cause either
an increase or a decrease, depending on the differences of the previous and the new structure. Both cases we, in fact, observed
already [2,10].

Similarly, like in the case of the electrical resistance, any
other property dependent on the electronic structure can serve as
an indicator for the phase transitions investigated. Of special
importance is the thermopower. A known temperature gradient has

to be created along the sample and the thermopower can be measured
by a high resistance potentiometer [12]. One advantage as compared
with the electrical resistance measurements has to be mentioned
here: Hydride phase transformations are often connected with the
appearance of irreversible macrodefects, like cracks and reduction
of the crystallite size. Its consequences are large irreversible
changes of the electrical resistance, which are, on the other hand,
mostly not reproducible from one sample to the other. This is due
to the change of the effective path of the current. A similar
difficulty does not appear during the measurement of the thermopower, where currentless conditions are realized. Therefore, this
property may be sometimes advantageous as compared with the electrical resistance measurements. It is of special importance if one
intends to investigate not only the formation but also the decomposition process and if a recycling procedure is planned.

     Other electronic properties like the thermal conductivity [13]
and the Hall effects [14] were already measured in some systems,
but as these methods are more complicated in details, they were not
used so far for direct indications of phase transitions in unknown
cases.

     Rather successful is the measurement of the magnetic moment
"in situ" conditions [15,16]. This method is especially to be considered in cases where radical changes of the magnetic order are
taking place during the hydride transformation. A typical example
here is nickel which loses completely its ferromagnetism when
going over to the hydride phase. But as this method will be treated
in detail in a separate article of this book, I don't go into
further details here.

     The discontinuous uptake of hydrogen which is connected with a
hydride formation can serve as a direct method for determination of
the transition pressure [at constant temperature] and the composition of the phase created. A sensitive indicator for this purpose
is the displacement of the piston in the device presented in Fig. 1,
if the resolution of this measurement is sufficient [7]. Thus, a
determination of the absorption and desorption isotherms is possible
Such measurements are very informative, especially in respect to
the quantitative thermodynamic interpretation. The disadvantage
of this method is the requirement of a sufficient amount of the
metal investigated and a suitable pretreatment of the sample, in
order to avoid sluggish and unreproducible exchange of hydrogen.
Usually a manifold recycling is desired before the final measurement
can be started. Thus, this method is to be proposed, first of all,
for cases where at least some preliminary knowledge concerning the
hydride formation is available.

Another direct method for the indication of phase transformations is the determination of the lattice parameter "in situ"
conditions. Here a transparent X-ray window is required. This was
realized in our laboratory in the lower pressure region by an
epoxy resin container [up to 700 bar] [17] and in the higher
pressure range [up to 10 kbar] by a beryllium container [18]. Long
exposition times are necessary in order to record the characteristic
lines by a photographic method. The increase of the hydrogen
pressure was carried out either by a mobile piston inside the
device [17] or by a separate multiplicator [18] from which a transmission capillary was constructed. As the measuring procedure is
rather complicated and tedious, this method is not a useful one
for a frequent application.

5. EXAMPLES OF HYDRIDES FORMED IN HIGH PRESSURE CONDITIONS

As it was mentioned above, the simplest measurement "in situ"
conditions can be carried out in respect to the electrical resistance. An additional preference of this property is, besides high
sensitivity to structural changes and high resolution of the single
measurement, the possibility of simultaneous registration of even
more than 10 different samples. Fig. 2 presents an example of the
course of the relative electrical resistance of palladium and two
palladium - platium alloys. All points were measured at stationary
conditions, that is, after the attainment of a stable equilibrium
between gaseous hydrogen and the metallic samples. The data presented in Fig. 2 correspond to an upper limit of pressure of gaseous
hydrogen around 10 kbar. The concentrations of hydrogen indicated
were calculated from low pressure results and exceed, as we now
know, the real uptake of hydrogen by these metals. For pure palladium, the maximal value of the electrical resistance is observed at
25°C if the metal is in contact with gaseous hydrogen of about 15
bar. Further increase of this pressure results in a continuous
decrease of the resistance which, around 15 kbar, coincides with
the initial value for pure palladium. This course is due to the
approach of palladium hydride to full stoichiometry which is
equivalent with a complete occupation of all octahedral interstitials in the fcc lattice of this metal. Thus the decrease of the
electrical resistance is due to the continuous transition from a
disordered to an ordered state in the metallic matrix. In other
words, the curves presented in Fig. 2 can be treated as an illus-
tration of Northeim's rule for two-component metallic alloys. From
the point of view of the electrical resistance we observe in the
palladium - hydrogen system a consequent continuity, especially if
we plot this quantity as a function of hydrogen concentration. The
composition around 0.6 atomic ratio, distinguished in this system
by rigid band arguments, if protonic character of the hydrogen
particles is assumed, is in no special way different from other
hydrogen contents. Much more involved seems the treatment of an

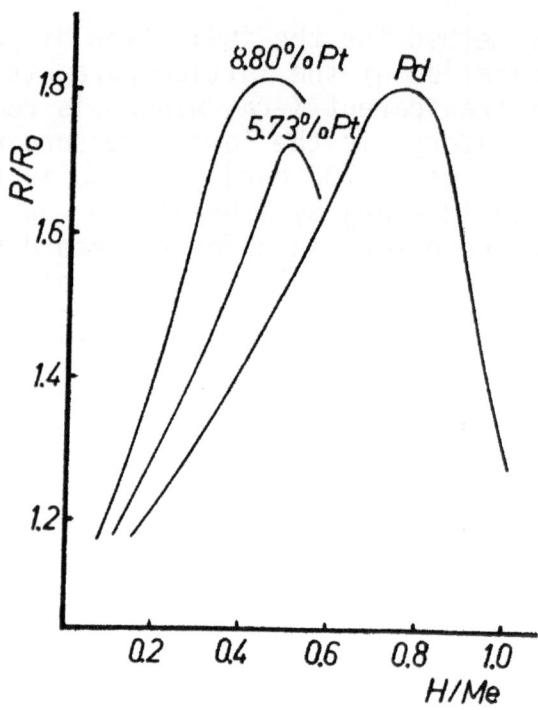

Fig. 2. Relative electrical resistance of palladium and two palladium - platinum alloys as function of hydrogen concentration at 25°C. R - resistance at the given high pressure
of gaseous hydrogen, R_0 - resistance at normal pressure in
absence of hydrogen.

overlapping of two ordered structures, namely, that of pure palladium and that of stoichiometric palladium hydride, whereby the last
requires a contact with high pressure gaseous hydrogen. At room
temperature, at least 10 kbar of this component is necessary to
achieve full stoichiometry of this hydride [19]. This approach to
full occupation of the octahedral vacancies of hydrogen particles
was the starting point for the later discovery of superconductivity
of palladium hydride [20] that is still a very active subject for
theoretical and experimental investigations [21]. It was preceded
by the observation of the disappearance of a resistance anomaly in
palladium hydride in the low temperature region if a sufficient
hydrogen-rich sample is investigated [22].

The above example concerns a system where the high pressure technique is suitable for an increase of hydrogen concentration inside a one-phase hydride. Of course, a similar influence of the high pressure gaseous hydrogen is to be expected in many other cases.

More interesting are systems in which new hydride phases are formed. Fig. 3 presents an example for the formation of chromium

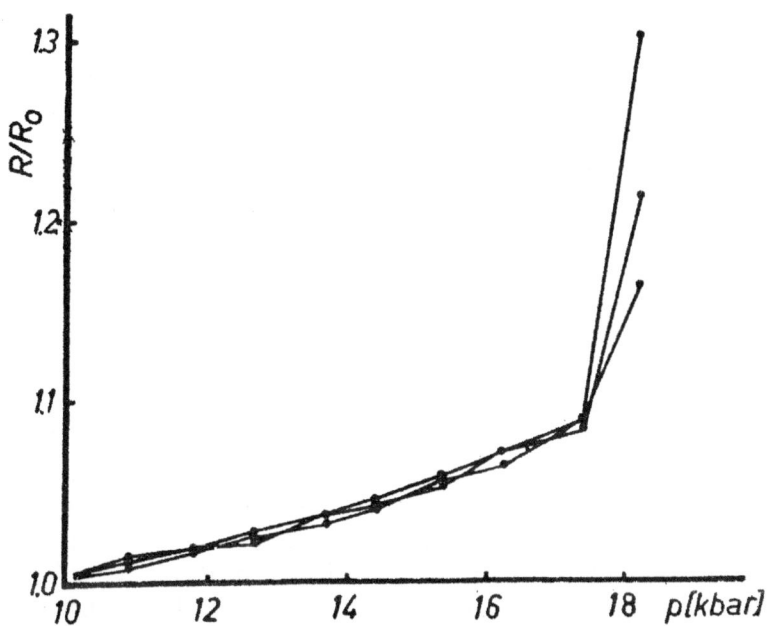

Fig. 3.   Relative electrical resistance of three different chromium
          samples kept at high pressure gaseous hydrogen at 150°C.
          Each point was measured after 30 minutes.

hydride [23]. Chromium foils of about 20μ thickness were exposed in gaseous hydrogen and the electrical resistance was measured after a time of 30 min. whereby constant pressure conditions were maintained. Of course, the results presented in Fig. 3 may differ from stationary values but the main feature is preserved. Up to the pressure of 17 kbar, a systematic increase of the resistance is noticed, but a discontinuity is observed first in the neighborhood of 18 kbar.   Here a remarkable increase of the resistance is noticed which can even lead to infinite values. This indicated that macroscopic cracks are created, which cause an interruption of the electrical circuit. Keeping the sample for several hours in the final conditions, cooling down the pressure vessel to temperatures below -30°C, reducing quickly the pressure and later keeping

the samples in liquid nitrogen, one can prove that a hydride phase
was formed. The atomic ratio hydrogen to chromium exceeds hereby
0,9 and the initial bcc lattice of chromium is reconstructed to a
hcp lattice of the new hydride. So far the same hydride phase
could be prepared by a special electro-chemical method only. An
extended study of the Cr-H system was later carried out in a wider
temperature and pressure range [24]. Hereby, it was shown that
the formation pressure is a complex function of temperature,
contrary to the decomposition pressures which are rising continuously
with the temperature.

Reconstructive in respect to the crystallographic lattice is
also the formation of manganese hydride which requires an even
higher temperature for the formation [25-27]. Recently, a stoichiometric molybdenum hydride was prepared [28] but a pressure of 65
kbar gaseous hydrogen and a temperature of 350°C were necessary.
Details concerning the phase diagram of this system in a more
extended temperature-pressure range were published more recently
[29]. A similarity with the Cr-H phase diagram was stated, but
much higher pressures are involved in the Mo-H system. In all
examples mentioned, the electrical resistance was taken as the
indicator for the formation and decomposition process.

Of course, the most direct way for the processes considered is
the measurement of the uptake of hydrogen as a function of hydrogen
pressure and temperature. In other words, the absorption and desorption isotherms are the most desirable data for a characteristic of
the hydride formation. The device presented in Fig. 1 can be used
for this purpose, as we mentioned shortly in Part 4. So far,
direct measurements were carried out in the Pd-H system [19] and
recently in the Ni-Cu alloy system [30]. Fig. 4 presents an
example of the curves taken up. It concerns a powder sample of
pure nickel, kept in contact with gaseous hydrogen at 25 and 65°C.
The metal used was pretreated by manifold cycling between formation
and decomposition of the hydride. Such a treatment activated the
surface of the metal, accelerating the kinetics of the process.
Let us discuss some details of Fig 4. At 25°C the increase of the
hydrogen pressure does not change the composition of the metallic
phase up to about 6 kbar. Around this pressure range a rapid
increase of the hydrogen concentration takes place. This discontinuity corresponds to the transition from the α-phase (solution
of hydrogen in the nickel) to the β-phase (nickel hydride phase).
The pressure plateau observed is typical for the two-phase region
of the diagram considered. It is well known that Sievert's law is
fulfilled in the nickel-hydrogen system at normal hydrogen pressures
The shape of the absorption isotherm just before the plateau is a
clear example for the violation of this law. Namely, we observe a
much larger increase of the hydrogen concentration inside the

metallic matrix than it could be expected from this law. A further proof for this behavior can be given as follows. If we calculate from the hydrogen solubility in nickel at normal conditions, the pressure of this gas required to achieve the atomic ratio H/Ni around 10^{-1}, we come to the value above 10 kbar, even then when we replace in Sievert's law the pressure by fugacity [31]. As Fig. 4 shows, the mentioned concentration is reached at a much lower pressure. In quite macroscopic terms we can explain this discrepancy by an attractive interaction of the hydrogen particles inside the metal. This attractive interaction is the very condition for the creation of a miscibility gap, that is equivalent in our case with the formation of a high concentrated hydride phase. At 25°C the pressure plateau achieves quite closely the stoichiometric

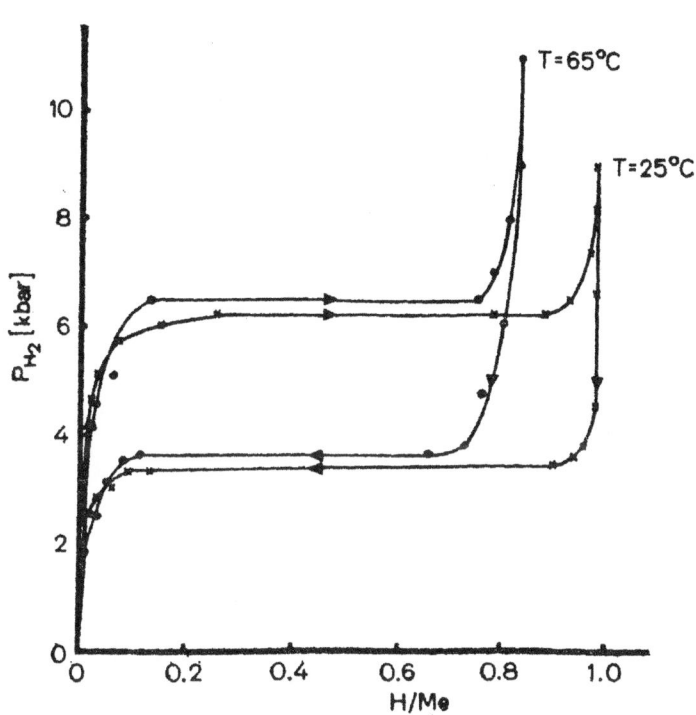

Fig. 4. Absorption and desorption isotherms of pure nickel (direction of pressure change indicated).

composition H/Ni = 1 which corresponds to a complete occupation of the octahedral vacancies in the nickel lattice. It was previously proved by neutronographic investigations that the tetragonal vacancies in the fcc lattice of nickel are not occupied by the hydrogen particles [32]. Further increase of the hydrogen pressure does not result in an uptake of hydrogen. Here we meet a radical difference between the Pd-H and Ni-H systems. As the first one changes its composition in a wide range of hydrogen pressure (from 10^{-2} to 10^{-4}

bar), the second system, as it is shown in Fig. 4, reaches the stoichiometric composition at 25°C in an extremely narrow pressure range. When going back with the pressure, a very large hysteresis is noticed. The hydride phase is maintained much below the formation plateau. The difference between the formation· and the decomposition pressures in the two temperatures exceed two kbar.

As palladium and nickel form a continuous set of fcc substitutional alloys, it was interesting to look for the possibility of hydride formation in these alloys [33]. The discontinuity of the thermoelectric power served as an indicator of the phase transition. The formation pressures of the corresponding hydrides form a continuous curve as is shown in Fig. 5. The results presented are the very beginning of a systematic study of thermodynamics of this system. Let us remark that the hydrides of the Ni-Pd alloys exhibit an unexpected behavior in respect to the superconductivity [34].

More interesting, are investigations of binary alloys where one component only is hydride forming and the second unknown in this respect. A careful interpretation of the results obtained

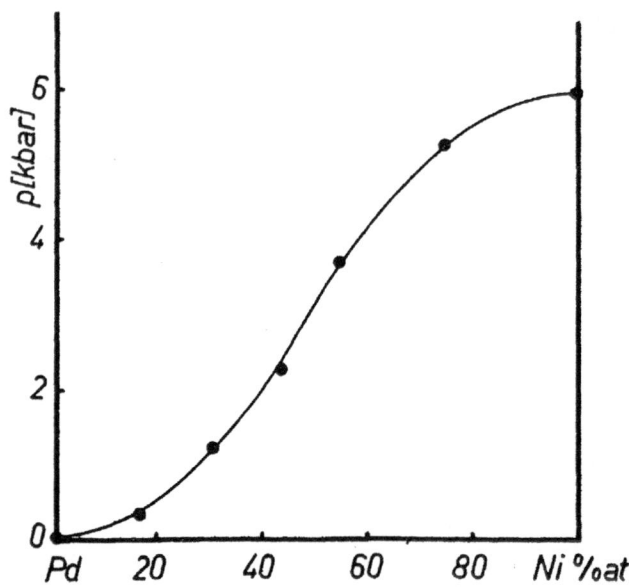

Fig. 5.   Formation pressure of the hydride as a function of composition in the Pd-Ni alloy system.

may be very useful in examining the probable conditions required for the formation of the unknown hydride. For instance, the very inactive behavior of platinum in alloys with palladium could be taken as an indication that below 30 kbar of gaseous hydrogen at room temperature, no possibility exists for the formation of platium hydride [35]. As an opposite example, the alloys Pd-Rh can be mentioned where an active contribution of rhodium in the uptake of hydrogen was stated [36]. Even a rough estimate of the conditions required for the rhodium hydride formation was made. Unfortunately, the pressure limit required was outside the region available in our laboratory. But some years later the Soviet group was successful in preparation of rhodium hydride in high pressure conditions [37]. A discontinuous increase of hydrogen solubility was observed at 250°C and a hydrogen pressure range of 40 ±5 kbar. The hydride formed was also of fcc type with an increase of the lattice parameter of 6% as compared with the original lattice of the metal. As could be expected from the formation conditions, the new hydride is unstable under normal pressure. Thus we meet here a similarity to the nickel hydride and in respect to the structure changes involved also to palladium hydride. The set of these three metals is a good example how the increase of hydrogen pressure can be successful in the formation of new hydrides.

In respect to the interaction with hydrogen under the transition metals, a special role is played by iron. As it is well known, a very small concentration of hydrogen in steels can sometimes cause very big troubles. An extended literature is devoted to the explanation of these effects, which are connected with large losses. One of the assumptions being considered is the formation of iron hydride which may be responsible for the phenomena observed. As we could show in a systematic study of the Ni-Fe alloy system up to the highest limit of 30 kbar at room temperature, no indication exists which could support the formation of this hydride [38,39,40]. Figs. 6 and 7 present some of the results obtained. The alloy with 17.2 atomic percent of iron exhibits a discontinuous increase of the electrical resistance above 15 kbar of gaseous deuterium. As could be expected, all alloys with smaller iron concentrations are characterized by formation pressures lying between pure nickel and the above mentioned value for the 17.2 alloy. The alloy with 25,8 atomic percent of iron can be prepared in two different states of the iron distribution, either with random distribution or in a superstructure. Fig. 6 shows that the statistically distributed iron atoms enhance the deuterium absorption as compared with the ordered alloy. This difference is clearly demonstrated by the relative electrical resistance presented. Quite an opposite behavior was previously observed in the Pd_3Fe system where the ordered alloy absorbs much more hydrogen than the disordered one [41]. The simplest explanation available would be that local effects are dominant in the Pd-H system whereby long range influence is determining the hydrogen absorption in the Ni-D system.

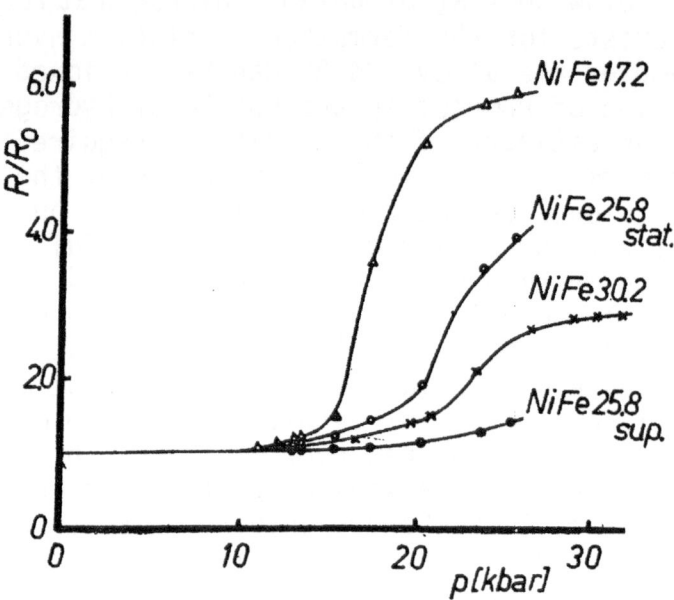

Fig. 6. Relative electrical resistance of three different Ni-Fe
         alloys as a function of deuterium pressure at 25°C. The
         alloy NiFe$_{25.8}$ is presented in the ordered state (sup)
         and with statistically distributed iron atoms (stat).

       Alloys with the highest concentrations of iron are presented
in Fig. 7. Here considerable changes of the electrical resistance
require pressure around and higher than 20 kbar of gaseous hydrogen.
The alloy with 65.9% of iron lies inside the invar region of this
system where many irregularities are known. All higher concentrated (in respect to Fe) alloys are of bcc structure. The alloy with
the highest iron content (87.6% Fe) changes its electrical resistance in a very insignificant extent, supporting the opinion
expressed above that no hope exists for the formation of iron hydride below the limit of 30 kbar of gaseous hydrogen at room
temperature.

       The very experience with different metallic systems in the
high pressure range of gaseous hydrogen is the rich variety of
behavior observed. Even very similar systems at normal pressure

conditions can diverge in a considerable way in the high pressure
range of gaseous hydrogen. An example is given in Fig. 8 [42] if
we compare it with Fig. 6. Similarity between Fe-Ni and Co-Ni
systems concerns the shift of the formation pressures to higher
values when increasing the concentration of the inert components.

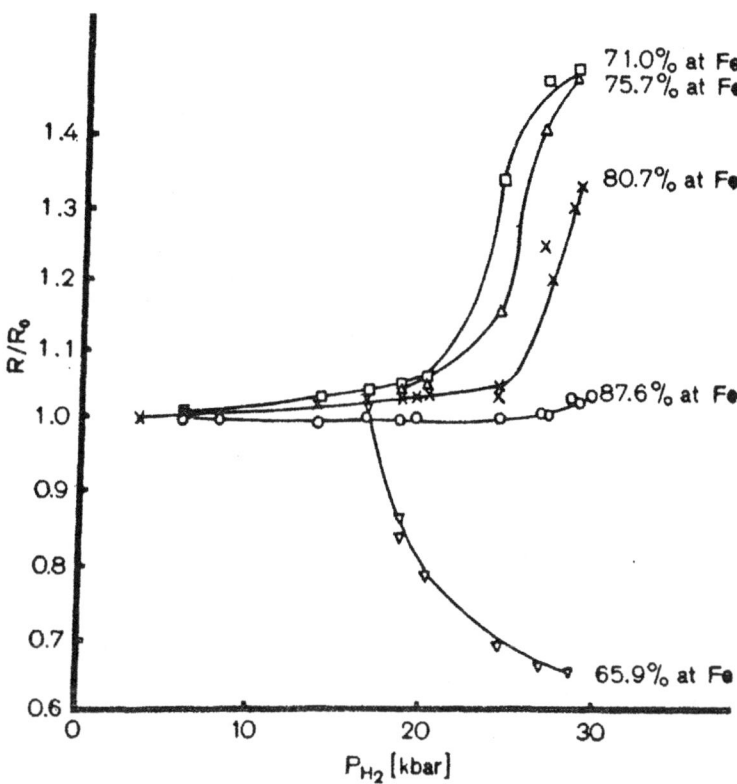

Fig. 7. Relative electrical resistance of different Ni-Fe alloys
         as a function of hydrogen pressure at 25°C.

But a quantitative comparison of the formation pressures in both
cases leads to the conclusion that it should be easier to prepare
a cobalt hydride than an iron hydride [16]. This expectation is
fulfilled in the results of the Soviet group where recently a cobalt
hydride was obtained [43]. A hydrogen pressure of 65 kbar at a
temperature of 225°C was required for this synthesis. A hexagonal
structure of the lattice was found for the new hydride phase.

     Well, there are still several transition metals which are
inert in respect to the formation of a hydride phase. It is a
question of time how far the high pressure technique can be useful
in extending the family of metallic hydrides to further systems.

6. SECONDARY PROPERTIES POSSIBLE TO BE INVESTIGATED UNDER HIGH
    PRESSURE OF GASEOUS HYDROGEN.

    The high pressure technique developed can be applied for the
investigation of properties not directly connected with the hydride
formation. Let us give here some examples which are not mentioned
in the above discussion. The phase transformations treated so far
are closely connected with the mobility of hydrogen particles in the

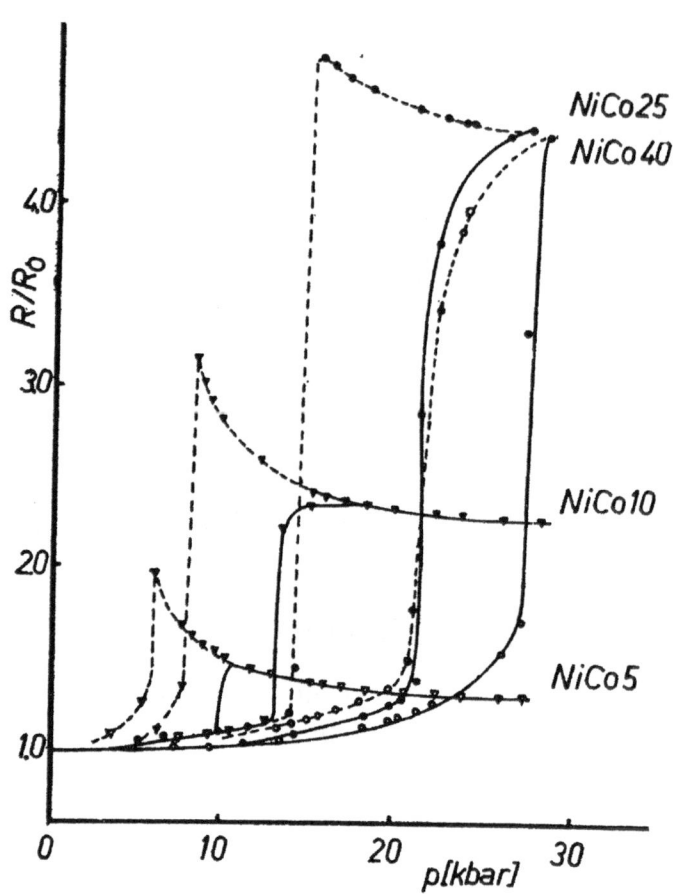

Fig. 8. Relative electrical resistance of different Ni-Co alloys
         as a function of deuterium pressure at 25°C.

metallic matrix. A quantitative measure of this property is the
diffusion coefficient. If the electrical conductivity can serve
as an indicator of the hydrogen concentration, its time behavior
can be correlated with the solution of II - Fick's law, if appropriate initial and boundary conditions are taken into account. The
most simple and realistic way is to impose hydrogen pressure jumps

and to follow the response of the system in respect to the hydrogen exchange between the gaseous and metallic phases. Such a relaxation method can finally lead to the evaluation of the diffusion coefficients involved. Some years ago this procedure was applied in Pd-H system [44] and recently a more improved technique was taken into account in the same system [45]. The increase of the hydrogen content is connected with a continuous decrease of the diffusion coefficient considered. An anomaly was found in the low temperature region which can either be interpreted as caused by a partial trapping of the hydrogen particles or by a correlated movement in these conditions. It disappears continuously when going over to higher temperatures. The same relaxation method can probably be applied to other systems too. In all cases a careful activation of the surface is required in order to avoid kinetic barriers during the transition to the bulk phase. In other words, a special care is necessary to be sure that bulk diffusion is the rate determining step of the kinetics investigated.

Another example for the usefulness of the high pressure technique could be the electrochemical measurements of a hydrogen electrode being in contact with the gaseous hydrogen. Taking palladium or platinum as the electrode material and an aqueous solution of hydrochloric acid, the silver/silver chloride electrode could serve as the second half-cell. The electromotive force of the mentioned cell was measured as a function of the hydrogen pressure up to the limit of freezing of the electrolyte solution (around 8 kbar at room temperature) [46]. The results obtained could be used for the evaluation of the partial molar volume of hydrogen in the electrolyte solution. These values could be compared with the corresponding results for pure gaseous hydrogen.

7. CONCLUSIONS

The examples described illustrate the high pressure technique with gaseous hydrogen as a very effective and successful method for preparation of metallic hydrides in cases where a high activity of hydrogen is required. Several new hydride phases could be discovered already and it seems quite probable that the list of possible examples is not closed. On the other hand, high pressure gaseous hydrogen can serve as a contact phase for the investigation of all properties "in situ" conditions if the concentration of hydrogen in the metallic phase can be changed to a suitable extent. The small working volume available in the high pressure conditions can be a serious limitation for such a purpose but here also the list of potential possibilities if far from being closed.

REFERENCES

1. B. Baranowski. "Nichtgleichgewichtsthermodynamik in der physik-
    alischen Chemie," VEB Deutscher Verlag für Grundstoffindustrie,
    Leipzig (1975).
2. B. Baranowski. Ber. Bunsenges. Physik. Chem. 76, 714 (1972).
3. B. Baranowski. Bull. Polon. Acad. Sci. 10, 451 (1962).
4. B. Baranowski, K. Bocheńska. Z. Physik. Chem. (N.F.) 45, 140
    (1965).
5. B. Baranowski, R. Wiśniewski. Bull. Acad. Polon. Sci. 14, 273
    (1965).
6. B. Baranowski, W. Bujnowski. Roczn. Chemii. 44, 2271 (1970).
7. B. Baranowski, W. Bujnowski, M. Tkacz. High Temperature - High
    Pressure 8, 656, (1976).
8. B. Baranowski. Plat. Met. Rev. 16, 10. (1972).
9. W. E. Antonov, F. T. Belash, W. F. Degtjareva, B. K. Ponomarev,
    E. G. Poniatowskij, W. G. Tissen. Solid State Physics 20,
    2680, (1978).
10. B. Baranowski. "Metal-Hydrogen Systems at High Hydrogen
    Pressures" in "Hydrogen Metals II", Ed. by G. Alefeld and J.
    Völkl (Springer-Verlag, Berlin-Heidelberg-New York, 1978),pp.
    157-200.
11. B. Baranowski, M. Śmialowski. J. Phys. Chem. Solids 12, 206
    (1959).
12. T. Skośkiewicz. Phys. Stat. Sol. (a) 6, 29 (1971).
13. A. W. Szafrański, B. Baranowski. J. Phys. E. 8, 823 (1975).
14. R. Wiśniewski, A. J. Rostocki. Phys. Rev. B 3, 251 (1971).
15. H. J. Bauer, B. Baranowski. Phys. Stat. Sol. (a) 40, K35
    (1977).
16. B. Baranowski. Z. Phys. Chem. (N.F.) 114, 71 (1979).
17. S. Majchrzak, B. Baranowski, W. Bujnowski, M. Krukowski.
    Roczn. Chemii 46, 1173 (1972).
18. S. Majchrzak. Roczn. Chemii 51, 1549 (1977).
19. M. Tkacz, B. Baranowski. Roczn. Chemii 50, 2159 (1976).
20. T. Skośkiewicz. Phys. Stat. Sol. (a) 11, K123 (1972).
21. B. Stritzker, H. Wühl. "Superconductivity in Metal-Hydrogen
    Systems" in "Hydrogen in Metals II", edited by G. Alefeld and
    J. Volkl (Springer-Verlag, Berlin-Heidelberg-New York, 1978),
    pp. 243-269.
22. T. Skośkiewicz, B. Baranowski. Phys. Stat. Sol. 30, K33
    (1968).
23. B. Baranowski, K. Bojarski. Roczn. Chemii 46, 525 (1972).
24. E. G. Poniatowskij, I. T. Belash. Dokl. Akad. Nauk SSSR,
    229, 1171 (1976).
25. M. Krukowski, B. Baranowski. Roczn. Chemii 49, 1183 (1975).
26. E. G. Poniatowskij, I. T. Belash. Dokl. Akad. Nauk SSSR 224,
    667 (1975).

27. M. Krukowski, B. Baranowski. J. Less - Commn. Metals 49, 385 (1976).
28. I. T. Belash, W. E. Antonov, E. G. Poniatowskij. Dokl. Akad. Nauk SSSR 235, 379 (1977).
29. W. E. Antonov, I. T. Belash, E. G. Poniatowskij. Dokl. Akad. Nauk SSSR 248, 635 (1979).
30. B. Baranowski, M. Tkacz. Polish J. Chem. 54, 819 (1980).
31. B. Baranowski, K. Bocheńska, S. Majchrzak. Roczn. Chemii 41, 2071 (1967).
32. E. O. Wollan, J. N. Cable, W. C. Koehler. J. Phys. Chem. Solids 24, 1141 (1963).
33. T. Skośkiewicz. Phys. Stat. Solidi (a) 48, K165 (1978).
34. B. Baranowski, T. Skośkiewicz. "High Hydrogen Pressures in Superconductivity" in "High-Pressure and Low Temperature Physics", Edited by C. N. Chu and J. A. Woollam (Plenum Press, New York and London, 1978), pp. 43-53.
35. B. Baranowski, F. A. Lewis, S. Majchrzak, R. Wiśniewski. J. Chem. Soc. Faraday Trans. I 68, 653 (1972).
36. B. Baranowski, S. Majchrzak, T. B. Flanagan. J. Phys. Chem. 77, 35 (1973).
37. W. E. Antonov, I. T. Belash, W. F. Degtjareva, E. G. Poniatowskij. Dokl. Akad. Nauk SSSR 239, 342 (1978).
38. B. Baranowski, S. Filipek: Roczn. Chemii 47, 2165 (1973).
39. S. Filipek, B. Baranowski. Roczn. Chemii 49, 1149 (1975).
40. S. Filipek. B. Baranowski. Polish J. Chem. 53, 951 (1979).
41. T. B. Flanagan, S. Majchrzak, B. Baranowski. Phil. Mag. 25, 257 (1972).
42. S. Filipek, B. Baranowski, M. Yoneda. Roczn. Chemii 51, 2243 (1977).
43. I. T. Belash, W. E. Antonov, E. G. Poniatowskij. Dokl. Akad. Nauk SSSR 235, 128 (1977).
44. M. Kuballa, B. Baranowski. Ber. Bunsenges. Physik. Chem. 78, 335 (1974).
45. B. Baranowski, S. Majorowski. (in preparation)
46. B. Baranowski, T. Szymczyk. Polish J. Chem. (in press)

ELECTRONIC STRUCTURE OF METAL HYDRIDES

D.A. Papaconstantopoulos

Naval Research Laboratory

Washington, DC   20375

ABSTRACT

Theoretical studies of the electronic structure of metal hydrides will be discussed from the point of view of ordinary band theory and from the view of disordered materials theories such as the coherent potential approximation. The presentation will cover an introduction to the methodology followed in such calculations and analysis of the results obtained. A comparison will be made between the band structure of the host metal and that of the corresponding hydride. Trends as a function of changing the element of the metal site, and as a function of hydrogen content will be examined.

I.   INTRODUCTION

In a series of papers[1-7] starting in 1971 Switendick established the foundation of the band theory approach for the study of the electronic properties of metal hydrides. Subsequently several other workers contributed to the understanding of the electronic structure of these systems. In this article we review the methodology followed in such calculations as well as the basic physical concepts which emerged. Both the methods and the results described here reflect more closely our own work.[8-15]

The paper is organized in two parts. The first part (Section II) deals with the theory and the computational techniques used in our calculations. The second part (Section III) discusses the results.

II.   THEORY AND COMPUTATIONAL METHODS

1.   The APW Method

The Augmented Plane Wave (APW)[16-19] method deals with the central problem of solid state physics, namely, the solution of the one-electron wave equation in momentum space. The wave equation to be solved has the form

$$H\Psi(r;k) = E(k)\Psi(r;k) \tag{1}$$

and is either the Schrödinger equation or, if relativistic effects are included, the Dirac equation. In the calculations to be discussed in this article the Dirac equation is solved with the simplification of omitting the spin-orbit interaction.[19-21] Therefore, the Hamiltonian H contains, besides the usual kinetic and potential energy terms, the relativistic components with mass-velocity and Darwin corrections. In the APW method the so-called muffin-tin approximation (MTA) is usually made which assumes a spherically symmetric potential within spheres surrounding each atom in the solid and a constant potential in the interstitial region. The MTA is a very good approximation for cubic structures like the fcc (studied in the present work), because only 0.26 of the unit cell volume is outside the spheres. Removing the MTA in those structures introduces a very small correction. The solutions of (1) are represented by linear combinations of the form

$$\Psi(r;k) = \sum_n v(k_n)\phi(r;k_n,E) \tag{2}$$

where $k_n = k+K_n$ with K_n the reciprocal lattice vectors. The wave vector k specifies the translational as well as the space group operations. For details about the implementation of group theory in the APW method, the reader is referred to the classic article by Mattheiss, Wood, and Switendick.[17] The so-called APW's ϕ of Eq. (2) have different functional forms in the two different regions. ϕ is a single relativistic plane wave in the region of constant potential. Inside the muffin-tin spheres, the wave function has the form:[19]

$$\phi(r; k_n, E) = \sum_{\lambda \mu} A(k_n)_{\lambda \mu} \begin{pmatrix} g_\lambda(r) \chi_{\lambda \mu} \\ if_\lambda(r) \chi_{-\lambda \mu} \end{pmatrix} \tag{3}$$

$\chi_{\lambda \mu}$ is a two-component spinor that is the relativistic equivalent of the spherical harmonics, and $A_{\lambda \mu}$ is a coefficient determined by matching the value of ϕ at the sphere radius to the plane wave value. The radial functions $g_\lambda(r)$ and $f_\lambda(r)$ satisfy the following set of coupled differential equations:[19-21]

$$f_\lambda'(r) = \frac{(\lambda-1)}{r} f_\lambda + \frac{(V(r)-E)}{c} g_\lambda(r) \tag{4}$$

$$g_\lambda'(r) = -\frac{\lambda+1}{r} g_\lambda(r) + \left(\frac{E-V(r)}{c^2} + 1 \right) cf_\lambda(r) \tag{5}$$

where $V(r)$ is the periodic crystal potential, E is the relativistic energy minus the rest mass energy, and λ is a relativistic quantum number related to the usual quantum numbers j and ℓ by:

$$\lambda = \ell \; ; \; j = \ell - \frac{1}{2} \quad \text{for} \quad \lambda > 0$$

and
$$\lambda = -\ell - 1; \; j = \ell + \frac{1}{2} \quad \text{for} \quad \lambda < 0$$

Now, if the value of f_λ from Eq. (5) is substituted in Eq. (4), one obtains a differential equation for g_λ in which the spin-orbit term can be easily identified and omitted.

2. Crystal Potential

We will now discuss how the starting crystal potential function $V(r)$ is constructed and how subsequent potentials are derived when the band calculation is carried out to self-consistency.

A starting crystal charge density $\rho(r)$ is found by a superposition of atomic charge densities $\rho_0(r)$ that are obtained from a relativistic Hartree-Fock-Slater atomic structure calculation.[22]

This superposition is done by the Löwdin α-expansion method,[23] which is an expansion in spherical harmonics about a central atom. This expansion is simplified by considering $\rho_0(r)$ to be a spherically symmetrical function. The result is[23]

$$\rho(r) = \rho_0(r) + \sum_i \frac{n_i}{2a_i r} \int_{a_i-r}^{a_i+r} r'\rho(r')dr' \tag{6}$$

where a_i is the distance from the central atom to the neighboring atom, and n_i is the number of atoms at the distance a_i. We have found that in the above summation we need not include more than five terms (i.e., $i_{max} = 5$). The resultant $\rho(r)$ is then used to solve Poisson's equation:[24]

$$\nabla^2 V_c(r) = -8\pi\rho(r) \tag{7}$$

and determine the coulombic potential $V_c(r)$ seen by one electron in the field of the nucleus and the other electrons. Since $\rho(r)$ is also a spherically symmetric function, Poisson's equation is reduced to the following expression[25] that gives $V_c(r)$ within the MT spheres,

$$V_c(r) = -\frac{2Z_s}{r} + \frac{2}{r}\int_0^r \sigma(r')dr' + 2\int_r^{R_s} \frac{\sigma(r')}{r'}dr' - 2\sum_{s'} W_{ss'}Q_{s'} - \frac{3\rho_0 V_s}{R_s} \tag{8}$$

where $\sigma(r) = 4\pi r^2 \rho(r)$, Z_s is the atomic number, R_s is the radius of the MT sphere,[26] $W_{ss'}$ are the Ewald constants, ρ_0 is the constant charge density in the interstitial region, V_s is the volume of a MT sphere, and Q_s is given by the expression:

$$Q_s = Z_s - q_s + \rho_0 V_s \tag{9}$$

where q_s is the charge inside a MT sphere. The index s labels the MT spheres for the case of a compound and we have suppressed it in our notation for $V_c(r)$ and $\sigma(r)$.

Outside the MT spheres, the coulombic potential is constant given by the following expression:[25]

$$V_c^O = \frac{1}{\Omega}\left[\frac{3}{5}\rho_o \sum_s \frac{V_s^2}{R_s} + 3 \sum_s \frac{V_s Q_s}{R_s} + 2 \sum_s V_s \sum_{s'} W_{ss'} Q_{s'}\right] \qquad (10)$$

where Ω is the outside-the-MT-spheres volume.

The total crystal potential $V(r)$ is found by adding the exchange potentials $V_{ex}(r)$ and V_{ex}^O to $V_c(r)$ and V_c^O respectively. In our hydride work these potentials were found by the following expressions using the so-called $X\alpha$ method:[27]

$$V_{ex}(r) = -6\alpha\left(\frac{3\rho(r)}{8\pi}\right)^{1/3} \qquad (11)$$

$$V_{ex}^O(r) = -6\alpha\left(\frac{3\rho_o}{8\pi}\right)^{1/3} \qquad (12)$$

where the coefficient α was determined[28] by the requirement that an atomic structure calculation performed with this form of exchange gives the same total energy of the atom as the Hartree-Fock method.

An alternate form of the exchange potential which contains correlation in a more transparent way than with the α of the $X\alpha$ scheme, was proposed by Hedin and Lundqvist.[29] This approach is supposed to be more appropriate for performing calculations of the total energy of the crystal. However, the two methods yield results which are in very close agreement for obtaining the energy bands and the densities of states.

3. Self-Consistent Energy Band Calculations

In order to carry out self-consistent APW calculations, a new spherically averaged radial charge density is constructed from the APW wave functions. This is given by the expression:[17]

$$\sigma_s(r) = \sum_{n,\ell} P_{n\ell s}^2(r) + \sum_{n,k} W_{nk}\sigma_{nks}(r) \qquad (13)$$

The first term of Eq. (13) involves only core states and is computed by using the crystal potential and a modified version of a relativistic atomic-structure computer code.[22] This procedure is known as the "soft-core" approximation. The second summation in Eq. (13) includes the "semi-core" states (for Pd 4s and 4p) and the occupied conduction-band states. This term samples a k-space mesh in the 1/48th of the cubic Brillouin zone (BZ). The coefficient W_{nk} is a weighting factor which depends on the k-point mesh and the symmetry of the corresponding k-points. As an example, we give in Table I the W_{nk}'s for the 6 point fcc mesh which we have used in our calculations for PdH.

Table I

K-(α/π)	$16 \times W_{nk}$
Γ(000)	1
Δ(100)	6
X(200)	3
Σ(110)	12
L(111)	4
W(210)	6

The charge density from Eq. (13) is averaged with the starting charge density to avoid divergence of the calculation.[30] With this averaged charge density, Poisson's equation is again solved to obtain the new crystal potential. This potential is then used to solve the radial wave equation and the procedure is repeated until the eigenvalues of two successive iterations satisfy the desirable convergence criterion. Convergence to within 3mRy is usually adequate when one is not interested in calculating the cohesive energy of the crystal. Such convergence is usually achieved after 5 to 7 iterations. Also, the above mentioned 6 k-point sampling is sufficient to give reliable results, especially when the Fermi level does fall in a flat region of the density of states graph.

Self-consistent calculations for compounds usually show substantial differences from the corresponding non-self-consistent ones. This is due to the uncertainties in constructing a sensible crystal potential. It appears, however, that in at least the case of monohydrides, the basic physical picture emerging from the band calculations is not particularly sensitive to self-consistency effects.[11]

4. Calculation of the Densities of States

The final self-consistent results are calculated on 89 k-points in the 1/48th of the fcc BZ. In order to obtain a meaningful density of states (DOS) from these calculations, an interpolation of the above results is needed. Several schemes for interpolating the results of the band structure calculations have been proposed. We have used the Quad,[31] the Tetrahedron,[32] and the Slater-Koster[33] methods. These approaches are outlined below:

a) Quad. In the fcc lattice, one constructs a cube that has twice the volume of the first BZ. Because of symmetry, only one octant of this cube needs to be considered. This octant is then subdivided into smaller cubic cells by choosing appropriate bisecting planes.[31] This bisection forms a sequence of 27 evenly-spaced points arranged as corners of cubes. Knowing the eigenvalues at these 27 k-points, one can determine by a least-squares procedure the constant coefficients E_0, $\vec{b}$, and $\vec{c}$ of the following quadratic expansion:

$$E(\vec{k}) = E_0 + \vec{b} \cdot \vec{k} + \vec{k} \cdot \vec{c} \cdot \vec{k} \tag{14}$$

This provides an analytic expression for the energy which is used to calculate the DOS. In our work we have modified the Quad scheme in order to evaluate the angular momentum components of the DOS given by:

$$n_\ell = \sum_n \int_S \frac{Q_{n,\ell}(\vec{k})}{|\nabla E_n(\vec{k})|} \, dS \tag{15}$$

where $Q_{n,\ell}(\vec{k})$ are the electronic charges inside the MT spheres as defined by Mattheiss et al.[17]

b) Tetrahedron Method. In this method the BZ is divided in tetrahedra of the same volume. The interpolation formula is now linear of the form:

$$E(\vec{k}) = E_0 + \vec{b} \cdot (\vec{k} - \vec{k}_0) \tag{16}$$

here the coefficients E_0 and $\vec{b}$ are determined exactly using energy values at the four corners of a tetrahedron. This approach has the advantage over the Quad that no least-squares fitting is necessary for the determination of the expansion coefficients.

Although the tetrahedron has been found to be a more

accurate method for calculating the DOS than the Quad, the differences are usually of the order of a few percent. So in the hydrides reported in this paper (especially in the cases where E_F is away from a peak in the DOS), the two methods give practically the same results. In most of our calculations we have used the Quad procedure to compute the DOS.

Recently, Boyer[34] proposed a method that interpolates the first principles APW results to a much finer mesh of k-points before one goes to the Quad or tetrahedron technique. Boyer's technique consists of a Fourier series expansion which has built in the compatibility relations between symmetry points and axes for the space group considered. The calculations presented here have not utilized this technique since it was not then available. This method, however, was successfully used in the calculations of the DOS of the A15 compounds[35] where high precision was needed. We believe that Boyer's method is probably the most accurate one to use for calculating DOS.

c) Slater-Koster Interpolation. The Slater-Koster (SK) interpolation is a tight-binding method in which the energy integrals are taken as parameters whose values are obtained by fitting to the results of first principles band calculations (in our case an APW calculation). In the case of PdH, we have used a basis set of 10 functions involving the s, p, and d orbitals of Pd, and the s orbital of H. We have set the overlap matrix equal to unity and included second neighbors of the same kind of atom for a total of 38 parameters. These parameters were determined by a non-linear least-squares procedure using symmetry to reduce the size of the 10x10 secular equation at each k-point. We have fit the energy values of the lower six bands at a uniform grid of 19 k-points. In order to obtain parameters which reflect correct wave function character, we also included in the fit energies from higher bands which correspond to the H-Pd antibonding s-states and the p-like Pd states. Having determined the parameters, we diagonalize the SK Hamiltonian for a large number of k-points (typically of the order of 400) and then use the tetrahedron method to find the DOS.

5. Coherent Potential Approximation

The coherent potential approximation (CPA)[36] is a mean field theory designed to calculate the electronic states of disordered materials. The basic idea of the CPA is that the electrons in the solid can be regarded as moving in an effective medium, the Hamiltonian of which is determined self-consistently by the condition that the average scattering of electrons by the atomic potentials relative to the medium is zero.

The CPA that we have used in PdH and AgH is based on the assumption that the metal-ion sublattice is perfectly periodic, while the non-metal sublattice can have vacant sites. The form of the CPA which we have employed here is the tight-binding CPA of Faulkner,[37] with the improvement of using a more elaborate SK Hamiltonian than that used in Ref. 37. Following Faulkner, we define a hydrogen scattering matrix,

$$\tilde{t}_H = (\tilde{\epsilon}_H - \tilde{\Sigma})[\tilde{I} - (\tilde{\epsilon}_H - \tilde{\Sigma})\tilde{G}]^{-1} \tag{17}$$

and a vacancy scattering matrix,

$$\tilde{t}_v = - [\tilde{G}]^{-1} \tag{18}$$

where $\tilde{\epsilon}_H$ is, in this case, simply the on-site diagonal SK parameter which describes the hydrogen-hydrogen interaction in PdH. The matrix $\tilde{\Sigma}$, called the CPA self-energy, is a complex quantity and since our SK Hamiltonian has only the s-hydrogen orbital, it is also a scalar. The Green's function G is given by the following integral over the BZ,

$$G(E) = \int_{BZ} \frac{d^3k}{z - H(k)} \tag{19}$$

where $z = E + i\delta$ is the complex energy, and H(k) the 10x10 SK Hamiltonian with the parameter ϵ_H replaced by the self-energy Σ. The vacancy matrix $\tilde{t}_v$ has the simple form of Eq. (18), since one sets the corresponding ϵ to infinity.

Now we write the so-called CPA condition for zero average scattering:

$$c\tilde{t}_v + (1-c)\tilde{t}_H = 0 \tag{20}$$

where c is the concentration of vacancies. Substituting·Eqs. (17) and (18) into (20) we obtain

$$\Sigma(E) = \epsilon_H - C/G_{10,10}(\Sigma, E) \tag{21}$$

where $G_{10,10}(\Sigma, E)$ is the 10th diagonal element of the Green's function matrix which corresonds to the hydrogen-hydrogen matrix element. Equations (19) and (21) are solved by iteration self-consistently, and then the DOS are evaluated from the expression:

$$n_\ell(E) = -\frac{1}{\pi} \lim_{z \to E^+} \mathrm{ImTrace} G_\ell(E) \tag{22}$$

where ℓ refers to the angular momentum components of the DOS.

III. RESULTS AND DISCUSSION

1. Band Structure of Transition Metal Monohydrides

The first band structure calculations on transition-metal hydrides were performed by Switendick[2,3] on PdH and NiH. Subsequently, several other groups[11,38-43] presented band calculations of PdH and studied various applications. In our work,[9-15] we carried out self-consistent band structure calculations for PdH, and we used the results to compare with several different experiments. In particular we calculated the superconducting properties of this material.[9-15]

In Figs. 1 and 2, we show the energy bands of Pd and PdH. As it was first pointed out by Switendick, hydrogen has a small effect on the d-states (these states are roughly centered at the symmetry points $\Gamma_{25'}$ and Γ_{12}). However, the introduction of hydrogen in the lattice makes the Fermi level E_F move to higher energy above the d-bands. On the other hand, the low-lying bonding levels (at Γ_1) are located approximately 2eV deeper in PdH than in Pd, and they have strong hydrogen s-like character. An entirely new band (high Γ_1), characterizing the antibonding states, appears at about 3eV above E_F. This band also has strong hydrogen s-like character. A dramatic effect of hydrogenation is seen at the point $L_{2'}$ (p-like symmetry) which is above E_F in Pd and falls by approximately 5eV in PdH. The bonding states were observed in photoemission experiments.[3] Our results are in good accord with those measurements.

Several calculations[2,40,43] of the energy bands of NiH show close similarity to PdH with the main difference being the position of the bonding states. Kulikov[43] has argued that hydrogenation leads to a reduced Stoner parameter and hence NiH is not ferromagnetic.

For the purpose of studying hydride formation and stability, Williams et al.[42] performed total energy calculations on Co, Ni, Cu, Rh, Pd, Ag, and the corresponding monohydrides. They

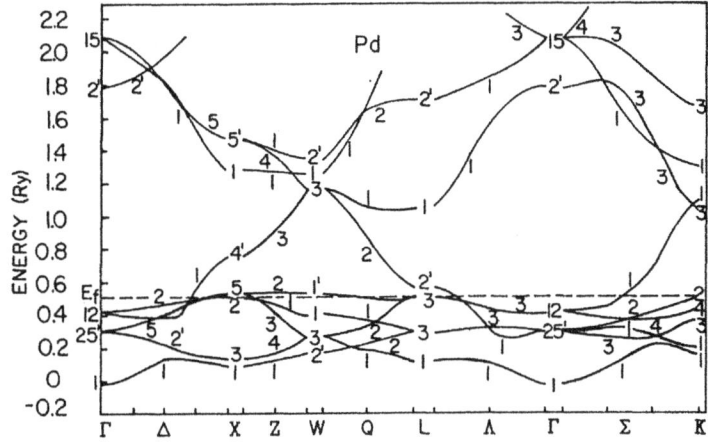

Figure 1

Energy bands of Pd.

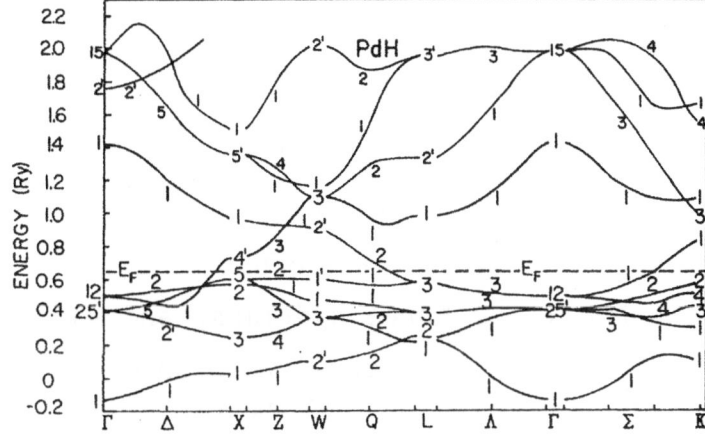

Figure 2

Energy bands of PdH$_{1.0}$.

conclude that hydrogen increases the volume of the unit cell and that the heat of solution is understood by considering the H proton and electron separately.

We also have studied the band structure of monohydrides which have not been made in crystalline phase.[14] Our calculations were made on RhH, AgH, and PtH, assuming the NaCl structure. In the case of PtH, we suggest that, on the basis of the very strong similarity to the band structure of PdH, platinum hydride is a good candidate for a high temperature superconductor.[15] On the other hand, our band-structure results for RhH and AgH provided us with the necessary justification to use rigid band and virtual crystal arguments in order to explain the superconducting proper-ties of the ternary systems $Pd_{1-y}Rh_yH_x$ and $Pd_{1-y}Ag_yH_x$. Indeed, an inspection of Fig. 3 shows that the DOS of $RhH_{1.0}$ and $PdH_{1.0}$

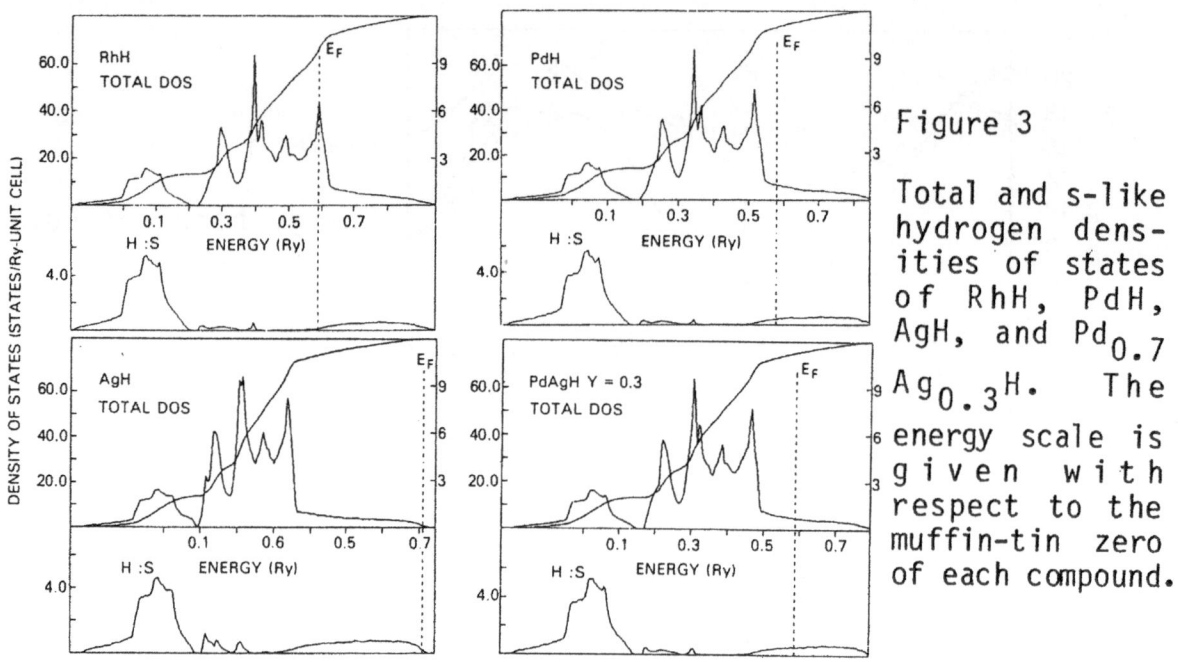

Figure 3

Total and s-like hydrogen densities of states of RhH, PdH, AgH, and $Pd_{0.7}Ag_{0.3}H$. The energy scale is given with respect to the muffin-tin zero of each compound.

are almost identical. The DOS of $AgH_{1.0}$ is also similar, but with the d-band complex being narrower by about 0.1 Ry. We have concluded that the rigid band approximation (RBA), and presumably the virtual crystal approximation (VCA), are safely applicable between $RhH_{1.0}$ and $PdH_{1.0}$. But it is questionable to extend them to include $AgH_{1.0}$. However, the values of the DOS just above the last peak are very close for these three hydrides; and since E_F falls in this region of the DOS, it appears reasonable to adopt the VCA for the purpose of obtaining the DOS at E_F in the $Pd_{1-y}Rh_yH_{1.0}$ and $Pd_{1-y}Ag_yH_{1.0}$ alloys. The fourth panel of Fig. 3 shows the DOS of a $Pd_{0.7}Ag_{0.3}H_{1.0}$ derived using the VCA. Calculations by Stocks et al.[44] for Pd-Ag alloys using the CPA, show a splitting of the Pd and Ag d- bands which, of course, the VCA cannot reproduce. However, at least for the Pd-rich alloys, the DOS at E_F are originating from the Pd component, and they have values close to those given by the VCA.

In these calculations we have used the RBA to approximate the effects of the vacancy disorder on the hydrogen sublattice and deduce the values of the DOS at E_F as a function of hydrogen content.

In all cases we have applied the RBA to the DOS of Fig. 3 from the last peak which corresponds to 10 electrons up towards the 12 electron mark in the flat structureless regions of the DOS graphs. From Fig. 3, one can see that to the right of the 10-

electron peak the DOS decreases while the H s-like component has the opposite behavior. This increase of the s-like hydrogen-site DOS has been proposed by us as the explanation of the increase of the superconducting temperature T_c.[10-15]

2. Band Structure of Transition Metal Dihydrides

The dihydrides have the fluorite structure (CaF_2) in which the space lattice is fcc with the hydrogen atoms occupying the tetrahedral interstitials. Switendick presented the first band structure calculations in these systems.[6] Using the TiH_2 and YH_2 systems as prototypes, he showed that the hydrogen-metal bonding states are significantly lower than in the host metal (as they do in the monohydrides). In addition, an antibonding band is formed below E_F which, according to Switendick, contributes to the hydride stability. Figure 4 shows the energy bands of YH_2

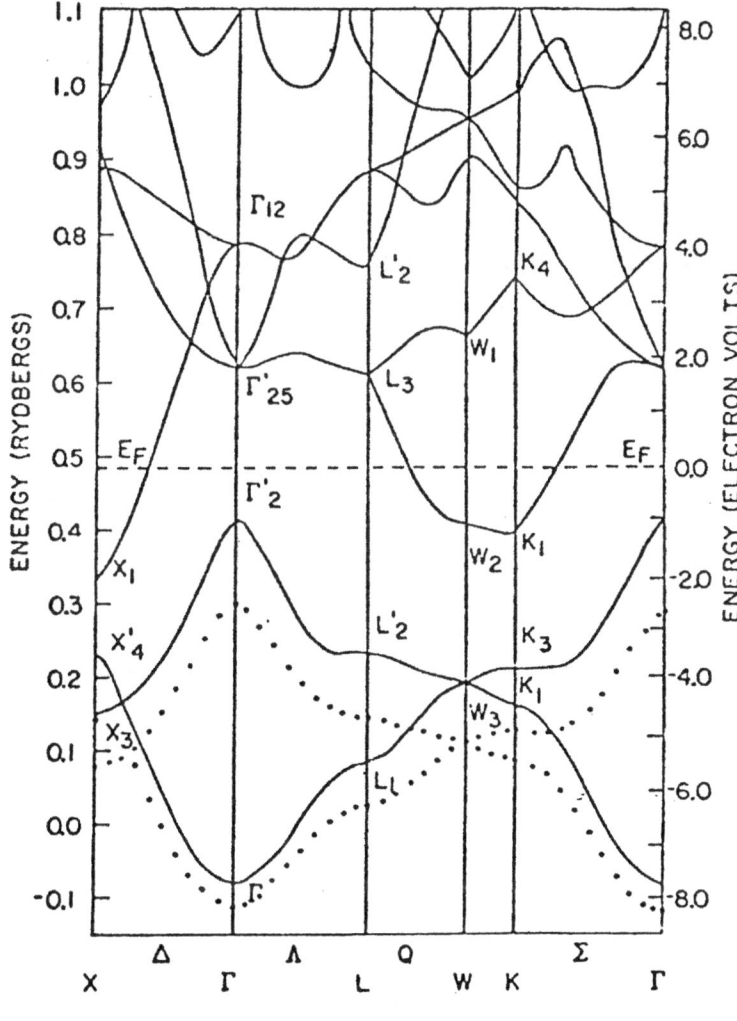

Figure 4

Energy bands of YH_2 after Peterman et al. The dotted line indicates the position of the non-self-consistent bands.

as calculated self-consistently by Peterman et al.[45]   These calculations display some quantitative differences from those of Switendick,[1] but the main characteristics of his results are preserved.   One can identify the bonding states centered at Γ_1 and the antibonding states at $\Gamma_{2'}$.   The position of the $\Gamma_{2'}$ with respect to E_F is the Switendick criterion of dihydride stability. For YH_2 $\Gamma_{2'}$ is well below E_F, while in ScH_2 (also calculated by Peterman et al.), $\Gamma_{2'}$ is just above E_F which should account for the relative instability of this system.

Band structure calculations for several dihydrides were performed by Kulikov[43] using a model Hamiltonian KKR approach, and by Gupta[40] using the APW method.   Kulikov's calculations differ from those discussed above in that they always locate the $\Gamma_{2'}$ state above E_F.   Both Kulikov and Gupta have calculated the superconducting properties of the dihydrides. We refer the reader to M. Gupta's article in this volume for details on this work.

3.   <u>Densities of States of Pd and PdH$_{1.0}$</u>

Using the Slater-Koster (SK) fit to the APW results, we described in Sec. II.4, we have calculated the density of electronic states for Pd and $PdH_{1.0}$ shown in Figs. 5 and 6.   We caution the reader that the energy scale is given with respect to different MT zero for Pd and $PdH_{1.0}$.   An approximate rescaling can be achieved by aligning the graphs at the last peak.   These DOS are decomposed per symmetry (s-, p-, T_{2g}- and Eg-like) and also per site.   We note that for Pd, E_F falls in a pronounced peak, while for $PdH_{1.0}$, E_F has moved well beyond the peak and gives a much smaller DOS value.   The low lying states are a mixture of s-, p-, and d-like symmetry for pure Pd. However, in $PdH_{1.0}$ the lowest peak of the DOS (H-Pd bonding states) consists of H and Pd s-states as well as Pd p-states, together with Pd Eg states.   It is interesting to note that H has pulled down to lower energies only the Eg-symmetry d-states.   At this point we should note that SK-DOS (which are calculated within the whole unit cell), and the Quad-APW-DOS[11] (which are calculated within the MT spheres), do not agree as to the relative magnitude of the s-like H and Pd contributions to the H-Pd bonding states in $PdH_{1.0}$.   This discrepancy is probably due to the somewhat arbitrary choice of the MT radii in the APW method.   If a smaller H radius and a larger Pd radius were used, then the H and Pd s-like DOS would come close to the SK values.

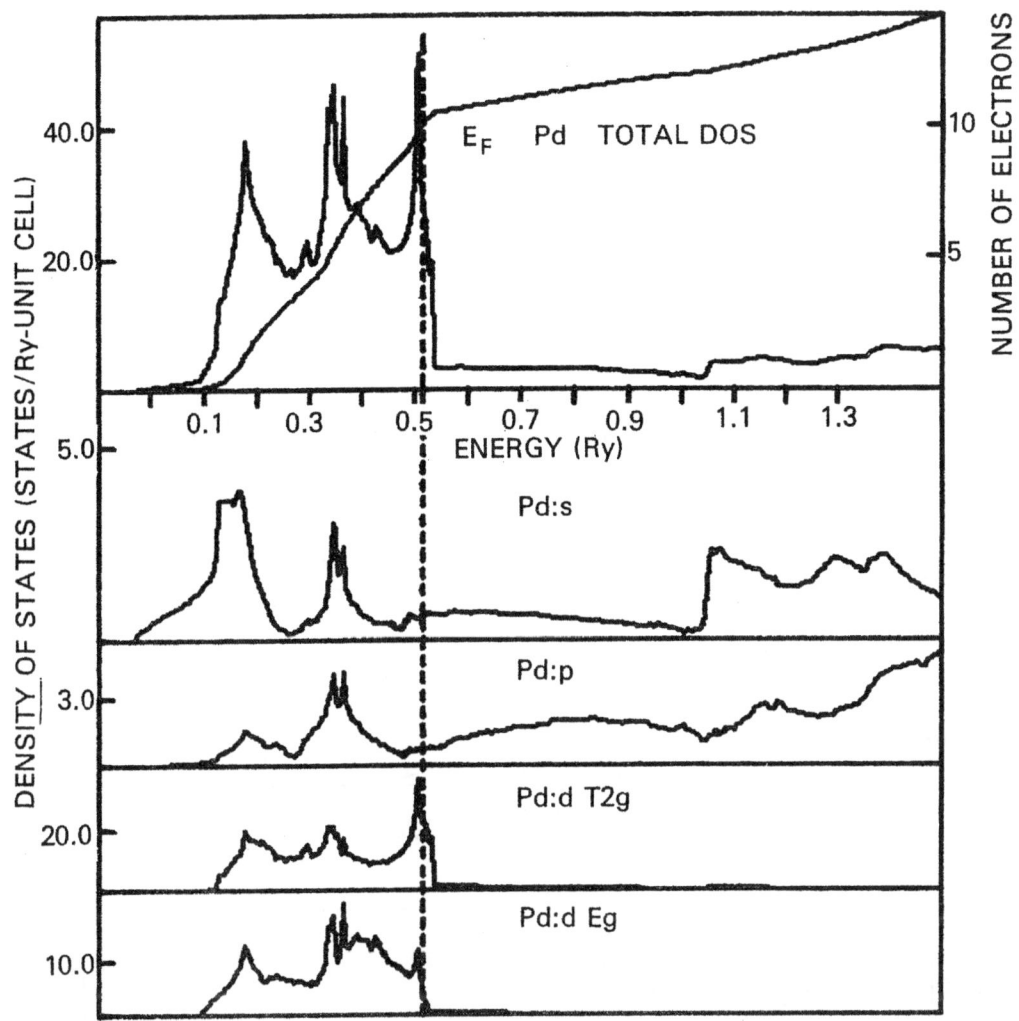

Figure 5. Slater-Koster Densities of States for Pd

On the other hand, there is some uncertainty attached to the SK
wavefunctions which could limit the accuracy of the decomposed
SK-DOS.   The above arguments are also affecting our discussion
of charge transfer.[11]   In Ref. 11, ignoring the uncertainty
introduced by the MT radii, we predicted charge transfer from Pd
to H. Our SK results, however, predict the opposite direction of
charge transfer.   The latter results we find more reliable.

     It should be stressed here that the foregoing discussion
does not pose any problem in the evaluation of the electron-phonon
interaction parameter η (discussed in Sec. III.5); because the
decomposed DOS enter the theory in a ratio which is independent of
the MT radius.   Finally, returning to Fig. 6, we point out the
appearance of H-Pd antibonding s-like states well above E_F.

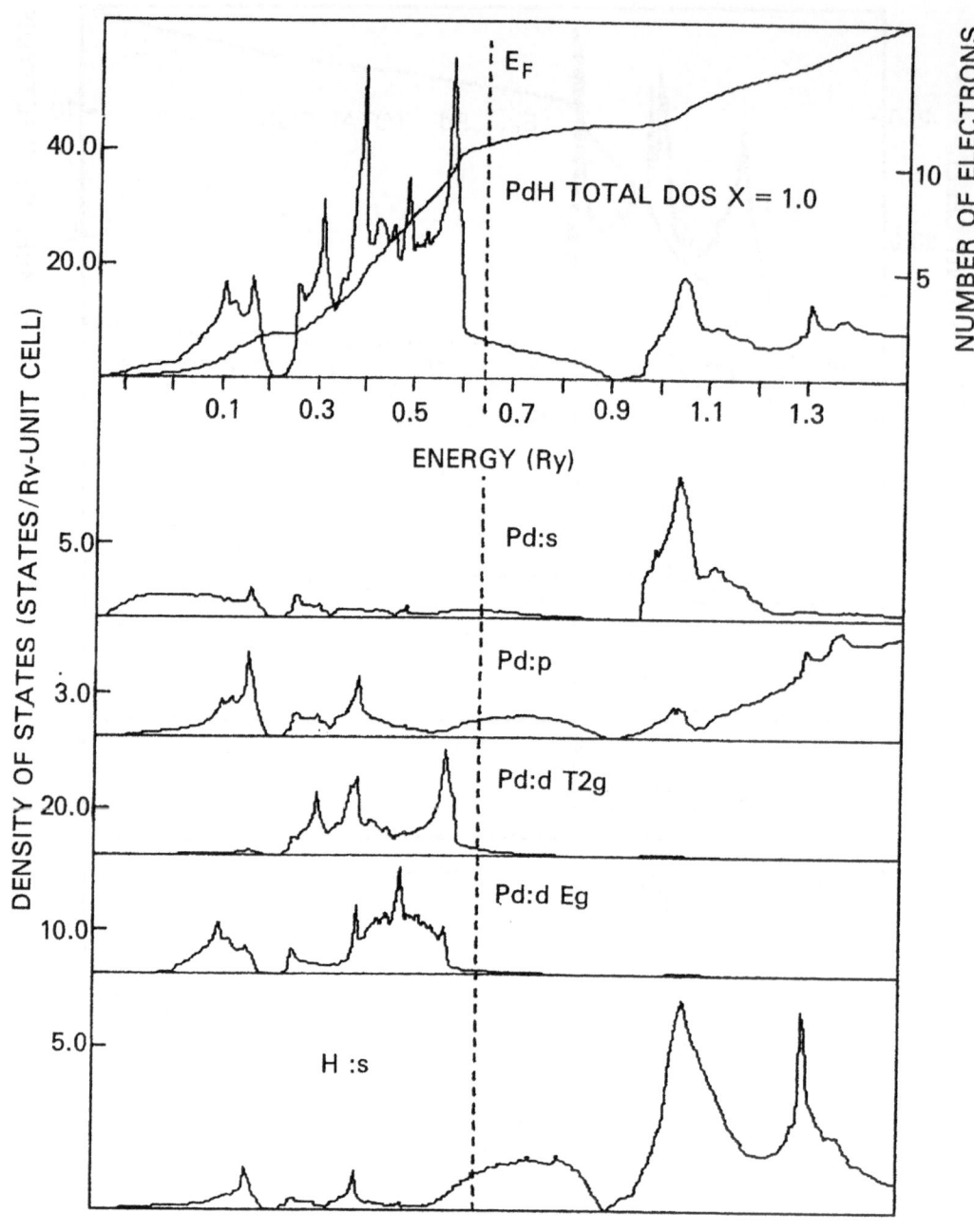

Figure 6. Slater-Koster Densities of States for PdH$_{1.0}$

In Table II, we give the total DOS at E$_F$, together with the Fermi velocity and plasmon energy for Pd and PdH$_{1.0}$.

<u>Table II</u>

	N(E$_F$) states/Ry/cell/spin	$\langle v^2(E_F)\rangle^{1/2}$ (10^8cm/sec)	v(E$_F$) (10^8cm/sec)	$\hbar\Omega_p$ (eV)
Pd	14.6	0.37	0.29	7.29
PdH$_{1.0}$	3.27	0.74	0.73	6.83

From Table II, we note the rather large difference between $\langle v^2(E_F) \rangle^{1/2}$ and $v(E_F)$ for Pd indicating the anisotropic nature of the Pd-Fermi surface (FS). In contrast, the same quantities are about equal for $PdH_{1.0}$ which suggest an isotropic character of its FS. Indeed, we have calculated the FS of $PdH_{1.0}$ and found it very similar to that of Cu or Ag, being nearly spherical with necks at the L point.

We conclude this section by noting that the plasmon frequencies Ω_p, listed in Table II, can be used in conjunction with resistivity measurements to obtain the electron-phonon coupling constant λ_{tr}. (For a more complete discussion see E.N. Economou in this volume.[46]) Using the measurements of MacLachlan et al.[47] and our Ω_p we find for $PdH_{1.0}$ λ_{tr} = 0.68. This value should be contrasted to the value of λ = 0.54 that we have reported[11] using neutron scattering data and the theory for the electron-phonon interaction discussed in Sec. III.5.

4. Electronic States of Substoichiometric PdH_x and AgH_x

We have used the CPA described in Sec. II.5 to calculate the DOS of substoichiometric PdH_x[12] and also of AgH_x. The Green's functions were determined by performing a weighted sum (Eq. 19) over 770 k points in the 1/48th of the BZ which is sufficient to achieve adequate convergence.[12] In Fig. 7, the PdH_x CPA-DOS for x=0.8 are shown. This figure does not contain the antibonding states. So when comparing it with the stoichiometric case (Fig. 6), one has to allow for a difference in the energy scale. It appears that qualitative differences exist only in the s-like DOS, while especially the d-DOS are insensitive to the reduction of H concentration. As we have discussed in Ref. 12, the Pd site DOS values at E_F have small variations while the H s-like component, $n_{SH}(E_F)$, reduces by a factor of two as we go from x=1.0 to x=0.8. According to our theory, the variation of $n_{SH}(E_F)$ with x controls the values of the electron-phonon coupling constant λ , which enters into the determination of the electronic specific heat coefficient γ and the superconducting temperature T_c. Our calculated value of γ for x=0.8 is 1.5 mj/moleK^2, which is in perfect agreement with the measured value.[48] In Ref. 12, we have argued that using the $PdH_{1.0}$ SK parameters to do CPA calculations for

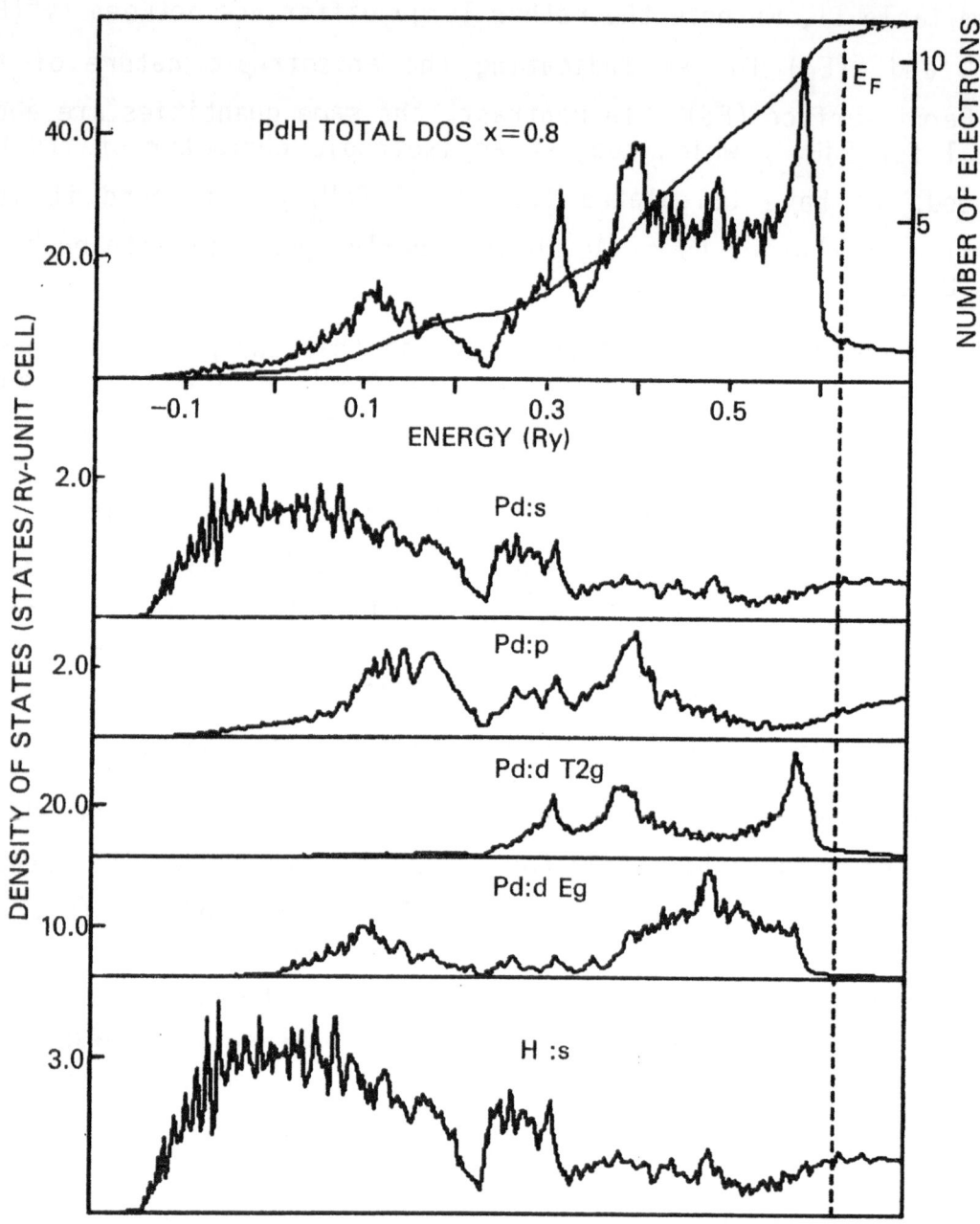

Figure 7.   CPA Densities of States for $PdH_{0.8}$

small values of x (x<0.6), is probably inaccurate. In the present
work we have performed a CPA calculation for x=0.1 by employing
the Pd-Pd SK parameters of pure Pd (also given in Ref. 12) and the
H-H and H-Pd parameters of $PdH_{1.0}$.   The DOS resulting from this
calculation are shown in Fig. 8. This figure shows a sharper peak
in the DOS associated with the hydrogen impurity, in contrast to
the broad peak seen for x=0.8 (Fig. 7). Apparently this band
broadens and moves down in energy for high hydrogen concentrations.

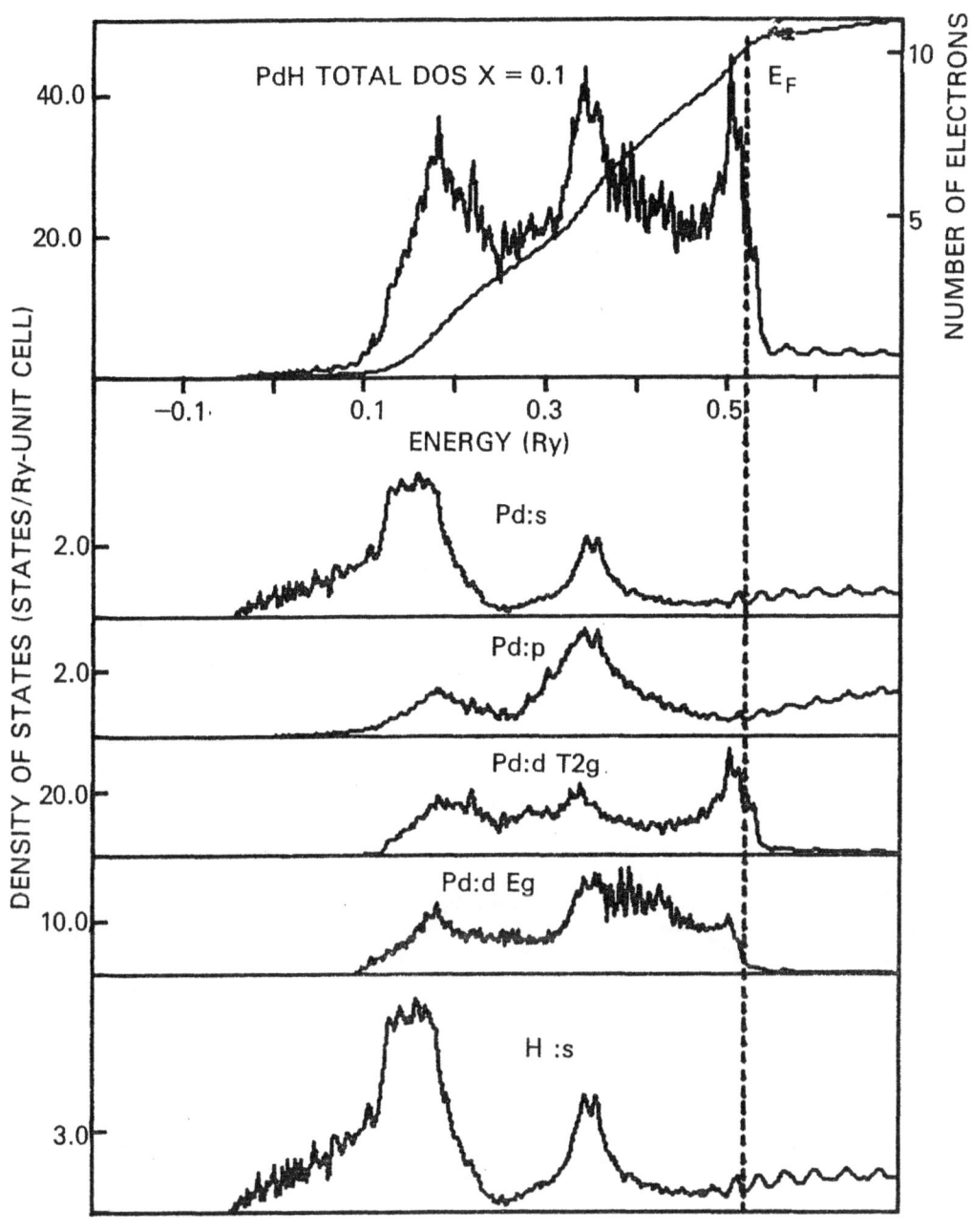

Figure 8.   CPA Densities of States for $PdH_{0.1}$

On the other hand, the DOS around E_F are not affected by the presence of small amounts of hydrogen, and look almost identical to those of Pd.   This is why we believe that for small x it is extremely difficult to obtain a reliable quantitative estimate of the Fermi surface areas needed to interpret the very interesting experiments of Venema et al.[49] Bansil et al.[50] using the average-t-matrix approximation have presented such calculations but, in our opinion, they are probably of limited accuracy.

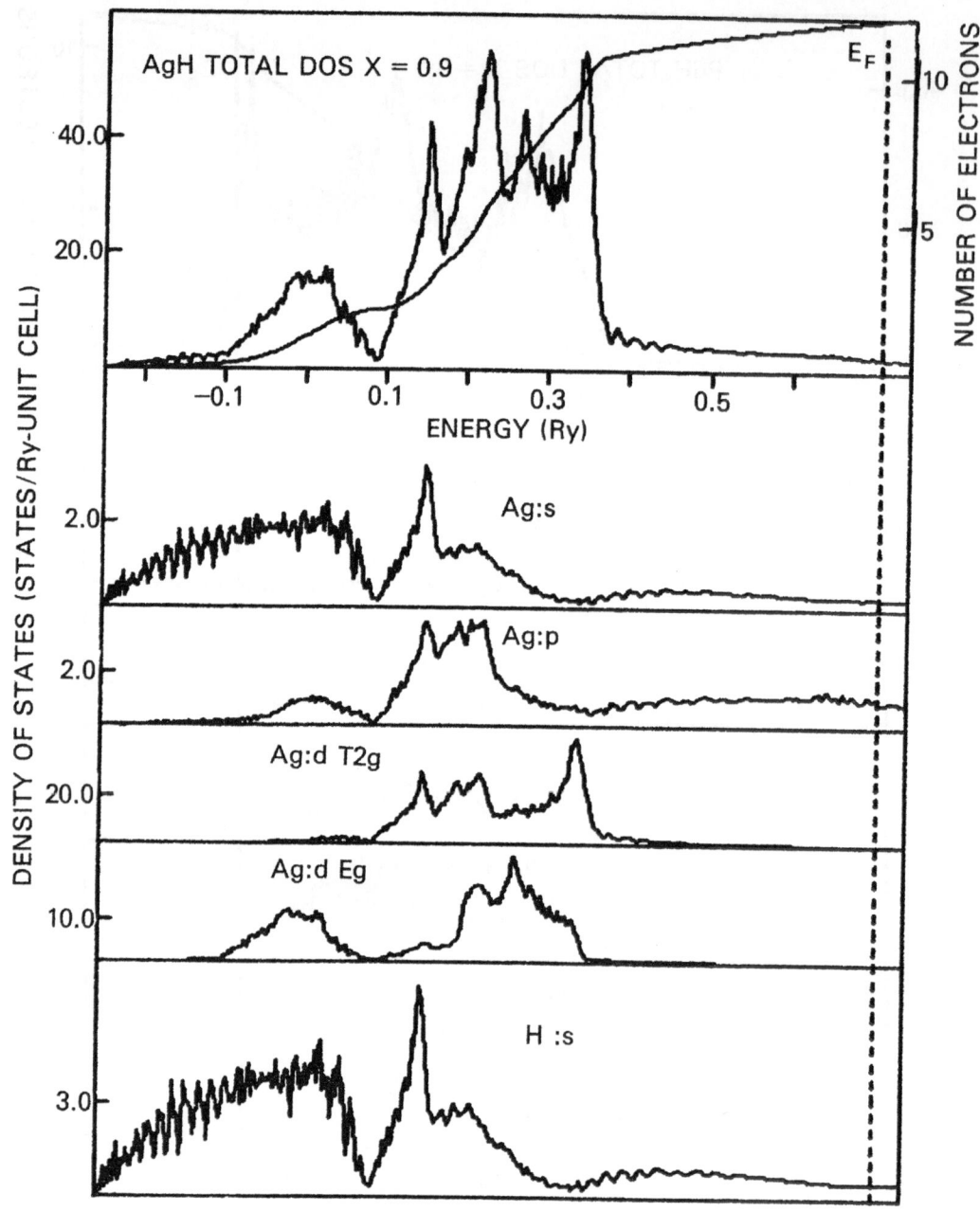

Figure 9.   CPA Densities of States for $AgH_{0.9}$

Finally, in Fig. 9 we show the DOS of AgH_x for x = 0.9. The main differences from the corresponding PdH DOS are that E_F is located further away from the d-bands, and that the d-band-complex is significantly narrower.

5.   Electron-Phonon Interaction

In a recent paper, Klein and Papaconstantopoulos[51] suggested a generalization to compounds[52] of the classic expression derived by McMillan[53] for the electron-phonon coupling constant λ. In the case of the monohydrides discussed in this article, this formula takes the form:

$$\lambda(x) = \frac{x \eta_H(x)}{M_H \overline{\omega_H^2}} + \frac{\eta_{Met}(x)}{M_{Met} \overline{\omega_{Met}^2}}$$

(23)

with $\eta_H(x) = N(x) I_H^2(x)$ and $\eta_{Met}(x) = N(x) I_{Met}^2(x)$

where x is the hydrogen content, $N(x)$ is the Fermi-level value of the DOS per spin, and $I_H^2(x)$, $I_{Met}^2(x)$ are mean square electron-ion matrix elements averaged over the Fermi surface. While the force constants $M_H \overline{\omega_H^2}$ and $M_{Met} \overline{\omega_{Met}^2}$ are determined from neutron scattering data,[54] $\eta_H(x)$ and $\eta_{Met}(x)$ are calculated from first principles using the elegant theory of Gaspari and Gyorffy.[55] These authors proposed that the local environment of a vibrating atom can be approximated by rigidly displacing the MT potential for that atom. This approach is now referred to as the rigid muffin-tin approximation. The formula of Gaspari and Gyorffy gives $\eta_{H,Met}(x)$ in terms of quantities found from band structure calculations. For the H-site, we have:

$$\eta_H(x) = \frac{E_F(x)}{\pi^2 N(x)} \sum_{\ell} 2(\ell+1) \sin^2(\delta_\ell^H(x) - \delta_{\ell+1}^H(x)) \gamma_\ell^H(x) \gamma_{\ell+1}^H$$

(24)

with an analogous equation for $\eta_{Met}(x)$. In Eq. (24), E_F is the Fermi level in rydbergs, δ_ℓ is the scattering phase shift of energy E_F at the MT radius, R_s, and γ_ℓ is the ratio of the angular-momentum-decomposed band structure DOS, n_ℓ, to the so-called "single scatterer" DOS which is given by:

$$N_\ell^{(1)} = \frac{\sqrt{E_F}}{\pi} (2\ell+1) \int_0^{R_s} dr \, r^2 u_\ell^2(r, E_F)$$

(25)

where u_ℓ is the radial wave function. In the case of a relativistic calculation u_ℓ is taken to be equal to the large component. In Fig. 10, we present a plot of the scattering phase shifts for $PdH_{1.0}$. One can see the Pd d-like phase shift approaching reson-

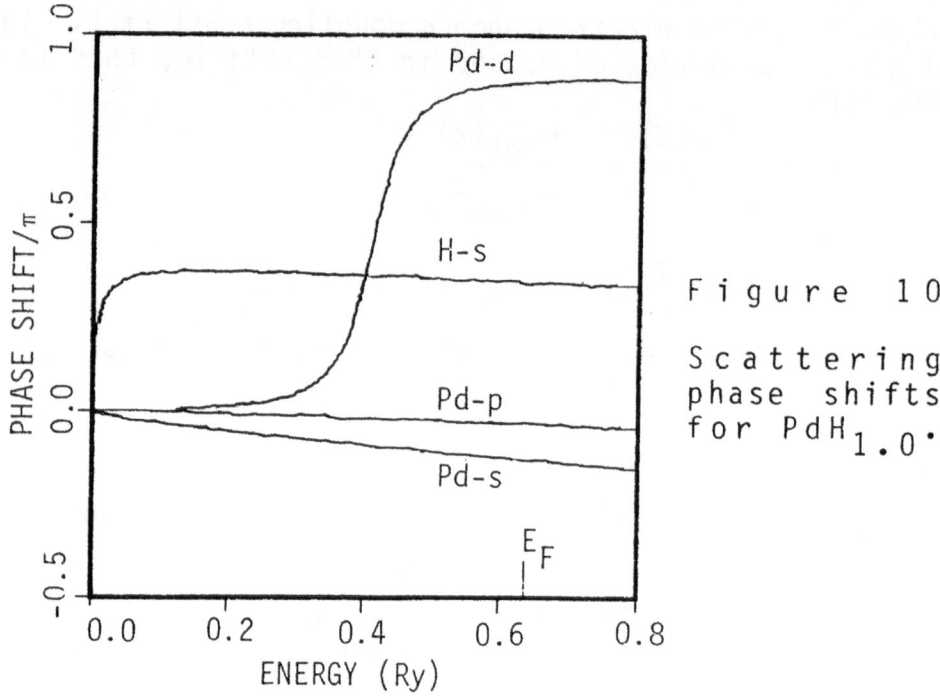

Figure 10

Scattering phase shifts for $PdH_{1.0}$.

ance in the energy range of the d-bands. Also it is interesting to note the fairly strong and constant H s-like phase shift.

One should stress here that the above formalism is rigorously applicable only to the MT-APW results. Therefore, for treating the non-stoichiometric alloys, we used, as discussed earlier,[11] the rigid band approximation applied in a range of approximately 0.7 eV around E_F. Our analysis has shown that in the Pd-based hydrides the important quantity which determines the variation of λ is the hydrogen component n_H. The metal component n_{Met} is nearly constant in the hydrogen-rich region where superconductivity occurs. We have traced the variation of n_H with x to the variation of the s-like H-DOS at E_F n_{SH} and concluded that $n_H(x)$ is, to a good approximation, proportional to $\frac{n_{SH}}{N}(x)$. Using this relationship, we have calculated $n_H(x)$ from our CPA results,[12] and we find good agreement with those of the RBA approach.[11] In Fig. 11, we show the quantity n_H plotted versus hydrogen content x for the various hydrides that we considered. One can conclude from this figure that: (i) addition of Rh decreases n_H; (ii) addition of Ag increases n_H; (iii) for a given Pd to Rh or Ag ratio, the addition of more H increases n_H; and (iv) for AgH_x,

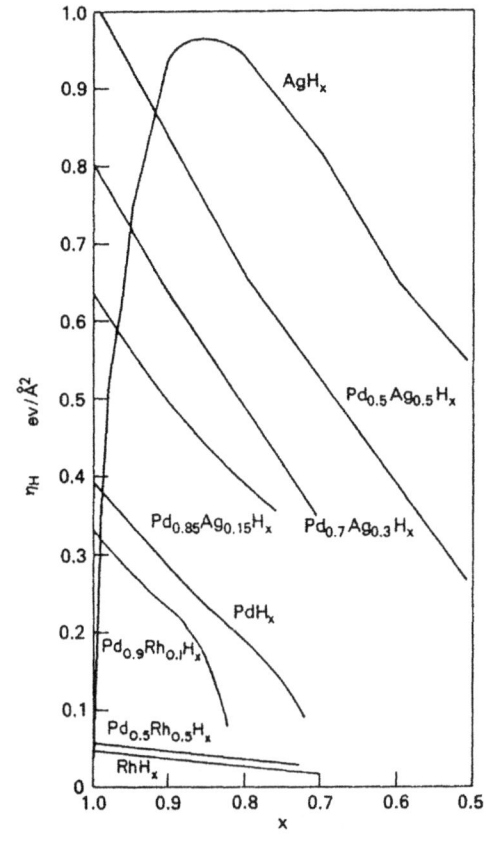

Figure 11

Hydrogen-site electron-phonon interaction η_H as a function of hydrogen content x.

η_H drops sharply as $x \to 1$.

In this volume, E.N. Economou's article[46] discusses the implications of these results in the calculation of T_c, the superconductivity mechanism, and presents a comparison with experiment. M. Gupta's article,[40] also in this volume, describes calculations of η for other hydride systems.

Acknowledgments

The author gratefully acknowledges collaborations with L.L. Boyer, E.N. Economou, J.S. Faulkner, and B.M. Klein. He also wishes to thank A.C. Switendick, W.E. Pickett, and W.M. Temmerman for several stimulating discussions.

References

1. A.C. Switendick, Electronic Band Structures of Metal Hydrides, Solid State Commun. 8, 1463 (1970); Metal Hydrides-Structure and Band Structure, Int. J. Quantum Chem. 5, 459 (1971).

2. A.C. Switendick, Electronic Energy Bands of Metal Hydrides-Palladium and Nickel Hydride, Ber. Bunsenges, Physik. Chemie 76, 535 (1972).

3. D.E. Eastman, J.K. Cashion, and A.C. Switendick, Photoemission Studies of Energy Levels in the Palladium-Hydrogen System, Phys. Rev. Lett. $\underline{27}$, 35 (1971).

4. A.C. Switendick, "Hydrogen in Metals - A New Theoretical Model", in Hydrogen Energy, Part B, ed. T.N. Veziroglou (Plenum Press, NY, 1975) pp 1029-1042.

5. A.C. Switendick, Influence of the Electronic Structure on the Titanium Vanadium-Hydrogen Phase Diagram, J. Less-Common Metals $\underline{49}$, 283 (1976).

6. A.C. Switendick, "The Change in Electronic Properties on Hydrogen Alloying and Hydride Formation", in Topics in Applied Physics, Vol. 28: Hydrogen in Metals I: Basic Properties, G. Alefeld and J. Völkl eds. (Springer Verlag, Berlin 1978) pp 101-129.

7. A.C. Switendick, Bandstructure Calculations for Metal Hydrogen Systems, Zeitschrift Physik. Chemie, Vol. 117, pp 89-112 (1979).

8. D.A. Papaconstantopoulos and B.M. Klein, Superconductivity in the Palladium-Hydrogen System, Phys. Rev. Lett. $\underline{35}$, 110 (1975); B.M. Klein and D.A. Papaconstantopoulos, Calculation of the Electron-Phonon Interaction and Superconductivity in the Palladium-Hydrogen System, in Proceedings of 14th Intern. Conf. on Low Temperature Physics, eds. M. Krusius and M. Vuorio (North Holland, Amsterdam, 1975) Vol. 2, pp 399-402.

9. B.M. Klein, D.A. Papaconstantopoulos, and L.L. Boyer, Calculations of the Superconducting Properties of Compounds: Refractory Carbides, PdH and V_3Si, in Proceedings of the 2nd Rochester Conf. on Superconductivity in d- and f-Band Metals ed. D.H. Douglass (Plenum Press, NY, 1976) pp 339-359.

10. B.M. Klein, E.N. Economou, and D.A. Papaconstantopoulos, On the Inverse Isotope Effect and the x-Depencence of the Superconducting Transition Temperature in PdH_x and PdD_x, Phys. Rev. Lett. $\underline{39}$, 574 (1977).

11. D.A. Papaconstantopoulos, B.M. Klein, E.N. Economou, and L.L. Boyer, Band Structure and Superconductivity of PdD_x and PdH_x, Phys. Rev. B$\underline{17}$, 141 (1978).

12. D.A. Papaconstantopoulos, B.M. Klein, J.S. Faulkner, and L.L. Boyer, Coherent-Potential-Approximation Calculations for PdH_x, Phys. Rev. B$\underline{18}$, 2784 (1978).

13.   D.A. Papaconstantopoulos, E.N. Economou, B.M. Klein, and L.L. Boyer, Superconductivity in Palladium-Based Hydrides, J. Physique 6, C 435 (1978).

14.   D.A. Papaconstantopoulos, E.N. Economou, B.M. Klein, and L.L. Boyer, Electronic Structure and Superconductivity in Pd-Ag-H and Pd-Rh-H Alloys, Phys. Rev. B20, 177 (1979).

15.   D.A. Papaconstantopoulos, Platinum Hydride: A Possible High Temperature Superconductor, J. Less-Common Metals 73, 305 (1980).

16.   J.C. Slater, Wave Functions in a Periodic Potential, Phys. Rev. 51, 846 (1937).

17.   L.F. Mattheiss, J.H. Wood, and A.C. Switendick, A Procedure for Calculating Electronic Energy Bands Using Symmetrized Augmented Plane Waves, in Methods in Computational Physics, Vol. 8, pp 63-147 (1968).

18.   T. Loucks, "Augmented Plane Wave Method", Benjamin, NY (1967).

19.   J.O. Dimmock, The Calculation of Electronic Energy Bands by the APW Method, Solid State Phys. 26, 103 (1971).

20.   L.F. Mattheiss, Band Structure and Fermi Surface for Rhenium, Phys. Rev. 151, 450 (1966).

21.   D.D. Koelling and B.N. Harmon, A Technique for Relativistic Spin-Polarized Calculations, J. Phys. C10, 3107 (1977).

22.   D.A. Liberman, D.T. Cromer and J.T. Waber, Relativistic Self-Consistent Field Program for Atoms and Ions, Comput. Phys. Commun. 2, 107 (1971).

23.   P.O. Löwdin, Quantum Theory of Cohesive Properties of Solids, Advan. Phys. 5, 1 (1956).

24.   In what has come to be known as the Mattheiss prescription (L.F. Mattheiss, Phys. Rev. 133, A1399 (1964)) a different procedure is followed. The difference is the fact that Eq. (6) is also used in order to calculate $V_c(r)$ as a super-position of atomic potentials $V_0(r)$ instead of solving Poisson's equation.

25.   S. Asano and J. Yamashita, On the Self-Consistent Potential of the Band Calculation, J. Phys. Soc. Japan, 30, 667 (1971); for a computer code see D.A. Papaconstantopoulos and W.R.

Slaughter, Calculation of Crystal Potentials, Comput. Phys. Commun. $\underline{7}$, 207 (1974); $\underline{13}$, 225 (1977).

26.   The MT sphere radius is usually taken equal to half the nearest neighbor distance for monatomic materials. For compounds we have chosen the radii by imposing the condition that the starting crystal potentials are equal at the point of contact of the MT spheres.[11]

27.   J.C. Slater, Statistical Exchange-Correlation in the Self-Consistent Field, in Advances in Quantum Chemistry, Vol. $\underline{6}$, pp 1-92, Academic Press (NY) 1972.

28.   K. Schwarz, Optimization of the Statistical Exchange Parameter α for the Free Atoms H to Nb, Phys. Rev. B$\underline{5}$, 2466 (1972); Optimized Statistical Exchange Parameter α for Atoms with Higher Z, Theor. Chim. Acta $\underline{34}$, 225 (1974).

29.   L. Hedin and B.I. Lundqvist, Explicit Local Exchange-Correlation Potentials, J. Phys. C$\underline{4}$, 2064 (1971).

30.   In our calculations we have used the expression:

$$\sigma(r) = F\sigma^{old}(r) + (1-F)\sigma^{new}(r)$$
where F = 0.75.

31.   F.M. Mueller, J.W. Garland, M.H. Cohen, and K.H. Bennemann, Quadratic Integration: Theory and Application to the Electronic Structure of Platinum, Ann. Phys. (NY) $\underline{67}$, 19 (1971).

32.   G. Lehmann and M. Taut, On the Numerical Calculation of the Density of States and Related Properties, Phys. Status Solidi (b)$\underline{54}$, 469 (1972); O. Jepsen, and O.K. Anderson, The Electronic Structure of hcp Ytterbium, Solid State Commun. $\underline{9}$, 1763 (1971).

33.   J.C. Slater and G.F. Koster, Simplified LCAO Method for the Periodic Potential Problem, Phys.Rev. $\underline{94}$, 1498 (1954).

34.   L.L. Boyer, Symmetrized Fourier Method for Interpolating Band Structure Results, Phys. Rev. B$\underline{19}$, 2824 (1979).

35.   B.M. Klein, L.L.Boyer, D.A. Papaconstantopoulos, and L.F. Mattheiss, Self-Consistent Augmented-Plane-Wave Electronic-Structure Calculations for the A15 Compounds V_3X and Nb_3X, X = Al, Ga, Si, Ge, and Sn, Phys. Rev. B$\underline{18}$, 6411 (1978).

36.   P. Soven, Coherent-Potential Model of Substitutional Disordered Alloys, Phys. Rev. $\underline{156}$, 809 (1967).

37. J.S. Faulkner, Electronic States of Substoichiometric Compounds and Application to Palladium Hydride, Phys. Rev. B13, 2391 (1976).

38. J. Zbasnik, and M. Mahnig, The Electronic Structure of Beta-Phase Palladium Hydride, Z. Phys. B23, 15 (1976).

39. M. Gupta and A.J. Freeman, Electronic Structure and Proton Spin-Lattice Relaxation in PdH, Phys. Rev. B17, 3029 (1978).

40. M. Gupta and J.P. Burger, Experimental and Theoretical Investigation of the Coupling of Electrons with Acoustical and Optical Phonons in Metal Hydrides Relationships with Superconductivity, this volume.

41. C.D. Gelatt, Jr., H. Ehrenreich, and J. Weiss, Transition Metal Hydrides: Electronic Structure and the Heats of Formation, Phys. Rev. B17, 1940 (1978).

42. A.R. Williams, J. Kubler, and C.D. Gelatt, Jr., Cohesive Properties of Metallic Compounds: Augmented-Spherical-Wave Calculations, Phys. Rev. B19, 6094 (1979).

43. N.I. Kulikov, Band Structure and Electronic Properties of Transition Metal Hydrides, Phys. Status Solidi (b)91, 753 (1979).

44. G.M. Stocks, R.W. Williams, and J.S. Faulkner, Electronic States in Ag-Pd Alloys, J. Phys. F3, 168 (1973); A.J. Pindor, W.M. Temmerman, B.L. Gyorffy, and G.M. Stocks, On the Electronic Structure of Ag_cPd_{1-c} Alloys, J. Phys. F (1980) to be published.

45. D.J. Peterman, B.N. Harmon, J. Marchiando, and J.H. Weaver, Electronic Structure of Metal Hydrides II: Band Theory of ScH_2 and YH_2, Phys. Rev. B19, 4867 (1979).

46. E.N. Economou, Superconductivity in Palladium-Based Hydrides, this volume.

47. D.S. MacLachlan, R. Mailfert, B. Souffaché, and J.P. Burger, Electrical Resistivity and Superconductivity in PdH, in Proc. of 14th Intern. Conf. on Low Temperature Physics, eds. M. Krusius and M. Vuorio (North Holland, Amsterdam 1975) Vol. 2 pp 40-43.

48. C.A. Mackliet, D.J. Gillespie, and A.I. Schindler, Specific Heat, Electrical Resistance, and Other Properties of Superconducting Pd-H Alloys, J. Phys. Chem. Solids 37, 379 (1976).

49.   W.J. Venema, R. Griessen, R.S. Sorbello, H.L.M. Bakker, and P.E.M. Mijnarends, Effect of Zero-Point-Motion on the Electronic Structure of Pd-H(D), Proc. of Physics of Transition Metals Conf., Leeds (1980).

50.   A. Bansil, R. Prasad, S. Bessendorf, L. Schwartz, W.J. Venema, R. Feenstra, F. Blom, and R. Griessen, Electronic States and Fermi Surface Properties of α-Phase PdH_x, Solid State Commun. $\underline{32}$, 1115 (1979).

51.   B.M. Klein and D.A. Papaconstantopoulos, On Calculating the Electron-Phonon Mass Enhancement λ for Compounds, J. Phys. F$\underline{6}$, 1135 (1976).

52.   This formula (Eq. 23) is approximate because of the ommission of the cross term but it is particularly accurate when the mass difference $M_{Met}-M_H$ is large as in the present case.

53.   W.L. McMillan, Transition Temperature of Strong-Coupled Superconductors, Phys. Rev. $\underline{167}$, 331 (1968).

54.   J.M. Rowe, J.J. Rush, H.G. Smith, M. Mostoller, and H.E. Flotow, Lattice Dynamics of a Single Crystal of $PhD_{0.63}$, Phys. Rev. Lett. $\underline{33}$, 1297 (1974).

55.   G.D. Gaspari and B.F. Gyorffy, Electron-Phonon Interaction d-Resonances and Superconductivity in Transition Metals, Phys. Rev. Lett. $\underline{28}$, 801 (1972).

ELECTRON-PHONON COUPLING AND SUPERCONDUCTIVITY

IN PALLADIUM HYDRIDES AND DEUTERIDES

J.P. BURGER

Laboratoire de Physique des Solides

Université Paris-Sud 91405 Orsay (France)

ABSTRACT

We discuss first the preparation, handling and characterization of bulk and thin films of PdHx and PdDx. We show how information concerning the electron-phonon coupling can be extracted from the temperature dependant normal state resistivity. Particular emphasis will be drawn on the way of separating the acoustical and optical contribution to this coupling. Experimental resistivity data concerning PdH_x and PdD_x, supplemented by some supraconducting tunneling data will be presented. Finally we will discuss the possible origin of the positive isotope effect observed for the superconducting Tc and the influence of magnetic impurities on Tc.

INTRODUCTION

The observation of high Tc superconductivity [1] [2] raised immediatly several intersting questions as it was somewhat unexpected. Indeed the band structure of PdH can be considered a priori to be rather similar to the one of Ag a well known non superconductor ; the strong decrease of the Fermi level density of states as one goes from Pd, a non superconducting transition metal, to the hydride may also be considered to be an unfavourable circunstance for the occurrence of superconductivity in the hydrides. But one was forced to recognize that the light atoms, H or D are involved when the reversed isotope effect, i.e. Tc(D) > Tc(H) was discovered [2]. It became thus evident that it was necessary to measure separately the contribution to superconductivity coming from the Pd (or acoustical) vibrations and from H (or optical) vibrations. Apart from the measurement of the superconducting Tc it is possible to obtain information on the electron-phonon coupling parameter λ through the normal state phonon resistivity and through the superconducting tunneling density of states.

Other interesting aspects of this system are the strong increase of Tc with the H concentration as it approaches x = 1, and the further increase when Pd is substituted by monovalent metals. If one reminds that $\lambda = \dfrac{\eta}{M<\omega^2>}$ depends through η on electronic parameters like the density of states and on atomic parameters through the phonon energies ω, it appears important to find out what parameters, electronic or atomic, may explain the observed trends.

EXPERIMENTAL

The loading of Pd foils ($\simeq$ 50 microns thick) or thin films ($\simeq$ 500 A° thick) can be done either by implantation, by pressure or by electrolysis. The first method perturbs heavily the structure of the films ; on the other hand for to reach a maximum concentration of x with the second method it is interesting to operate at a low temperature, the equilibrium concentration for a given pressure of H_2 increasing with decreasing temperature ; but the handling of high pressure and the poor kinetics of the dissociation of the H_2 molecule at the surface makes this technique more difficult than electrolysis. With this last method it is possible to reach for pure Pd the concentration x = 1 at temperatures of the order of 180 to 200°K rather easily using low current densities of the order of 1 mA/cm^2 ; the electrolyte is a mixture of $C_2H_5OH(D) + SO_4H(D)_2$. The method works as well for bulk samples thin films and also for $Al/Al_2O_3/Pd$ tunneling junctions. The sample prepared this way are stable below 200°K if they are bulk and only below $\sim$120°K for thin films ; to avoid hydrogen loss when we cool down the samples we must in this last case continue to apply a voltage between the electrodes, one being the Pd film, the other a platinum wire. During the electrolysis, we monitor continously the electrical resistance R of the sample and Fig. 1 shows the typical shape of R(t) where t is the electrolysis time [3]:R goes through a maximum near x $\simeq$ 0,75 and later on decreases when x approaches the stoichiometric concentration x = 1. This decrease is related to the decrease of the number of vacancies in the hydrogen lattice . A careful measurement of R(x) for bulk specimens permits us now to determine the concentration x even in thin films where it would be difficult to make a volumetric determination of the absorbed amount of H.

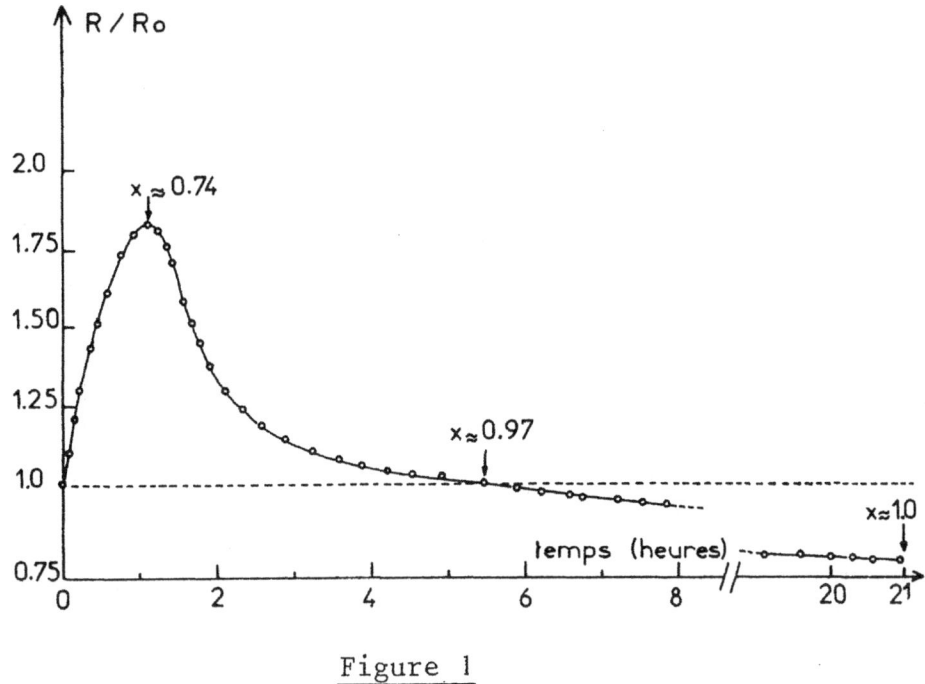

Figure 1

Resistance versus electrolysis time (in hours) for a 40 μ thick Pd foil.

ELECTRON-PHONON COUPLING AND PHONON RESISTIVITY

The phonon electrical resistivity $\rho(T)$ and the superconducting T_c can be written in the following way [4] :

$$\rho(T) = C \int \alpha^2_t \, F(\omega) \, f(\frac{\hbar\omega}{kT}) \, d\omega \tag{1}$$

$$T_c = <\omega>\exp(-\frac{1}{g}), \quad g = \frac{\lambda}{1+\lambda} - \mu^*, \quad \lambda = 2\int \frac{\alpha^2 F(\omega)}{\omega} \, d\omega \tag{2}$$

C is a constant involving the effective mass and the density per unit volume of conduction electrons, $F(\omega)$ is the phonon density of states, α^2_t is a transport electron-phonon coupling matrix element which differs from the α^2 appearing in λ by a factor involving the scattering angle, $f(x) = \frac{x}{(e^x-1)(1-e^{-x})}$ is a statistical factor. The expression for T_c is the standard McMillan type formula.

At temperatures high enough $f(x) \simeq \frac{1}{x}$ and $f(T) \simeq AT$ where A is proportional to $\lambda_t = 2 \int \frac{\alpha^2_t \, F(\omega)}{\omega} \, d\omega \simeq \lambda$.

Thus, it is possible to extract λ from measured data of $\rho(T)$. However, for compounds like $PdH(D)_x$ we have to consider contributions to λ coming from acoustic and optic phonons, i.e. :

$$\lambda = \lambda_{ac} + \lambda_{op} \tag{3}$$

Both contribution depend now on the masses M and m of the Pd and $H(D)$ ions, even if the total $\lambda = \lambda_{ac} + \lambda_{opt}$ is mass independant. This mass dependance, which is due to the simultaneous motion of both ions in each phonon mode can be shown to be negligeable [5] if the masses M and m are very different which is the case here.

The resistivity $\rho(T)$ can in a similar way be separated into two contributions, i.e. :

$$\rho(T) = \rho_{ac}(T) + \rho_{op}(T) \tag{4}$$

the second term $\rho_{op}(T)$ depends now drastically on the mass m through the statistical factor $f(x)$.

It is possible to separate experimentally the two contributions because at low enough temperature ($T < 50$ K) only acoustical phonons are thermally excited, so that $\rho(T) = \rho_{ac}(T)$. Approximating in this temperature range $\rho(T)$ by a Grüneisen function

$\rho_{ac}(T) = \rho_G(\frac{T}{\Theta ac}, A_{ac})$ and assuming that this approximation is valid at higher temperatures as well, it is possible to isolate the optical phonon contribution in the whole temperature interval considered i.e. $\rho_{op}(T) = \rho(T) - \rho_{ac}(T)$. The optical contribution is then analysed by approximating the optical phonon spectrum

with an Einstein spectrum i.e. $\rho_{op}(T) = \rho_E(\frac{T}{\Theta_{op}}, A_{op})$. A best fit of $\rho_{ac}(T)$ and $\rho_{op}(T)$ with ρ_G and ρ_E permits us to determine Θ_{ac} (the Debye temperature), Θ_{op} (the Einstein temperature), A_{ac} and A_{op}. Since the constant C is not know precisely, one can determine

only the ratio $\frac{\lambda_{op}}{\lambda_{ac}} = \frac{A_{op}}{A_{ac}}$. Absolute values of λ_{op} and λ_{ac} can

nevertheless be obtained by using this ratio together with relation (2) and taking $\mu^* = 0,1$, a value generally admitted in non transition metals for the electron-electron repulsive interaction parameter and

$$<\omega> = (\Theta_{ac})^{\frac{\lambda_{ac}}{\lambda}} (\Theta_{op})^{\frac{\lambda_{op}}{\lambda}}$$

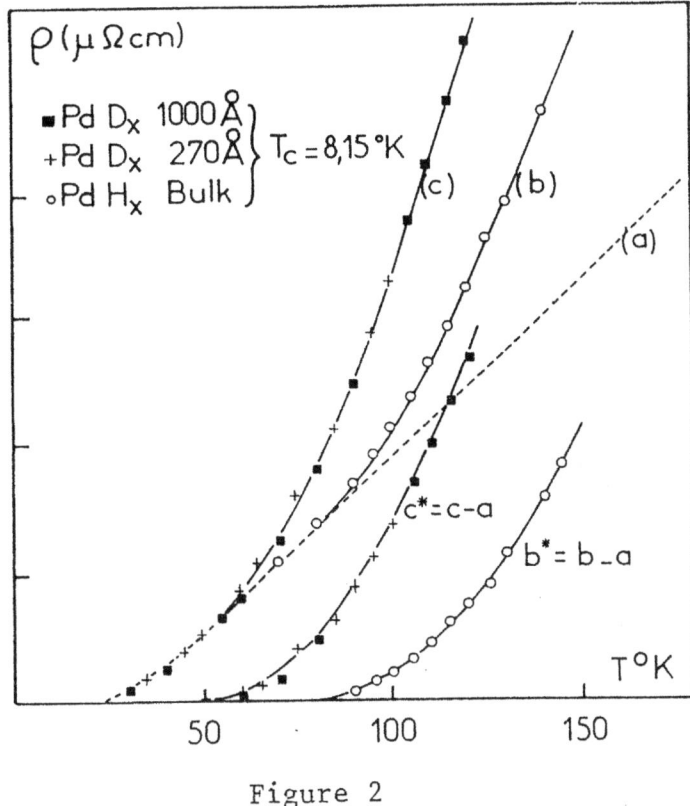

<u>Figure 2</u>

Experimental phonon resistivity for PdH_x (curve b) and PdD_x (curve c). The dashed curve (a) represents a theoretical Grüneisen law fitted to the low température acoustical resistivities. Curves b* and c* are best fit Einstein laws through the experimental points obtained by substracting (a) from (b) and (c).

Fig. 2 shows the measured resistivities of different hydrided or deuterated samples having all roughly the same T_c (8,15 for D, 9,4° K for H). In the range $T < 20°$K we observe a T^5 law and it is possible to fit the experimental data up to 50°K (or up to 80°K for H) with a Grüneisen law. For higher temperatures up to the limiting temperature of stability of the films, $\rho(T)$ increases faster than the just mentioned Grüneisen law. It is interesting to note that the resistivities for both isotopes overlap for $T < 50°$K (when only acoustic phonons contribute) and that the optical phonon contribution sets in at a lower temperature for D than for H, a behaviour which agrees with the expected decrease of the optical phonon energy with isotope mass [6].

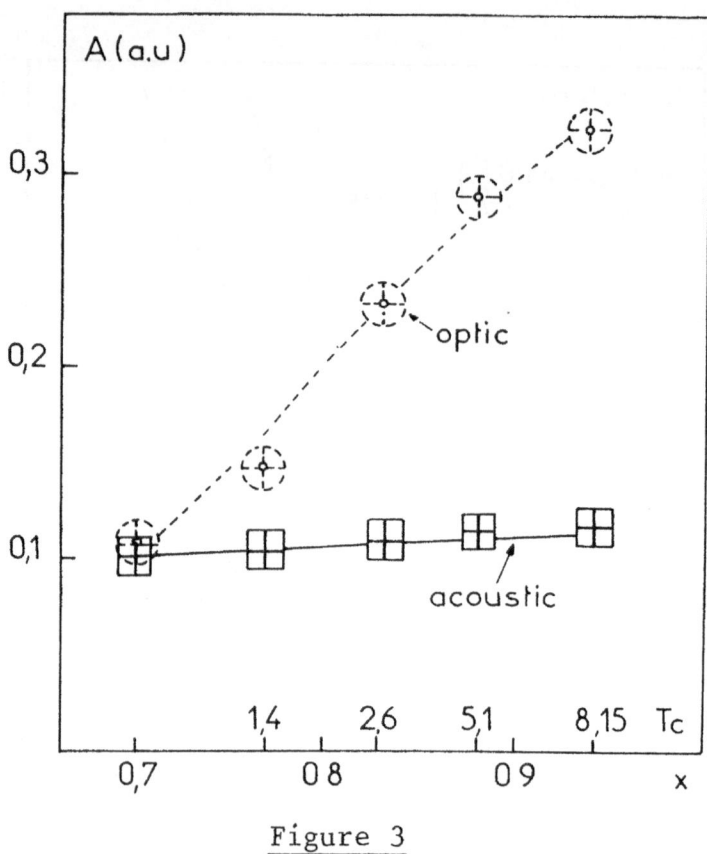

<u>Figure 3</u>

Variation of A_{ac} and A_{op} with x (or with superconducting T_c).
A_{ac} and A_{op} are the high temperature limits, divided by T,
of the acoustical or optical phonon resistivity i.e. $\rho = AT$

Some results concerning several PdD_x films with concentration
ranging from x $\simeq$ 0.7 to x $\simeq$ 0.94 (or with superconducting T_c ranging from below 1°K to 8.15°K) are reported on Figure 3.

The values of A_{ac}, A_{op} and also of Θ_D and Θ_E are obtained
through a best fit of the experimental $\rho(T)$ data with a theoretical law of the form :

$$\rho(T) = \rho_{ac} \left(A_{ac}, \frac{\Theta_{ac}}{T} \right) + \rho_{op} \left(A_{op}, \frac{\Theta_{op}}{T} \right)$$

This fit show clearly the following features :

− the value of A_{op} increases steadily with the deuterium concentration x

- A_{ac} is in first approximation independent of x

- A_{ac} is always smaller or at best equal to A_{op}

- the value of the Debye temperature Θ_{ac} is also independent of x and equal to 210° ± 10°K while the Einstein temperature Θ_{op} decreases from 520 ± 40 (near x ≃ 0.7) to 470 ± 10 (for x > 0.85).

From these results we may conclude that the conducting electrons of PdD_y couple mainly to the optical phonons.

It remains to understand why A_{op} (and through it λ_{op} and T_c) depends so critically on the concentration x. We know that the electron-phonon coupling parameter λ is also proportional to the electronic density of states, so that we must inquire how this quantity changes with x. Theoretical calculations [7] [8] and spin lattice relaxation time measurements [9] show indeed that the local, H site, density of states increases steadily with x, in agreement with the rigid band model : it means that the electrons at the Fermi level have increasing (with x) probability to be found around the D(H) site and couple more and more to the D(H) vibration. This is of course not enough to explain the occurrence of super-conductivity, we must also explain the relatively high value of λ_{op} needed to obtain a T_c in the 10°K range ; one can remark for this that Θ_{op} is only about 2 times Θ_{ac} despite the high value (≃ 50) of the ion mass ratio of Pd to D, so that one is forced to consider that these optical phonons are relatively soft, a factor probably decisive for the absolute value of λ_{op}.

To conclute this part, we can say that T_c is mainly determined by the optical contribution

$$\lambda_{op} = \frac{\eta_{op}}{m<\omega^2>_{op}}$$

The absolute value which is in the $\lambda_{op} \simeq 0,4$ range is mainly determined by the low values of the phonon energies but the trend of variation of λ_{op} with x comes from the numerator η_{op}.

SUPERCONDUCTING TUNNELING DATA

The energy (or voltage) dependent conductivity of a supercon-ducting tunneling junction is given by

$$\frac{\sigma_s(V)}{\sigma_N(V)} = \text{Re} \frac{V}{(V^2 - \Delta^2)^{1/2}}$$

(σ_N is the normal state conductivity, Δ the complex gap parameter). The real part $\Delta'(V)$ of Δ measures the strengtht of the electron-phonon coupling, the imaginary part $\Delta''(V)$ the life time of an excited electron.

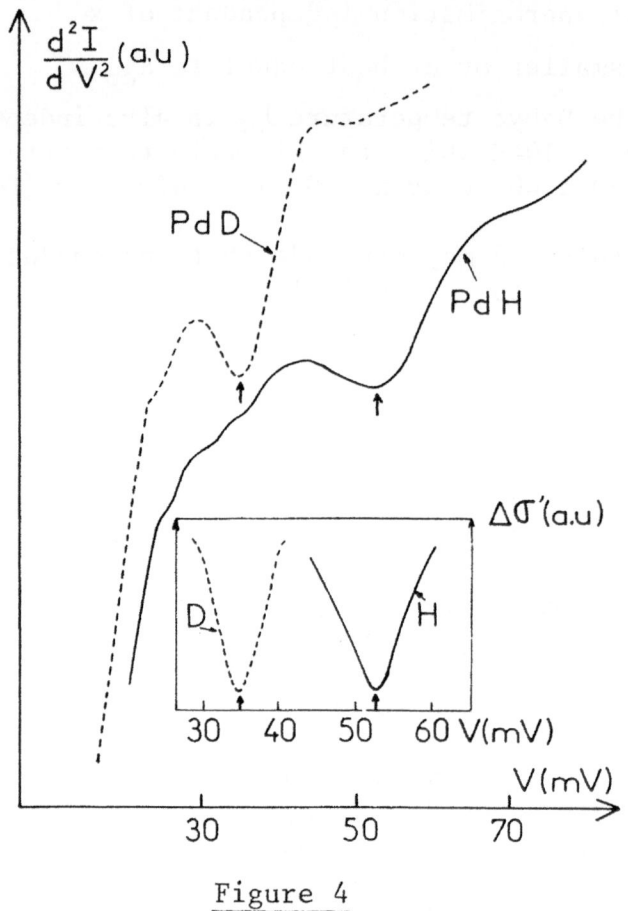

Figure 4

Experimental $\dfrac{d^2V}{dI^2}$ (V) data at 1.7°K for a deuterated (T_c=7.7°K)
and a hydrogenated (T_c = 5.3°K) tunneling junction. The inset
shows the sharp optical phonon anomaly if one substracts the
normal state background.

Both Δ' and Δ'' are directly related to $\alpha^2 F(\omega)$ which determines
also the superconducting T_c and the resistivity $\rho(T)$. For instance
any sharp peak in the phonon density of states $F(\omega)$ should give
a corresponding anomaly in $\sigma_S(V)$ when $eV = \hbar\omega$: it should be
possible this way to locate the optical phonon energy on the ω or
voltage scale. It is also possible in principle, using a deconvo-
lution program, to measure directly $\alpha^2 F(\omega)$ and so λ, but this is
a more difficult task, because of the small conductance change,
of the order of 10^{-4} involved by these anomalies. We report here
results on three PdH_x and one PdD_x compounds with T_c ranging from

5.3 to 8°K. Our main results are displayed on the second derivatives of the tunneling current I versus bias voltage V, $\frac{d^2I}{dV^2}$ (V). They are given in Figure 4 for $PdH_x(T_c = 5.3$ K) and

PdD_x (T_c = 7.7 K). These two samples have been prepared using the same parameters for electrolysis so that similar values of x are expected and the difference in T_c is probably related to the above mentioned isotope effect. Outside of the gap the structure of the

d^2I/dV^2 curve mainly reflect the phonon density of states. With the present accuracy we find no variation in these structures for the three PdH_x samples having T_c ranging between 5.3 and 7.5 K so that we may suppose to have nearly obtained the tunneling spectrum of the stoichiometric compounds PdH and PdD.

Figure 4 shows a well pronounced minimum around 35 mV for PdD. This structure, first observed by Eichler and al. [10] has been attributed to the optical phonon mode in agreement with neutron diffraction results. Figure 4 shows that this structure is missing in PdH and is replaced by the somewhat broader anomaly near 52 mV. This is the behaviour one expects for both the position and the width of the optical phonon induced anomalies. Other anomalies are observed both in PdH and PdD at energies between 10 and 24 meV which may be related to the acoustic phonons of the Pd lattice. However, these structures take place in an energy range where the BCS density of states deviates notably from a constant value and this makes it difficult to evaluate the contribution of these phonons with good enough accuracy. Anyway their amplitude appears to be weak, indicating again that the electrons couple less to these phonons. The contribution to the BCS density of states to d^2I/dV^2 can be neglected in first approximation in the energy range of optical phonons modes (V > 30 meV). In order to display accurately the deviation due to phonons on the second derivative, we calculate :

$$\Delta\sigma'(V) = (\frac{d^2I}{dV^2}_N)(V = \Delta_o) - (\frac{d^2I}{dV^2})_N(V) \quad (\Delta_o \text{ is the energy gap})$$

where $(\frac{d^2I}{dV^2})_{S,N}$ is the second derivative at 1.7 K in the super-conducting (S) and normal (N) state. The normal state in PdH(D) is obtained by quenching the superconductivity in a large enough magnetic field. In both cases the first Al electrode is normal and the modulation signal is the same so that $(\frac{d^2I}{dV^2})_S$ and $(\frac{d^2I}{dV^2})_N$ coincide at high energy (V > 70 meV). In this way we obtain $\Delta\sigma'(V)$ as shown on the inset of Figure 3 : the anomalies are now rather sharp and occur at $\omega_{op}(D) = 34.7 \pm 0.5$ mV and at $\omega_{op}(H) = 52.5 \pm 0.5$ mV.

ANHARMONICITY AND ISOTOPE EFFECT

One expects, in the harmonic approximation, that $\dfrac{\omega_{op}(H)}{\omega_{op}(D)} = \sqrt{2}$.

Our tunneling data indicate a greater ratio, of 1.51 ± 0.04. This value is confirmed by the resistivity data which give $\dfrac{\Theta_{op}(H)}{\Theta_{op}(D)} = 1.53 \pm 0.05$ for $x \simeq 1$. Both set of measurements overlap reasonably well, so that one may say, with a good degree of confidence, that the effect is outside experimental error. Two explanations have been proposed up to now. The more classical approach [11][5] considers the vibration of the H(D) ion inside the cage formed by the neighbouring heavy Pd ions ; the H ion has a larger vibration amplitude compared to D so that it senses more the steeper part of the H (or D) - Pd interaction potential, giving rise to an effective mass dependent force constant. Miller and Satterthwaite [12] argue on the other hand that the hydrogen, because of its larger vibration amplitude, spends more time on the neighbouring Pd atoms than does D, an effect which gives rise to a mass dependent H(D) - Pd potential through the variation in the overlap of H (or D) and Pd wave function. It is not possible, from our data, to favour one or other point of view. We would just mention that this relatively weak anharmonicity which has also been observed by neutron diffraction on non stoichiometric $PdH(D)_x$ [13] is strong enough to explain the reversed isotope effect. If we take $\lambda_{op} \simeq 3\lambda_{ac}$ and $\mu^{*} \simeq 0.1$ (μ^{*} is the repulsive electron-electron interaction parameter entering the expression of T_c), we obtain [5] for $x \simeq 1$:

$$\Delta T_c = T_c(D) - T_c(H) = (\Delta T_c)_h + (\Delta T_c)_{ah} \simeq + 3°K$$

where $(\Delta T_c)_h = -0.4°K$ is the harmonic contribution and $(\Delta T_c)_{ah} \simeq 3.4$ is the anharmonic contribution to ΔT_c. The calculated ΔT_c appears to be even larger than the observed ΔT_c of $\simeq 2°$ K [2], a discrepancy which reflects probably our poor knowledge of μ^{*}.

MAGNETIC IMPURITIES

As is usual, low concentration of d type impurities, forming local magnetic moments depress, considerably the superconducting T_c. The depression is larger than $100°K/at$ % for Fe or Cr [14]. But we would like to discuss more the case of Ni where the depression is less steep, below $1°$ K/at %. One can question if this is a magnetic effect at all, or if there is some other reason. Small Pt impurities, which are isoelectronic to Ni and Pd do not alter for instance T_c. Resistivity measurements [15] show indeed that there is a small decrease of the electron-optical phonon coupling which may probably be explained by variation of the electronic band structure of PdH if we substitute Pd by Ni. Recent theoretical band structure calculations [16] done on pure PdH and NiH compounds show that η_{op} is essentially sensitive to the H site Fermi level density of states. One of the main differences between PdH and NiH is that the Fermi level is much closer to the top of the d band for NiH ; this results in a lower s type density of states at the H site for NiH compared to PdH. The calculated η_{op} is more than two times smaller for NiH and this may well be one of the main reasons for the observed decrease of λ_{op} with nickel concentration. But one can show that this cannot explain the total decrease of T_c. A residual magnetic effect, despite Ni does not form a local magnetic moment in PdH, is probably present which may explain the difference in behaviour of Ni and Pt impurities inside PdH.

Acknowledgements : I would like to thank L. Dumoulin, P. Nedellec and B. Souffaché for many valuable discussions.

[1]   T. Skoskiewicz, Phys. Stat. Solid, 11K, 123 (1972)

[2]   B. Stitzker, W. Bückel, Zeits.f. Phys. 257, 1, (1972)

[3]   B. Souffaché, D.S. Mac Lachlan, J.P. Burger
       Rev. Phys. Appl. 14, 749 (1979)

[4]   A. Gorska, A.M. Gorski, J. Igalson, A.J. Pindor, L. Sniadower
       Proc. Int. Conf. on Hydrogen in Metals, Paris (1977)

[5]   J.P. Burger, D.S. Mac Lachlan, J. Phys. 37, 1227 (1976)

[6]   C. Arzoumanian, J.P. Burger, L. Dumoulin, P. Nedellec
       Zeits. f. Phys. Chem. 116, 117 (1979)

[7]   D.A. Papaconstantopoulos, E.N. Economou, B.M. Klein, L.L. Boyer
       J. Phys. Suppl. 39 C6-435 (1978)

[8]   M. Gupta, A.J. Freeman, Phys. Rev. 17, 3029 (1978)

[9]   C.L. Wiley, G. Cinader, F.Y. Fradin
       Bull. Am. Phys. Soc. 21, 404 (1976)

[10]   A. Eichler, H. Wühl, B. Stritzker, Sol. St. Com. 17,213 (1975)

[11]   B.W. Ganguly, Zeits. f. Phys. 265, 433 (1975)

[12]   R.J. Miller, C.B. Satterthwaite
       Phys. Rev. Lett. 34, 144 (1975)

[13]   A. Rohman, K. Sköld, C. Petizzari, S.K. Sinka
       Phys. Rev. B14, 3630 (1976)

[14]   J.C.M. Van Dongen, J.A. Mydosh, Zeits.f.Phys.Chem. 116,149
                                                         (1979)

[15]   L. Sniadower, L. Dumoulin, P. Nédellec, J.P. Burger
       to be published
[16]   M. Gupta, J.P. Burger, Journ. Phys. F (in press)

ELECTRONIC STRUCTURE AND ELECTRON-PHONON COUPLING

CONSTANT OF SOME METAL HYDRIDES

Michèle Gupta

Centre de Mécanique Ondulatoire Appliquée du CNRS
23, rue du Maroc - 75019 Paris
and Laboratoire de Physique des Solides, Bât. 510
Université Paris-Sud 91405 Orsay (France)

With the aim of explaining on theoretical grounds some experimental results on the superconducting transition temperatures, T_c, of metal hydrides and possibly make predictions on the values of T_c, the electron-phonon coupling parameter λ is evaluated for several simple, transition and non-magnetic rare earth metal hydrides. The electronic parameter η defined by Mc Millan is obtained within the rigid muffin-tin model from our augmented plane wave band structure results, while available experimental data are used to evaluate the 'phonon contribution'. The following general trends in the variation of η emerge from these calculations : (1) The magnitude of the electron-optical phonon matrix element which is dominated by the s-p scattering is found to be small for the early transition metal (TM) dihydrides and LaH_2 ; it is somewhat larger for the other TM dihydrides of the 4d series. This is in contrast to PdH which has a significantly larger value of η_H. Large values of η_H are also obtained for AlH_2 and AlH (which can be prepared by ion implantation) making these compounds good candidates for superconductivity. (2) As the metal site η_{metal} is essentially determined by the d-f scattering for the late members of the TM series while the p-d mechanism is important for the early members of the TM series, for LaH_2 and particularly for the AlH_x system. In most cases, a reduction of η_{metal} from its value in the pure metal is obtained for the cubic metal hydrides studied here. The essential features of the electronic structure of the stoichiometric metal hydrides under study are also underlined.

255

INTRODUCTION

The discovery of superconductivity with fairly high values of the superconducting critical temperature T_c in Th_4H_{15}[1] and in the palladium hydride and deuteride systems[2] gave an impetus to the search for superconductivity in other metal hydrides. An appealing feature of these systems is the possibility of varying the average number of valence electrons per atom ratio since most of the hydrides exist over a wide range of composition. The search for high T_c superconductors has been successful in a few cases ($T_c^{max} \sim 17$ K for $H/Pd_{55}Cu_{45} = 0.7$). For samples prepared by the ion implantation technique such as the AlH_x system, an increase of T_c from 1.6 K for pure Al to 6.8 K for hydrogen rich compounds has been observed[3]. This increase in T_c cannot be ascribed to the effect of lattice disorder only. Other examples can be found in the recent review of Stitzker and Wühl[4]. Nevertheless, most of the transition metal hydrides investigated to date do not show any superconducting behaviour down to 1 K : this is the case of hydrides of lanthanum[5] with hydrogen to metal ratios ranging from 1.8 to 2.36, while the pure La metal in the fec phase has a T_c of 6 K. Similarly, the hydrides of the high T_c group V transition metals such as TaH, VH_2 and NbH_2 are not superconducting down to 1 K. Other dihydrides of the early members of the transition metal (TM) series[7] TiH_2 and ZrH_2 as well as monohydrides of the end of the TM series such as NiH are not superconductors[8] above 1.5 K.

In order to explain some of these puzzling experimental results, we have calculated the electron-phonon coupling constant λ for the early cubic transition metal dihydrides of group IV and V ZrH_2 and NbH_2 ; LaH_2 is also investigated as an example of trivalent dihydride. We also studied the monohydrides with filled metal d bands PdH and NiH in order to understand why these isoelectronic compounds which are both diamagnetic do not have the same superconducting properties. The aluminium hydrides are also investigated and since the stability of the octahedral versus tetrahedral occupancy of the hydrogen in the Al lattice has not yet fully been ruled out[9], we have studied the monohydride AlH and the dihydride AlH_2 in which H occupies respectively the octahedral and tetrahedral interstices of the fcc metal lattice.

In his study of transition metals, McMillan[10] has shown that the electron-phonon coupling constant λ which essentially determines T_c can be written as follows in terms of the ratio of an 'electronic' contribution η to a 'phonon' contribution. For compounds with a large mass ratio between the two constituent atoms such as metal hydrides, an approximate expression of λ has been proposed[11] :

$$\lambda \simeq \frac{\eta_{metal}}{M_{metal}<\omega^2>_{acoustic}} + \frac{\eta_H}{M_H<\omega^2>_{optic}} = \lambda_{metal} + \lambda_H \qquad (1)$$

where M is the atomic mass and $<\omega^2>$ is the second moment of the renormalized phonon frequencies as defined by Mc Millan [10]. A study of the 'interference' term neglected in this approximate expression has been done in the case of PdH [12]. Using the rigid-ion approximation and expanding the Bloch functions into their angular momentum representation, Gaspari and Gyorffy [11] showed that the mean square of the electron-phonon matrix element $<I^2>$ can be conveniently expressed in terms of quantities obtained from ab-initio band structure calculation ; they have shown that :

$$\eta \sim \sum_K \frac{E_F}{N_\uparrow(E_F)\Pi^2} \sum_\ell 2(\ell+1)\sin^2(\delta^K_{\ell+1}-\delta^K_\ell) \frac{n^K_\ell(E_F)\, n^K_{\ell+1}(E_F)}{n^{K(1)}_\ell(E_F)\, n^{K(1)}_{\ell+1}(E_F)} \qquad (2)$$

where the summation on K runs on all the atoms in the unit cell, δ^K_ℓ is the single site scatterer phase shift at the Fermi energy E_F, n^K_ℓ the partial density of states (DOS) of angular character ℓ at the K site ; $n^{K(1)}_\ell$ is the corresponding partial DOS of a free scatterer and $N_\uparrow(E_F)$ the total DOS per spin at E_F.

Since the value of η is determined by the characteristics of the electronic states at the Fermi energy : phase shifts and partial DOS, we shall first discuss the essential modifications of the electronic structure of the pure metal upon formation of a hydride, found in the present and in previous work [13].

ELECTRONIC STRUCTURE

1) Monohydrides PdH and NiH

The main differences between the band structure and DOS of the monohydrides of the end of the TM series and those of the corresponding pure TM are characterized (i) by the formation in the monohydrides of low energy metal-hydrogen bonding states and (ii) by the filling of the pure metal d band holes, the Fermi energy of the monohydrides being in the metal s-p band. A wavefunction analysis reveals that the metal states which are strongly modified by the H potential are those having an s symmetry at the interstitial H site ; these state can have a s, p or d symmetry around the metal site ; they are drastically lowered in

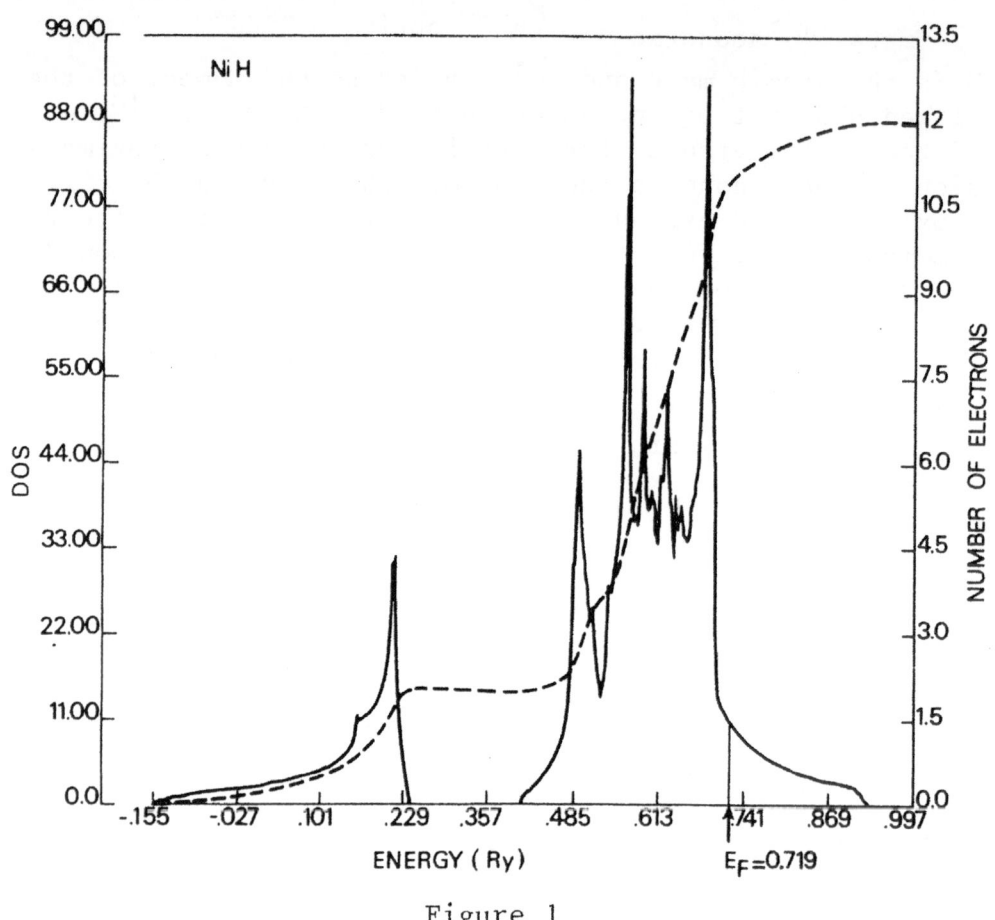

Figure 1

The DOS (states of both spin per Ryd per unit cell) of
NiH is represented by the full curve and the left-hand
side scale. The total number of electrons is represented
by the broken curve and the right hand side scale. The
arrow indicates E_F = 0.719 Ryd.

energy by the metal-hydrogen interaction. The low-lying metal-
hydrogen bonding band is responsible for the structure seen in
the DOS of NiH plotted in Fig. 1 ; these H induced states can be
observed in photoemission spectra. A partial DOS analysis into
angular momentum components of s,p,d symmetry around the metal
and the H sites shows [14] that the low-lying structure in the DOS
is due not only to H-s states but also to metal d and to a lesser
extent metal s and p states. This analysis shows already that the
oversimplified H^- anionic model is not correct. Direct information
on this partial wave analysis of the DOS can be obtained from a
study of the metal X-ray emission spectra since the electronic
transitions are governed by the dipole selection rule. In the
case of PdH and NiH, most of the states forming the metal-hydrogen

bonding band were already filled in the pure metal, in which they
were found at the bottom of the metal s-p band and in the lowest
lying metal d states. Nevertheless, it is interesting to notice
that a branch of states in the [111] direction which was unfilled
in the pure metal is found also in the low-lying metal-hydrogen
bonding band of the hydride. This feature has important consequences
for the position of the Fermi level and for the modification of the
Fermi surface of the pure metal upon formation of the hydride.
Since most of the states forming the low-lying band were already
filled in the pure metal, the extra electron brought by the H atom
cannot be fully accomodated in the low energy states and thus the
Fermi level of the metal is raised. But, as we previously discussed, less than one electron will be added at the Fermi level of the
metal since the metal hydrogen bonding band contains also states
previously unfilled in the pure metal. This discussion shows that
the protonic H^+ model is not correct. Its partial success is due
to the fact that in the monohydrides the metal d holes are filled
and the Fermi energy E_F lies in the metal s-p band, like in the
corresponding noble metal, although E_F remains closer to the top
of the d bands in the hydride. The large decrease in the DOS at
E_F from the pure metal to the hydride is in agreement with electronic specific heat and magnetic susceptibility data.

The main differences observed between the electronic structure of PdH and NiH concern the position of the metal hydrogen
bonding states and the position of the Fermi level. In the case
of NiH, E_F is closer to the top of the d bands than in PdH ; this
is due essentially to the fact that the number of holes in the
d bands is larger in Ni (0.6 holes) than in Pd (0.36 holes). This
feature has important consequences on the composition of the
electronic states at E_F and thus will affect the value of the
electronic contribution η of the electron-phonon coupling as we
shall see in the next section.

2) Dihydrides of early TM and lanthanides

As it can be seen in Fig. 2, the DOS of the cubic dihydrides is characterized by a structure at low energy which is due to
two bands : a metal-hydrogen bonding band, already found in the
case of monohydrides and, in addition, a second band which is due
to the interaction of the two H atoms in the unit cell. At the
Brillouin zone (BZ) center, this band is formed of hydrogen
hydrogen antibonding states [13,15,16]. The width and position of
the two low-lying bands depend critically upon the metal-hydrogen
distance. Thus in the lanthanides which have large lattice constants, the two low-lying bands do not overlap the metal d states
which lie at higher energies, while in the TM dihydrides such as
TiH_2 or NbH_2, a partial overlap is observed. The position of the
antibonding band at low energies plays an important role for the
stability of the compound. An empirical correlation has been

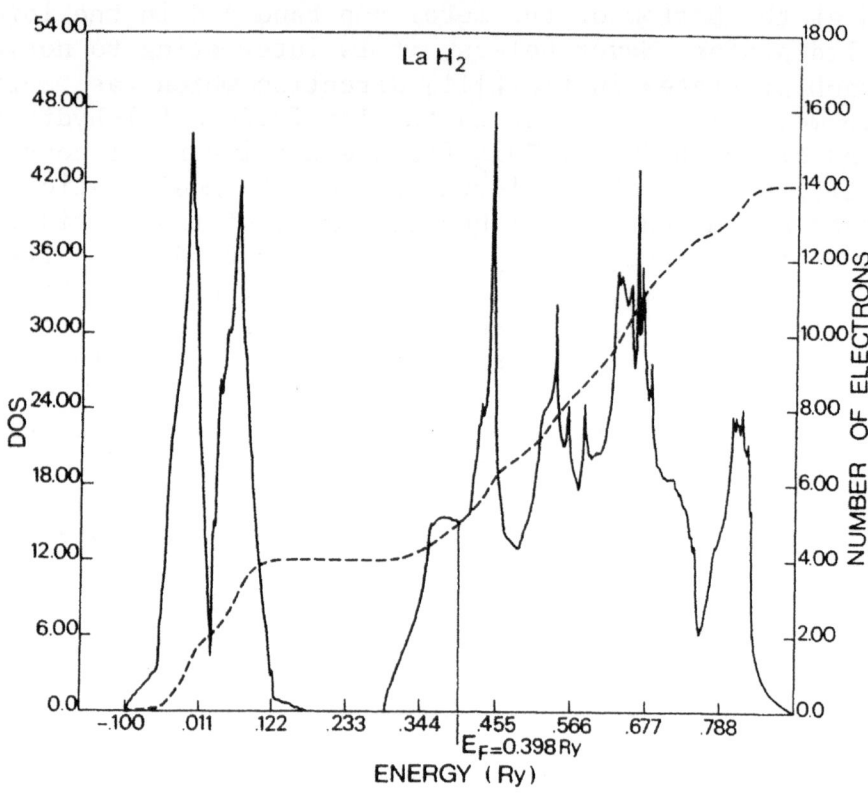

Figure 2

Total density of states of LaH_2 (full line curve and left-hand side scale) units are states of both spin/Ryd unit cell. The number of electrons (dashed line curve and right-hand side scale).

established [13] between the position of the H-H antibonding band above the Fermi level and the non-existence of the dihydride phase for the latest members of the TM series.

For the dihydrides of trivalent metal such as LaH_2, the two low-lying metal hydrogen bonding bands are filled and only one electron occupies the bottom of the metal d bands as it can be seen in Fig. 2. Schematically, because of the lowering of some metal states due to the metal-hydrogen interaction, the net result can be described as a depopulation of the metal d states in favor of the two low-lying bands, the d bands of LaH_2 accomodate only one electron. This should not however be considered as a depopulation of the pure metal bands in a rigid band model sense, since we have emphasized the deformation of the d bands and the hybridization of the low-lying metal hydrogen bonding states. The decrease of the d like DOS at E_F from the pure metal to the dihydride can

be invoked to explain the better conductivity of the dihydrides[15];
the depopulation of the d bands and also the large expansion of the
lattice lead to a decrease of the Ruderman-Kittel-Kasuya-Yoshida
indirect exchange interaction of the rare earth (RE) local moments
mediated by the conduction electrons in the magnetic RE dihydrides.
This can explain why the magnetic transition temperatures are far
lower for the dihydrides than for the pure RE metals [15]. In a
dihydride of tetravalent metal like TiH_2, two electrons are accomodated in the metal d bands ; the Fermi energy falls in a peak of
the DOS[15]. It is interesting to notice that a similar peak is also
observed in Fig. 2 in the d bands of LaH_2,which corresponds to the
filling of the d bands by one more electron. Indeed our ab-initio
studies of several dihydrides of group III, IV and V show that
the main structures of the DOS of the metal states of the dihydrides are consistently observed in all of these cubic compounds.
Of course, the width of the metal d bands and thus the exact values
of the DOS depend upon the compound under study through both the
lattice parameter and the position of the metal atom in the periodic table. The high value of the DOS at E_F in the group IV dihy-
dirdes has been invoked to interpret the cubic to tetragonal distortion observed in these compounds, in terms of a Jahn-Teller
type effect [15]. The degeneracy of the flat band responsible for
the large value of the DOS is lifted by the tetragonal distortion
and the total energy of the system is lowered. This interpretation
is in agreement with several experimental data such as the electronic specific heat, the magnetic susceptibility, the thermoelectric power data [17] and the sign change of the Hall coefficient
accompagnying the distortion.

3) Simple Metal Hydrides AlH and AlH_2

The parabolic s-p bands of pure Al are drastically modified by the metal-hydrogen interaction [18]. The bands of AlH are
shown in Fig. 3 ; we observe, as discussed previously for the
rocksalt structure monohydrides, a metal hydrogen bonding band,
at low energy. Three bands cut the Fermi level. The lowest one
which contains the majority of the two electrons to be accomodated,
has a strong admixture of H-s states. Indeed, at the BZ center,
the corresponding state is an antibonding combination of metal
and H-s states. For the dihydride [18], in addition to the low-
lying metal-hydrogen bonding band, the interaction of the two H
atoms lead to a second band at low energy. The two bands are filled
and the Fermi energy falls in the metal s-p band, which is also
hybridized with H-s states but to a lesser extent than in the
case of the monohydride.

The large hybridization with H-s states at E_F, which is particularly strong in AlH, is a new feature of the simple metal
hydrides. Such hybridization is not observed for the TM hydrides ;
this H-s character of the states at E_F has important consequences

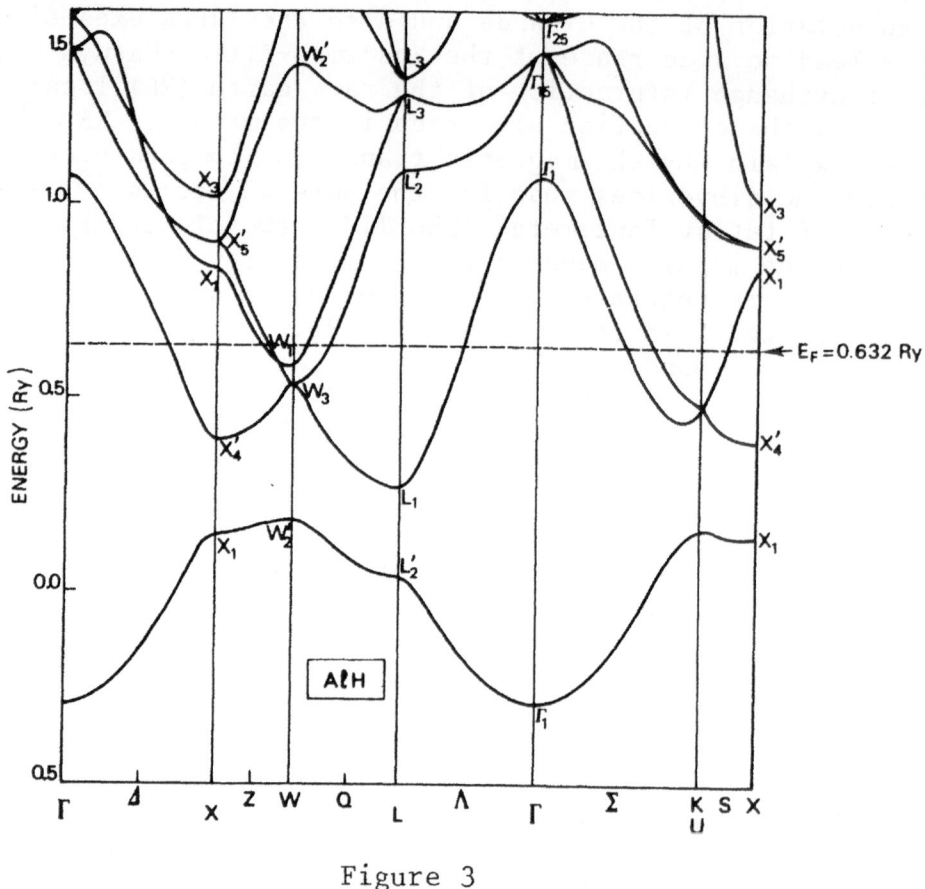

Figure 3

The energy bands of rocksalt structure AlH along several symmetry directions. Energies are in Rydberg.

on the value of the electron-optical phonon coupling, as we shall discuss in the next section.

ELECTRON-PHONON COUPLING

1) Trends in the variation of λ_H

We shall first examine the trends in the variation of λ_H in the metal hydrides under study. For the PdH_x system, which has been thoroughly investigated both from the experimental and the theoretical point of view, it is now well established that, in contrast to most of the superconducting transition metal compounds, the electron-optical phonon coupling plays a dominant role. The importance of low energy optical phonons ($\hbar\omega \simeq 50$ meV) has been revealed by superconducting tunneling experiments [19], neutron scattering data [20] and the study of the temperature dependence of the electrical resistivity [21] which shows the onset of the optical phonon scattering at $T \simeq 140$ K in PdH. Theoretical estimates of the electron-phonon coupling constant have shown[12,22], in agreement with the prediction of Ganguly[23], that the largest contribution to λ is provided by the electron-optical coupling. In spite of a general lowering of the acoustic phonon frequencies[20] by 20 to 30 % from the pure metal to $PdD_{0.63}$, the electron-acoustic phonon coupling is not for this system the essential mechanism which leads to a high value of T_c. It is thus interesting to find out whether the strength of the electron-optical phonon coupling (or its weakness) is responsible for the high value of T_c (or the lack of superconductivity) in other metal hydrides. In the case of Th_4H_{15} for which $T_c \sim 8$ K, the role of the electron-optical phonon coupling has not yet been clearly assessed ; nevertheless, the presence of optic modes of low energy[24] extending down to 50 meV, which are absent in the phonon DOS of the nonsuperconducting ThH_2 seem to indicate the possibility of a mechanism similar to that evidenced for PdH. This is corro-[25] borated by the fact that the acoustic modes of Th_4H_{15} are harder than those of pure Th, in spite of a lattice expansion, and thus less favourable for superconductivity. For the metal hydrides studied here, the essential parameters defined in Eqs(1) and (2) are listed in Tables I and II. We shall first study the variation of the electronic contribution at the hydrogen site η_H. The results listed in Table II show that the hydrogen potential scatters strongly the s waves ; the $s(\ell = 0)$ scattering phase shift δ_0^H is large at the Fermi energy for all the metal hydrides ;

δ_0^H is always close to a resonance $\delta_0^H \sim \frac{\pi}{2}$. Fig. 4 shows a typical

plot of δ_0^H in NiH as a function of energy. In the energy range

spanned by the metal d bands, the δ_0^H phase shift remains large

and is a very slowly varying function of energy ; the same beha- viour is observed for dihydrides. This indicates that all metal hydrides of the beginning as well as of the middle and of the end of the series will have large values of δ_0 no matter where the Fermi energy lies in the metal d bands. The results listed in Table II show that for all the hydrides, the phase shifts of higher angular momentum components at the H site are very small ; consequently, the value of η_H is dominated by the s-p scattering mechanism and its variation across a series of metal hydrides is controlled by the magnitude of the partial s and p DOS at E_F, rela- tive to the value of the total DOS. The value of n_s at the H site is found to be sizeable for the transition metal hydrides at the end of the series especially for PdH, and for the simple metal hydrides. This feature can be understood from our discussion of the electronic structure.

Table I

The partial wave analysis n_ℓ of the DOS inside the muffin-tin and hydrogen spheres at the Fermi energy. $N_\uparrow(E_F)$ is the total DOS at E_F. Units are states of one spin/Ry unit cell.

		n_s	n_p	n_d	n_f	$N_\uparrow(E_F)$
AlH	Al	0.621	0.757	0.217	0.037	3.203
	H	0.325	0.168	0.011	0.000_5	
AlH$_2$	Al	0.684	0.520	0.124	0.016	3.155
	1xH	0.073	0.107	0.005	0.000_3	
LaH$_2$	La	0.017	0.168	3.250	0.014	7.550
	1xH	0.017	0.154	0.0168	0.000_7	
ZrH$_2$	Zr	0.004	0.073	10.805	0.035	16.460
	1xH	0.008	0.432	0.021	0.000_8	
NbH$_2$	Nb	0.004	0.051	4.369	0.034	6.440
	1xH	0.016	0.119	0.010	0.001_2	
PdH	Pd	0.058	0.163	2.618	0.012	3.405
	H	0.255	0.029	0.001_6	0.001	
NiH	Ni	0.048	0.133	4.741	0.009	5.390
	H	0.161	0.035	0.002_5	0.001	

As pointed out previously, the H-s states in the TM hydrides are found essentially in the low-lying bands and since for the early transition and rare earth metal hydrides the Fermi level falls at the bottom of the metal d bands, it is not surprising to find small values of n_s^H in this energy range. For PdH and NiH, the d bands are filled and the values of n_s^H at E_F become sizeable. The Fermi level is not high enough to fall in the antibonding metal hydrogen band ; nevertheless, in this energy range some metal states of s symmetry at the H site have been lowered by the H potential, but not enough to fall below the d bands. The difference in the values of n_s^H between PdH and NiH is due to the fact that for NiH, the Fermi energy is closer to the top of the d bands than in PdH. The increase of n_s^H as a function of energy if

Table II

Values of the various parameters entering the calculation of
λ. Symbols are defined in Eqs(1) and (2). The angular momentum
dependent phase shifts δ_ℓ are given in radians.
(a) Theoretical values obtained by Butler Ref.31 for the
corresponding transition metals.

		δ_0	δ_1	δ_2	δ_3	$\eta_\circ$ $(eV/\mathring{A}^2)$	$\langle\omega^2\rangle$ $(eV/\mathring{A}^2)$	λ
AlH	Al	0.3365	014014	0.0528	0.0015	0.294		
	H	1.1856	0.0404	0.0012	0.0	2.292		
AlH$_2$	Al	0.2536	0.3714	0.0556	0.0018	0.224		
	1xH	1.1738	0.0379	0.0010	0.0	0.744		
LaH$_2$	La	-1.1006	-0.4556	0.5326	0.0006	0.753	7.35	0.103
	1xH	1.5290	0.0400	0.0009	0.0	0.043	3.35	0.013
ZrH$_2$	Zr	-0.9504	-0.3662	0.9042	0.0049	2.352 $(3.87)^a$		
	1xH	1.2462	0.0373	0.0009	0.0	0.088		
NbH$_2$	Nb	-1.0219	-0.4075	1.3749	0.0067	2.975 $(7.39)^a$		
	1xH	1.1593	0.0408	0.0011	0.0	0.102		
PdH	Pd	-0.5115	-0.1094	2.8066	0.0030	0.886	5.971	0.15
	H	1.1931	0.0280	0.0006	0.0	0.641	1.062	0.60
NiH	Ni	-0.3857	-0.0235	2.7916	0.0025	0.810	10.0	0.08
	H	1.0604	0.0318	0.0008	0.00	0.275	3.44	0.08

the rigid band model is applied to PdH has been invoked [26] to
explain (in part) the increase of T_c in the Pd-noble metal-H_x
systems. We remind the reader that this is not the only factor
which explains the increase of T_c upon alloying since there is
also experimental evidence [27] of a softening of the acoustic
phonons.

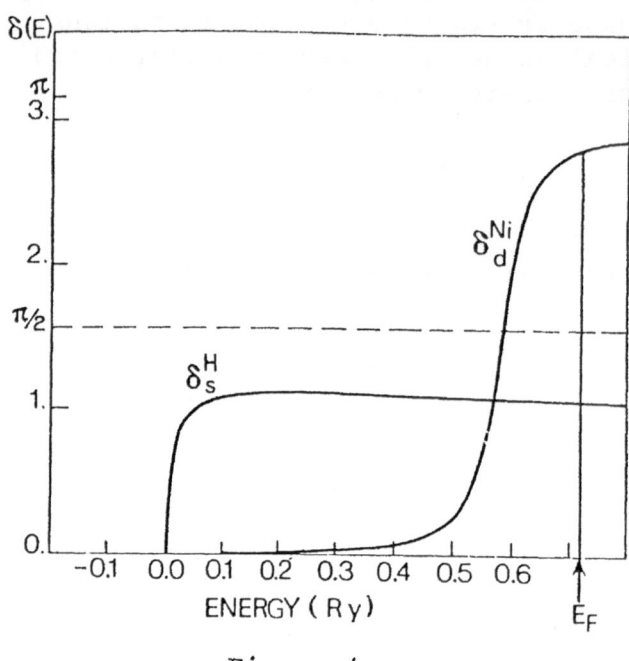

Figure 4

The d phase shift of the Ni potential and the s phase shift of
the H potential in NiH as a function of energy. Phase shifts
are in radians. The metal d bands in NiH are in the energy
range 0.4 Ry < E < 0.7 Ry.

For the simple metal hydrides, in contrast to the early
transition and rare earth (RE) metal hydrides, the values of n_s^H
at E_F is rather large. This is due to a larger hybridization
of the s-p metal states with the H-s states ; in the case of AlH
the presence of a metal-hydrogen antibonding band at E_F is responsible for the large value of n_s^H.

The DOS of p type ($\ell = 1$) at the H site is vanishingly small
for hydrides of the end of the TM series such as PdH and NiH
while hydrides of beginning and the middle of the series for which
n_s^H is very small, have larger values of n_p^H. This contribution to
the partial DOS arises from the metal d states having a p symmetry
at the H site, which have not been perturbed by the H potential
and remain almost unperturbed from the pure metal to the hydride.

Thus the s-p scattering mechanism which essentially determines the electronic contribution η_H in the angular momentum representation presently used should not be viewed as an intra atomic effect since the partial DOS of p type at the H site arises from the tails of the metal d states. We can conclude from the results listed in Table II that, except for the simple metal hydrides which have large values of n_H and for the hydrides of the end of the TM series such as PdH, the values of η_H per hydrogen site remain small for most of the metal hydrides studied here.

Besides the variation in the value of η_H, the magnitude of λ_H is expected to be very sensitive to the average energy of the optic phonons as it can be seen from Eq(1). To date, ab-initio predictions of the position of the optic modes in metal hydrides have not yet been made ; nevertheless some neutron scattering data are available for these compounds. A compilation of the experimental results [28] shows that the occupation of the octahedral sites by the H atoms such as in the fcc Pd metal or in the b.c.t. β $VH_{0.4}$ seems to lead to lower optic modes ($\hbar\omega \sim 50$ meV) than the occupation of the tetrahedral sites. In the cubic CaF_2 structure dihydrides or in the b.c.c. metals the energies of the optic modes is of the order of $\hbar\omega \sim 120$ meV. The difference in the position of the optic phonons is due to the size of the interstices (which is larger for the octahedral than for the tetrahedral holes) and to the differences in the electron-phonon interaction of electronic origin. Only detailed microscopic calculations can give an answer for the trends in the variation of the second factor. For compounds having essentially the same electronic structure, like all the rare earth dihydrides, it is the distance metal-hydrogen which appears to control the position of the optical phonon [29]. In the case of NiH, although H occupies the octahedral interstices, there is some experimental evidence that the optic modes have a much higher energy than in PdH. A study of the temperature dependence of the electrical resistivity [8] of NiH shows that the optical phonon contribution occurs at higher temperature than in PdH. The Einstein temperature of the optic modes of NiH has been estimated [8] to be at least a factor of 1.8 larger than that [21] of PdH. This feature is probably due in part to the fact that the metal-hydrogen distance is smaller in NiH than in PdH and certainly to differences in the electron-phonon interaction of electronic origin between two compounds of different rows of the periodic table. The values given in Table II for the phonon contribution $M_H<\omega^2>$ are taken from realistic neutron scattering data [20] in the case of PdH. In view of the lack of data for the other hydrides, we used a scaling factor derived from the average position of the optic modes when the corresponding data are available.

2) <u>Trends in the variation of</u> λ_{Metal}

We shall first summarize the main results obtained for η_{Metal} in the hydrides. From the results listed in Table II we can see that for the transition metal hydrides, the s and p phase shifts at the metal site are negative ; this indicates a repulsive character of the metal potential for the s and p waves, due to the orthogonalization conditions to the corresponding core states. The d wave phase shifts at the Fermi energy are positive and increase with the filling of the metal d bands thus δ_2^{Metal} is small for the hydrides whose Fermi energy falls at the bottom of the metal d bands like LaH$_2$; it increases and reaches a resonance, $\delta_2^{Metal} \sim \Pi/2$ in the middle of the d bands, the sharpness of the resonance being related to the width of the bands. When the d bands are filled like for PdH and NiH, the d wave phase shift is large and close to the value of Π. A typical energy dependence of δ_2^{Metal} for a transition metal hydride is illustrated by the plot of the d wave phase shift of Ni in NiH given in Fig. 4. A study of the partial wave decomposition of the DOS at E_F shows that for all the transition metal hydrides and LaH$_2$ the 'd' character is dominant at the metal site as it can be seen from the results listed in Table I. From the trends obtained in the values of the phase shifts and of the partial DOS at the metal site we found that the value of η_{Metal} for the transition metal hydrides of the middle and of the end of the series is dominated by the d-f scattering mechanism while the p-d mechanism is important also in the early transition metals and LaH$_2$. As an example, 60 % of the contribution of η_{La} in LaH$_2$ arises from the p-d scattering term and 37 % from the d-f contribution while for transition metal hydrides of the end of the series such as NiH and PdH, more than 80 % of the value of η_{Metal} is provided by the d-f scattering mechanism. We wish at this point to remind the reader that in the angular momentum representation used here, the d-f scattering should not be considered as an intra-atomic effect since the partial DOS of f type at one metal site is provided by the tails of the metal d functions of the neighboring sites ; thus the physical origin of this term should rather be understood in terms of a metal d-d interaction in a tight binding picture [30].

For the simple metal hydrides AlH and AlH$_2$, $\sim$ 80 % or more of the total value of η_{Al} is provided by the p-d scattering term ; the d-f mechanism is negligible since the d phase shifts and the partial DOS of d and f type are small for simple metals. The s-p scattering term is small in spite of the importance of the partial DOS of s and p type at E_F because the s and p phase shifts are nearly equal and lead to a cancelation of the phase shift dependent term in Eq.(2). A comparison of the values of η_{Metal} in the hydrides with the values of the corresponding pure metals shows, in most of the cases, an important decrease of η_{Metal} upon formation of the hydride. The theoretical values listed in parenthesis in Table II for the pure transition metals have been taken from the work of Butler [31]. From the results obtained for η_{Metal} in the metal hydrides under study it thus appears that the electronic contribution at the metal site will not lead to an enhancement of the superconductivity of the pure metal upon formation of the hydride. Of course, even when the electronic contribution decreases, an enhancement of λ_{Metal} could be still obtained if the acoustic modes become soft enough in the hydride. Unfortunately, experimental data on the acoustic phonon frequencies of all the hydrides investigated here are not always available. For PdH, the values of $M_{Pd}\langle\omega^2\rangle_{acoustic}$ listed in Table II has been obtained from the experimental phonon DOS of Rowe et al. [14]. Since similar data are not available for other hydrides, we have thus used the approximation

$$M_{Metal}\langle\omega^2\rangle_{ac} \sim \frac{1}{2} M_{Metal} \, \Theta_D^2$$

for the hydrides whose Debye temperature is known.

For LaH$_2$, for example, the increase of Θ_D from the pure metal [32] to the dihydride [33] leads to a reduction of λ_{Metal} ; since we have seen in the study of λ_H that the electron-optical phonon coupling provides a negligible contribution, we find a drastic decrease in the total value of λ upon formation of the dihydride which leads to a vanishingly small value of T_c. Thus, according to our theoretical estimate, LaH$_2$ should not be a superconductor ; this result is in agreement with the experimental investigation of Merriam and Schreiber [5]. For the bcc metals of group V, there exists some experimental evidence of a hardening of the acoustic phonons [34] with hydrogenation and this factor is of course not favourable for an enhancement of λ in the corresponding hydrides.

CONCLUSION

From the results obtained in the present work we find that the values of the electronic parameter η_H are sizeable for transition metal hydrides with filled d states, especially when the Fermi energy is not too close to the top of the d bands. η_H is especially large for simple metal hydrides, these compounds appear to be particularly favourable for a large electron-optical phonon coupling, provided that the energies of the optic phonons, which are not known experimentally, are not too high. The dihydrides of the beginning of the series and LaH$_2$ have small values of η_H ; the values of η_H increase for TM dihydrides of the middle of the series ; however, for these compounds, the energies of the optic phonons are high and this leads to small values of the electron-optical phonon coupling. At the metal site, we obtained in most cases a sizeable decrease of the electronic parameter from its value in the pure metal ; this decrease is not compensated by a

softening of the acoustic modes in most of the hydrides for which
experimental data are available and thus for these compounds, there
is no enhancement in the superconducting properties of the pure
metal due to the electron-acoustic phonon coupling. The theoretical results obtained here are in satisfactory agreement with
available experimental results on the search for superconductivity
in metal hydrides.

REFERENCES

1.   C.B. Satterthwaite and I.L. Toepke,
      Phys. Rev. Lett. 25, 741 (1970)

2.   T. Skoskiewicz
      Phys. Stat. Sol. (a) 11, K 123 (1972)

      B. Strizker and W. Bückel
      J. Phys. 257, 1 (1972)

3.   A.M. Lamoise, J. Chaumont, F. Meunier and H. Bernas
      J. de Physique Lettres, 36, L271 (1975)

4.   B. Stritzker and H. Wühl
      in Topics in Applied Physics, Vol. 28 : Hydrogen in
      Metals, edited by G. Alefeld and J. Völkl,
      Springer Verlag, Berlin, Heidelber, New-York 1968, p. 243

      M.F. Merriam and D.S. Schreiber
      J. Phys. Chem. Solids, 24, 1375 (1963)

6.   C.B. Satterthwaite and D.T. Peterson
      J. Less-Common Metals, 26, 361 (1972)

7.   B.T. Matthias, T.H. Geballe, V.B. Compton
      Rev. Mod. Phys. 35, 1 (1963)

8.   D.S. McLachlan, I. Papadopoulos and T.B. Doyle
      J. de Phys. Paris C 6, 430 (1978)

9.   J.P. Bugeat and E. Ligeon
      Phys. Lett. 71A, 93 (1979)

10.   W.L. McMillan
      Phys. Rev. 167, 331 (1968)

11.   I.R. Gomersall and B.L. Gyorffy
      J. Phys. F 4, 1204 (1974)

      G.D. Gaspari and B.L. Gyorffy
      Phys. Rev. Lett. 29, 801 (1972)

12.   H. Rietschel
      Z-Phys. B 22, 133 (1975)

      J.P. Burger and D.S. McLachlan
      J. de Phys. 37, 1227 (1976)

13. A.C. Switendick in Topics in Applied Physics
     Vol. 28 : Hydrogen in Metals I, ed. by G. Alefeld and
     J. Völkl, Springer Verlag, Berlin, New-York (1978)
     and references there in.

14. M. Gupta and A.J. Freeman
     Phys. Rev. 17, 3029 (1978)

     M. Gupta and J.P. Burger
     J. Phys. F 10 (1980) in press

15. M. Gupta
     Solid State Comm. 27, 1355 (1978)
     ibid. 29, 47 (1979)

16. M. Gupta and J.P. Burger
     Phys. Rev. B 22 (1980) in press

17. F. Ducastelle, R. Caudron and P. Costa
     J. de Phys. 31, 57 (1970)

18. M. Gupta and J.P. Burger
     J. de Phys. 41, 1009 (1980)

19. A. Eichler, H. Wühl and B. Stritzker
     Solid State Comm. 17, 213 (1975)

     P. Nedellec, L. Dumoulin, C. Arzoumanian and J.P. Burger
     J. de Physique C 6, 432 (1978)

20. J.M. Rowe, J.J. Rush, H.G. Smith, M.Mostoller and H.E. Flotow
     Phys. Rev. Lett. 33, 1297 (1974)

21. D.S. McLachlan, R. Mailfert, J.P. Burger and B. Souffaché
     Solid State Comm. 17, 281 (1975)

22. D.A. Papaconstantopoulos and B.M. Klein
     Phys. Rev. Lett. 35, 110 (1975)

23. B.N. Ganguly
     Z-Phys. 265, 433 (1975)

24. M. Dietrich, W. Reichardt and H. Riestchel
     Solid State Comm. 21, 603 (1977)

25. J.F. Miller, R.H. Caton and C.B. Satterthwaite
     Phys. Rev. B 14, 2795 (1976)

26. D.A. Papaconstantopoulos, E.N. Economou, B.M. Klein and
     L.L. Boyer
     Phys. Rev. B 20, 177 (1979)

27. M.R. Chowdhury and D.K. Ross
     Solid State Comm. 13, 229 (1973)

     M.R. Chowdhury
     J. Phys. F 4, 1657 (1974)

28.   T. Springer, in Topics in Applied Physics,
      Vol. 28, Hydrogen in Metals, ed. by G. Alefeld and
      J. Völkl, Springer-Verlag, Berlin, New-York (1978), p. 75

29.   D.G. Hunt and D.K. Ross
      J. of Less-Common Metals, 49 169 (1976)

30.   S. Barisic, J. Labbé and J. Friedel
      Phys. Rev. Lett. 25, 919 (1970)

31.   W.H. Butler,
      Phys. Rev. B 15, 5267 (1977)

32.   D.L. Johnson and D.K. Finnemore,
      Phys. Rev. 158, 376 (1967)

33.   Z. Bieganski, D. Gonzalez-Alvarez and F.W. Klaaysen
      Physica (Utrecht) 37, 153 (1967)

34.   J.M. Rowe, N. Vagelatos, J.J. Rush and H.E. Flotow
      Phys. Rev. B 12, 2959 (1975)

HYDROGEN IN V, Nb, AND Ta: MAGNETISM IN Fe, Co, AND Ni

G. C. Abell

Mound Facility*
Monsanto Research Corporation
Miamisburg, Ohio 45342

ABSTRACT

The concept of the four-atom Jahn-Teller resonance molecule. in transition metals is described, and its origin in the study of the Vb hydrides is retraced. The concept is schematically related to the theory of resonance localization of magnetic impurities. A model study that demonstrates plausibility and suggests generality of the four-atom resonance is reviewed. The nature of the electron correlation responsible for the resonance is revealed by a simple analogy to involve formation of a covalent bond. Generalization of the model Hamiltonian suggests the possibility of molecular magnetic moments and leads directly to a simple and successful description of magnetism in the Fe group.

INTRODUCTION

A recent study[1] of the properties of hydrogen in V, Nb and Ta proposes a molecular state made up of metal d-orbitals and localized near hydrogen, which qualitatively explains the unusual behavior of hydrogen in the Vb metals. In a subsequent study, a simple model Hamiltonian was developed for the purpose of quantitative calculations.[2] The model study shows that the basic idea is reasonable and, more significantly, that the localized state depends on hydrogen in a nonessential way. The latter fact raised the possibility that the localized molecular state might be an intrinsic feature -- not just of the Vb metals -- but of transition metals in

*Mound Facility is operated by Monsanto Research Corporation for the U. S. Department of Energy under Contract No. DE-AC04-76-DP00053.

general. This line of reasoning led eventually to a simple microscopic description that offers a quantitative explanation for the
very old puzzle of magnetism in Fe.[3] The present article retraces
that development. First comes a description of the localized molecular d-state and its origin in the Vb hydrides. This is followed
by a brief discussion of the problem of localization in metals.
Next there is a description of the simple model Hamiltonian developed for quantitative calculations. It is shown how this model
demonstrates plausibility and suggests generality. Then the model
is generalized to pure transition metals, and its behavior for a
more than half-filled band is described. The result is provocative
and leads directly to the final topic -- magnetism.

ORIGIN AND DESCRIPTION OF MOLECULAR STATE

    The group Vb metals (V, Nb, Ta) have the bcc structure. Hydro-
gen occupies certain interstitial sites in the lattice, and each
hydrogen site has four nearest neighbor metal atoms. (We emphasize
here that magnetic Fe is also bcc.) The properties of the Vb hydrides include the following paradox: The symmetry of the hydrogen
site is tetragonal, and the hydrogen vibrations, observed by neutron diffraction, show this symmetry. But the distortion of nearby
lattice atoms is observed to be cubic. The simplest ideas about the
interaction of hydrogen with the lattice tells us that these two
properties should at least have the same symmetry, preferably tetragonal. This paradox suggests the possibility that these two properties originate not in the same interaction but derive from distinct
force fields. This key idea suggests the possibility of molecular
d-states localized on the cluster of metal atoms closest to hydrogen.

    That cluster can be viewed as a separate molecule. There are
two atomic d-orbitals per cluster atom that have large molecular interactions within the cluster and these can be combined to give a
variety of molecular orbitals. One of these orbitals has the possibility of a strong Jahn-Teller (JT) effect, and is shown in Fig.
1a. It has bonding overlap along <101> directions, antibonding
overlap along <10$\bar{1}$> directions, and is nonbonding overall. It has
a degenerate orbital partner whose bonds and nodes are opposite to
those shown in Fig. 1a. The JT theorem states that a molecule with
orbital degeneracy can be stabilized by some distortion that breaks
that degeneracy. The particular distortion shown in Fig. 1b is one
possibility, and the reason for choosing it is simply that it re-
solves the paradox. Two of the metal atoms are displaced along a
<111> bond, giving a cubic lattice distortion. The hydrogen vibra-
tions, on the other hand, depend on the direct interaction of the
hydrogen atomic s-state with fully symmetric cluster molecular orbits.
The JT effect thus accomplishes the separation of force fields that
determine lattice distortion on the one hand and hydrogen vibrations
on the other.

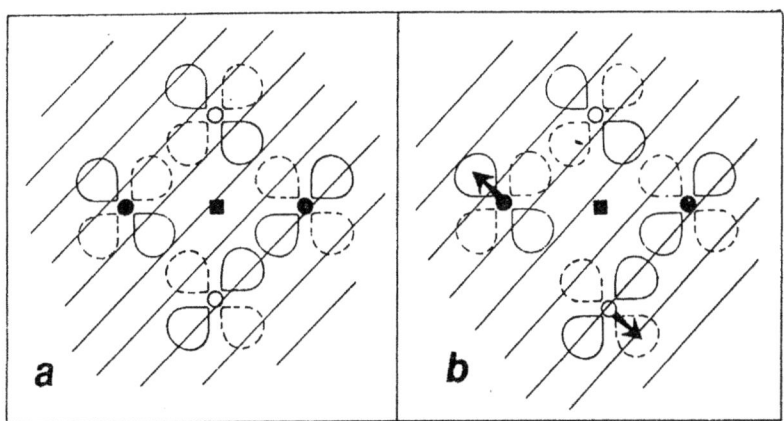

Fig. 1 Proposed Jahn-Teller state in Vb metal hydrides: (a) undis-
        torted degenerate state; (b) distorted state. View is
        along <010>. Filled and open circles represent M atoms at
        Z = a/4 and −a/4, respectively. The square represents an
        H atom at Z=0. The striping symbolizes band states.

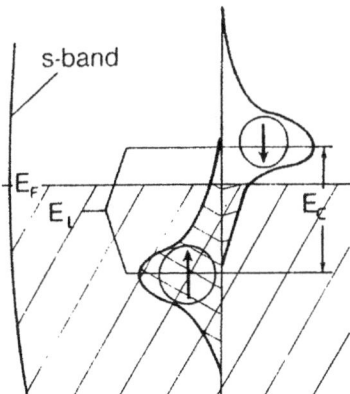

Fig. 2 Coulombic spin polarization of a magnetic impurity.

In fact this simple idea explains at least qualitatively most of the unusual behavior of the Vb hydrides as described in the original article.[1] One particularly interesting result is an unusual mechanism for hydrogen diffusion in which hydrogen revolves about the JT distortion axis on hexagonal rings. Occasionally this distortion axis undergoes a diffusive hop, and hydrogen whirls about the displaced axis on a new ring in a sort of dance through the crystal lattice. Another highlight is the explanation of certain blocking effects as being due to the molecular integrity of the JT state at high concentrations.

LOCALIZATION IN METALS

In spite of this apparent success, there is a serious difficulty in terms of fundamental theory. It is generally accepted that the important electron states in metals form a continuum in which electrons are not tied down in a given small region, but wander about more or less freely throughout the crystal. In other words, they are itinerant electrons -- not localized electrons. Localization in metals can occur, but it is somewhat pathological, requiring special circumstances. The JT state is a localized state, and so it has no place in a metal unless those special circumstances exist.

To establish that, we begin with a brief review of the Anderson Hamiltonian describing local magnetic moments in metals.[4]

$$H = H_F + H_L + H_C + H_{LF}$$

Here the total Hamiltonian is broken down into a term H_F giving the free electron energies, a term H_L giving the energy of the localized state, a term H_C describing correlation between localized electrons, and a term H_{LF} giving the coupling between localized state and free states. Anderson developed an approximate solution to this problem that was both physically reasonable and mathematically simple. Today his basic approach is widely accepted. Fig. 2 illustrates the concept explaining the existence of magnetic moments localized on certain transition metal impurities dissolved in simple metals (e.g., Fe in Cu). The vertical axis corresponds to increasing energy. In this example, the localized state is an impurity orbital with energy E_L. The two electrons that occupy this orbital are correlated in the sense that if the spin-up state is occupied, then it will cost an energy E_C to add a spin-down electron, due to coulombic repulsion. The stick diagram represents the effect for an isolated atom. When dissolved in Cu, though, the atomic orbital interacts with the conduction band states causing the discrete impurity levels to broaden out -- the localized states take on some band character. In this example the spin-up state is almost fully occupied while the spin-down state is nearly empty. The excess of spin-up over spin-down electron density due to the coulombic polarization gives a magnetic moment, μ, localized on the impurity atom. The existence of a local moment requires that the correlation energy be somewhat greater than the impurity-conduction band coupling energy.

Fig. 3 illustrates the analogy of JT localization to the preceding example. The correlation energy here is due to the molecular JT interaction, driven by the tendency of the four-atom molecule to maximize covalent bonding. Notice that the free electrons in Fig. 3 belong mostly to the narrow d-band characteristic of transition metals. The observable effect of JT orbital polarization is a static distortion localized on the four-atom molecular cluster as in Fig. 1. We must now demonstrate that the JT correlation energy is big enough to overcome the coupling energy.

SIMPLE JT MODEL

Rather than try to attack the problem of hydrogen in the Vb metals directly, I have constructed a simple model that is analogous to the real problem. The advantages of this are numerical and physical simplicity, at the risk of excluding an important feature. Fig. 4 shows the model: hydrogen is on a (100) surface of a simple cubic lattice. The lattice is described by an s-band tight-binding (TB) Hamiltonian, which reproduces the gross features of transition metal d-bands. The dashed circles represent the atomic orbitals localized on the cluster atoms. An important feature of the model is its isomorphism to the real problem. Symmetrized molecular orbitals constructed by LCAO-MO theory are as follows:

$$|A_1> = \tfrac{1}{2}(|1> + |2> + |3> + |4>), \quad \varepsilon(A_1) = -2T,$$

$$|B_2> = \tfrac{1}{2}(|1> - |2> + |3> - |4>), \quad \varepsilon(B_2) = 2T,$$

and $\quad |E_{\pm}> = \tfrac{1}{2}(|1> \overset{+}{\underset{-}{}} |2> - |3> \overset{+}{\underset{-}{}} |4>), \quad \varepsilon(E_{\pm}) = 0,$

where $|A_1>$ is a fully symmetric bonding state, $|E_+>$ is a double degenerate nonbonding state, and $|B_2>$ is an anti-bonding state. The ε are the corresponding orbital energies in terms of the TB parameter, T, which is essentially a measure of the overlap integral between neighboring metal orbitals. Hydrogen interacts only with the fully symmetric state, and causes a depletion of electron density on the metal cluster.[2] Effectively, the metal cluster is left with a positive charge -- the system must respond to restore local electroneutrality. It will be shown later that JT localization accomplishes that and, as a result, the JT state is indirectly tied down to the hydrogen site. We turn now to the JT interaction, which affects only the degenerate state, $|E_+>$.

The molecular JT problem is illustrated by Fig. 5. The degenerate electronic state couples to a tetragonal (B_1) vibration

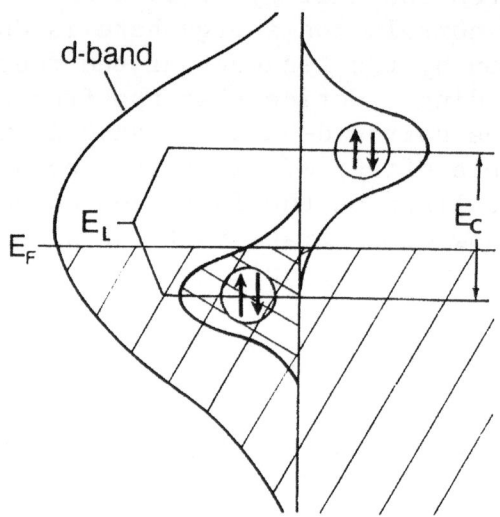

Fig. 3   Jahn-Teller orbital polarization in Vb metal hydrides.

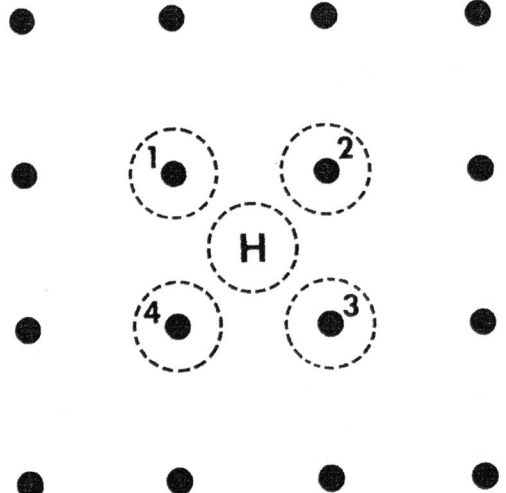

Fig. 4   Hydrogen on (100) surface of a simple cubic lattice.

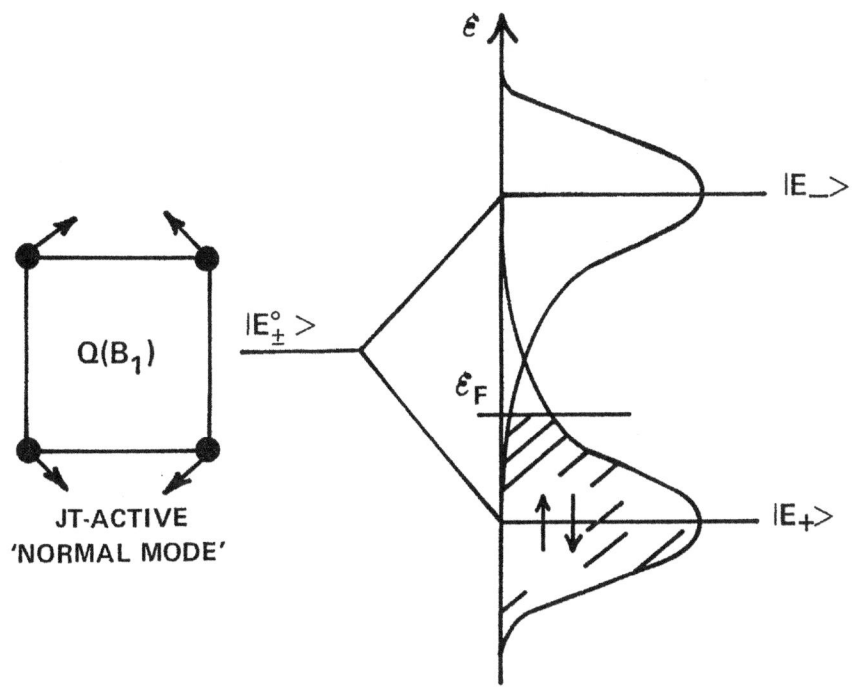

Fig. 5 The molecular Jahn-Teller electron-phonon interaction.

corresponding to the generalized displacement coordinate, Q. The
JT interaction is linear in Q and drives the distortion. There is
also an elastic restoring energy quadratic in Q that resists distortion. The molecule will be stabilized by a finite static distortion. The energy of stabilization (the correlation energy, E_C)
is a function of the linear coupling term, of the elastic force constant, and of the occupancy of the two JT energy levels.[2] In the
example of Fig. 5, the lower JT orbital is fully occupied and the
upper one empty. Interaction with band states broadens the JT levels
as depicted. It is appropriate at this point to introduce an important analogy which clearly reveals the nature of the JT correlation in terms of the inherent instability of symmetrical four-atom
rings with respect to formation of diatomic valence bonds. The
molecular JT problem ($H_{LF} = 0$) described above is essentially equivalent to the four-center exchange reaction

$$H_2 + D_2 \underset{\leftarrow}{\rightarrow} \begin{matrix} H \cdots D \\ \vdots \quad \vdots \\ H \cdots D \end{matrix} \underset{\leftarrow}{\rightarrow} 2HD,$$

For this reaction, it has been deduced from symmetry considerations
alone[5] that the activation energy must be roughly given by the energy required to break a covalent bond. A similar argument explains

the nonexistence of aromatic C_4 rings.[5] It is therefore reasonable to claim that the symmetrical four-atom ring exists at the expense of a covalent bond, and that E_C may well be on the order of several eV. In contrast to the well-known Coulombic correlation, we have here a particular example of covalency correlation. The latter has been described in very general terms as being responsible for the existence of "negative U" centers,[6] which apparently dominate the behavior of amorphous insulators and semiconductors.[6]

The Hartree-Fock (HF) description of the full JT problem, including interaction with band states, is given below.

$$\varepsilon_{\pm} = \pm (n_{+} - n_{-}) \varepsilon_{JT}, \tag{1}$$

$$\rho_{\pm} = \rho_0 [(1 - \varepsilon_{\pm} \mathrm{Re}G_0)^2 + (\pi \varepsilon_{\pm} \rho_0)^2]^{-1}, \tag{2}$$

$$n_{\pm} = 2 \int_{-\infty}^{\varepsilon_F} \rho_{\pm}(\varepsilon) d\varepsilon, \tag{3}$$

with a critical condition for polarization:

$$4\varepsilon_{JT} > \rho_0^{-1}(\varepsilon_F). \tag{4}$$

Equation (1) gives the effective one-electron energies of the two JT orbitals, $|E_+>$ and $|E_->$: n_+ and n_- are the occupation numbers of the two orbitals; ε_{JT} is the square of the linear coupling term divided by the force constant associated with the normal mode Q. Equation (2) gives the effect of band-state coupling on the localized states: ρ_0 is the LDOS (local density of states) of the JT state when the correlation energy vanishes; and G_0 is the Green's function for the JT state in the same limit. The occupation numbers, which have a maximum value of two, are given in Equation (3) by the integral of ρ over all energies up to E_F (Fermi level). The quantities ρ_0 and G_0 are known for the simple cubic lattice so that the HF solution to the above equations is easily obtained. The only unknown quantity is ε_{JT}, which from the above discussion should be several eV. Note that there is a critical condition: if the relation in Equation (4) is not satisfied, there is no orbital polarization. In other words, n_+ and n_- are the same and E_C vanishes. These results are similar to Anderson's treatment of magnetic impurities.[4]

Fig. 6 shows HF solutions for ρ for different values of ε_{JT}. The LDOS curves are shown as a function of energy. ε_F is fixed to correspond to the Vb metals. The dashed curve is the solution for zero correlation, with $\varepsilon_{JT} > 1.2T$ the critical condition for nonzero correlation. The solid curves represent solutions for two values of ε_{JT} in the vicinity of 2eV, which has been established

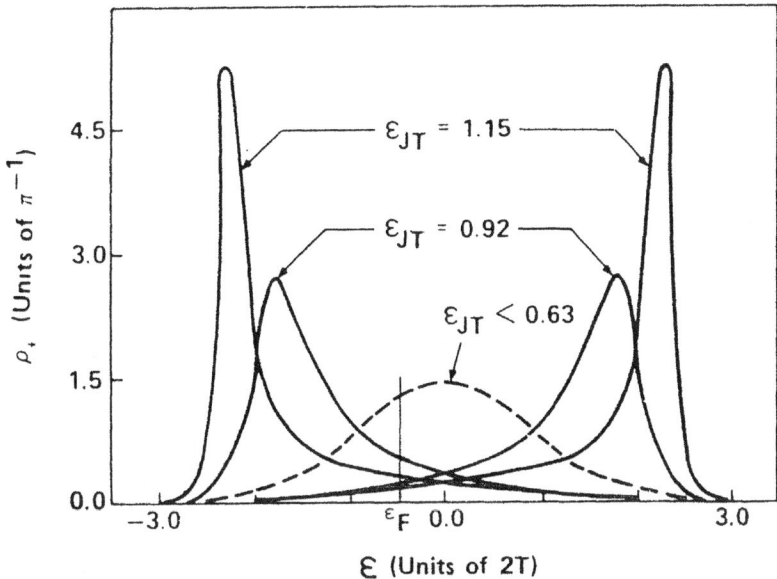

Fig. 6 Jahn-Teller resonance states for several values of ε_{JT}. All energies are in units of 2T.

as the probable magnitude for Nb.[2] This simple model study thus demonstrates the plausibility of a localized JT state. But how is this state connected to the hydrogen site? The hydrogen direct interaction creates an effective + charge on the metal clusters. It is a simple matter to show that JT localization, as represented by the solid curves in Fig. 6, pulls sufficient electron density below E_F to approximately neutralize the + charge. The JT state is thus ionically pinned to the hydrogen site.

GENERALIZED JT MODEL

The indirect nature of this attachment suggests that if JT states exist, they are intrinsic to the host metal. Moreover, as far as the model is concerned, there is nothing special about the Vb metals. Any transition metal can be represented simply by adjusting ε_F. To pursue this line of reasoning, we have generalized the model by discarding hydrogen and by allowing for the possibility of coulombic spin polarization in the upper JT state. The following equations show the generalized HF theory of JT states:

$$\varepsilon_+ = -\varepsilon_{JT}\langle n_+-n_-\rangle,$$

$$\varepsilon_-^\uparrow = \varepsilon_{JT}\langle n_+-n_-\rangle + U\langle n\downarrow\rangle,$$

$$\varepsilon_-^\downarrow = \varepsilon_{JT}\langle n_+-n_-\rangle + U\langle n\uparrow\rangle,$$

$$\rho_\pm^\sigma = \rho_\pm^o \left[(1-\varepsilon_\pm^\sigma \mathrm{Re}G^o)^2 + (\varepsilon_\pm^\sigma \mathrm{Im}G^o)^2\right]^{-1},$$

$$\langle n_\pm^\sigma\rangle = \int_{-\infty}^{\varepsilon_F}\rho_\pm^\sigma d\varepsilon.$$

In these equations, U represents the correlation due to coulombic repulsion between two electrons in the upper JT state. It is this term that gives the possibility of spin polarization in the upper state. There should be a similar term affecting the lower JT state, but it is assumed to be unimportant.[3] There are now three parameters: ε_F, ε_{JT}, and U. The generalized model shows a variety of solutions including both orbital and spin polarization, with the latter occurring only for a more than half-filled band.

Fig. 7 is an exact numerical result showing a spin-polarized upper state. The location of ε_F corresponds approximately to Fe. The dashed curve is the uncorrelated LDOS. The upper half of Fig. 7 shows LDOS curves for spin-down electrons, the lower half for spin-up electrons. Note that the lower JT state is the same for both spins. The orbital polarization is due to the covalency correlation. Spin polarization of the upper JT orbital results directly from the coulombic correlation, but does not exist apart from the covalency correlation. The net result is a magnetic moment localized on a distorted four-atom 'molecule'. Why should this result be interesting?

MAGNETISM IN Fe

The famous Heisenberg description of ferromagnetism has elementary magnets on individual atoms. The atomic moments are due to electron spin $\underline{S}$, and are coupled by an exchange interaction $\underline{J}$:

$$H = -\sum_{kk'} J_{kk'}\tilde{S}_k\cdot\tilde{S}_{k'}. \tag{5}$$

Equation (5) is the Heisenberg Hamiltonian, expressed as a sum of two-body interactions in which J is usually important only for near neighbors. There are two very old questions here which have never been satisfactorily answered: "What is the origin of the local moment?" and "Why is $\underline{J}$ positive?" The simple model of the preceding section shows that for Fe, molecular covalency correlations can result in a spin-polarized upper state and therefore a local moment. The magnetic moment was not localized on individual atoms, but on

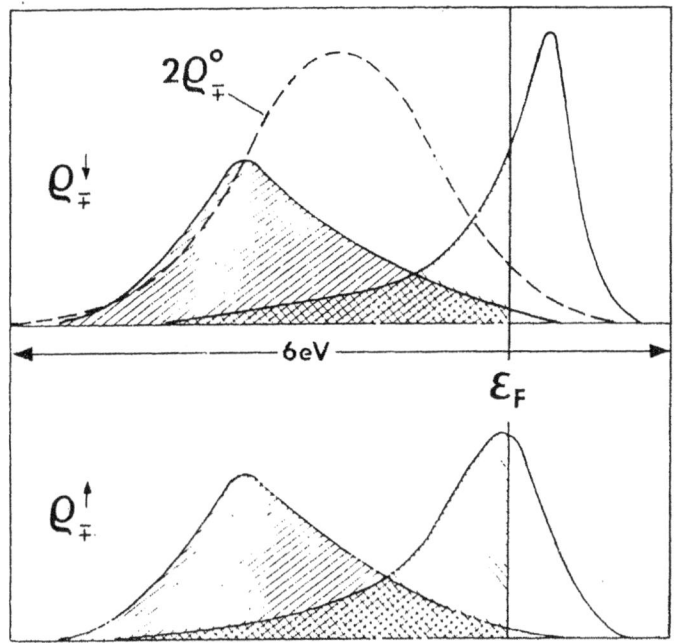

Fig. 7 Jahn-Teller stabilized resonances with a spin-polarized
        upper state. ε_{JT} = 0.5 eV, U = 1.5 eV.

a four-atom molecular cluster. Is it possible that magnetism in Fe
is due to an array of tetra-atomic molecular JT magnets?

Fig. 8 shows such an array in bcc Fe. The individual moments
are symbolized by μ, which also locates the geometric centers of the
tetragonally distorted JT quasimolecules. The molecular magnetic
state is explicitly displayed for one of the JT-centers. The constituent atomic d-orbitals are shown in perspective. The lobes are
at an oblique angle relative to the figure plane, and are oriented
approximately along nearest neighbor <111> directions. This JT
state is similar to that proposed for the Vb hydrides. The arrows
indicate the tetragonal distortion of the JT centers. The dashed
grid represents the statically distorted lattice and shows tetragonal JT centers alternating with regions of symmetric compression
and dilation along <110> directions.

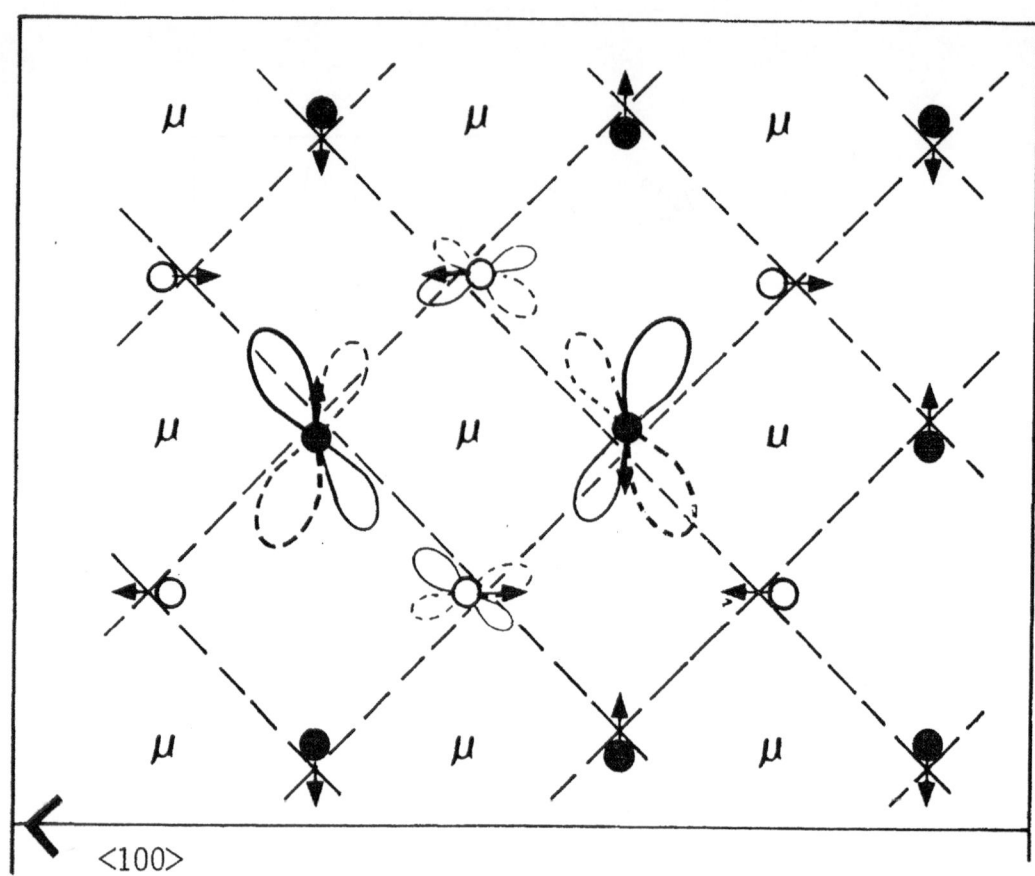

Fig. 8 bcc array of JT magnetic moments. View along <001>; closed
circles = $-a/4$, open circles = $a/4$, $\mu = 0$. (See text for
full description.)

The array of Fig. 8 was constructed to give the maximum concentration of JT centers, subject to the constraint that nearest neighbor
centers be orthogonal. The reason for this orthogonality constraint
is that non-orthogonal nearest neighbor JT centers add electrons to
the upper JT state, which is energetically unfavorable. Note that
nearest neighbor JT centers have a common atom, but they use different d-orbitals on that atom and are indeed orthogonal. This can be
visualized by superimposing their
corresponding d-orbitals onto the
common atom. In the array of Fig.
8 each JT magnet has four nearest
neighbor magnets, and each nearest
neighbor pair has a common atom.
Consider for now a single pair:

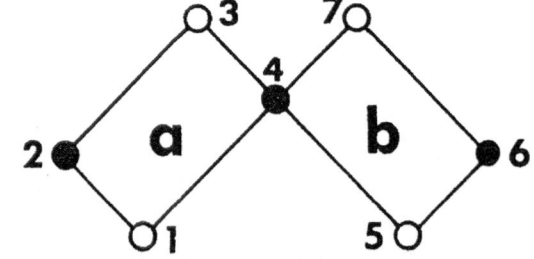

The individual JT centers are labeled a and b; the corresponding magnetic states are shown below as particular linear combinations of atomic d-orbitals:

$$\phi_a = \tfrac{1}{2}(|1\rangle - |2\rangle - |3\rangle + |4\rangle),$$

$$\phi_b = \tfrac{1}{2}(|\overline{4}\rangle + |5\rangle - |6\rangle - |7\rangle). \tag{6}$$

The d-orbitals are represented by numbers 1, 2, 3, and 4 for center a, and $\overline{4}$, 5, 6, and 7 for center b. The fact that the two JT centers use distinct d-orbitals on atom 4 is indicated by using the superior bar. For the sake of simplicity, we assume that ϕ_a and ϕ_b are singly occupied. This is the maximum spin polarization for each center and gives an upper limit for the pair interaction. This assumption seems to require that the various molecular centers be effectively localized, not so much with respect to resonance broadening through a hopping interaction with band states, but with respect to hopping between different JT centers. This latter question can probably be most effectively addressed by developing and studying an appropriate Hubbard Hamiltonian.[7] The exchange interaction between a and b gives a symmetric two-electron state, Φ_S; and an antisymmetric state, Φ_A:

$$\Phi_S(r_1, r_2) = N[\phi_a(r_1)\phi_b(r_2) + \phi_a(r_2)\phi_b(r_1)],$$

$$\Phi_A(r_1, r_2) = N[\phi_a(r_1)\phi_b(r_2) - \phi_a(r_2)\phi_b(r_1)],$$

where r_1 and r_2 are the electron coordinates and N is a normalization constant. The electron spins are parallel in the antisymmetric state, corresponding to ferromagnetic coupling. The exchange energy, J, is the energy difference between Φ_A and Φ_S, given by the two-electron exchange integral:

$$J = \langle\phi_a(r_1)\phi_b(r_2)\,|\frac{2e^2}{r_1 - r_2}|\,\phi_a(r_2)\phi_b(r_1)\rangle.$$

To calculate J, we replace ϕ_a and ϕ_b by their atomic orbital expansions (Equation 6). We will assume that of the resulting 256 terms, only that term with both electrons on the shared atom (the on-site term) is important, with the result:

$$J \cong \langle 4\overline{4}|\frac{2e^2}{r_1 - r_2}|4\overline{4}\rangle/16 = J_{at}/16. \tag{7}$$

The quantity in brackets is the well-known atomic exchange constant J_{at}. The factor of 1/16 is $(1/2)^4$ from the expansion of ϕ_a and ϕ_b. The justification for ignoring the remaining 255 terms is: (i) the on-site term is by far the largest; (ii) the next largest terms tend to cancel because of the symmetry; and (iii) in any event, the remaining interatomic terms should be effectively screened out by the conduction electrons. Thus, Equation (7) should be a good approximation. The quantity J_{at} is the Hund's rule intra-atomic exchange coupling, by which isolated transition metal atoms prefer states with parallel spins. Thus, the present model predicts ferromagnetism -- that is, J > 0.

The Curie temperature, T_c, is the temperature at which thermal energy is about equal to the energy required to go from a parallel arrangement of molecular spins, corresponding to the magnetized state, to an arrangement in which a given JT magnet is antiparallel to its four nearest neighbors. The energy cost is thus four times the pairwise exchange energy of Equation (7), but we must take only half of this so as not to count the pair energy twice. The result is $kT_c \simeq J_{at}/8$. For the Fe group, J_{at} is about 1eV giving $T_c \lesssim 1500°C$. This result is an upper limit because it was calculated assuming maximum spin polarization of the upper JT state. The agreement with the experimental T_c's is excellent, given the longstanding inability to understand even the magnitude of this quantity.

Finally, the present model gives a very simple explanation for magnetic anisotropy. The localized magnets tend to align parallel to an external magnetic field. The total energy depends on the crystallographic orientation of the field. For Fe, the <100> orientation is lowest in energy and so the <100> axis is called the easy axis of magnetization. This effect is thought to be due to spin-orbit coupling. The spin, $\underline{S}$, which is responsible for the localized magnets, couples to the orbital angular momentum, $\underline{L}$, which in turn couples to the lattice. The proposed JT states have a non-zero component of $\underline{L}$ only along the molecular symmetry axis, η_{JT}, and $\underline{LS}$ coupling gives lowest energy when $\underline{S}$ is parallel to η. Table 1 compares the observed easy axes for Fe, Co, and Ni with the symmetry axes for the corresponding tetra-atomic clusters. A rough calculation for Fe gives order of magnitude agreement with the observed anisotropy energy.[3]

The preceding discussion of magnetism in the Fe group has tried to show: that spin polarization of the upper JT state can explain

Table 1. Comparison of Observed Easy Axis with
Molecular Symmetry Axis

Element	Easy Axis	η_{JT}
Fe	<100>	<100>
Co	hex.	hex.
Ni	<111>	<111>

the existence of localized moments in Fe; that ferromagnetism is due to Hund's rule intra-atomic exchange coupling between tetra-atomic JT magnetic states having a common atom; and, finally, that magnetic anisotropy is determined by the symmetry axis of the JT molecular clusters.

SUMMARY AND CONCLUSION

The study of hydrogen in the Vb metals suggests the existence of a tetra-atomic JT resonance state localized near the hydrogen site. Solution of a simple model Hamiltonian demonstrates the plausibility of this idea and reveals that the JT state is a "negative U" center[6] created by covalency correlations. Of greater significance, the model study suggests that the JT state might be a general feature of transition metals, essentially independent of hydrogen. Generalization of the simple model reveals the possibility of a spin-polarized JT state for Fe. This leads directly to a microscopic description of ferromagnetism that has a number of attractive features, including simplicity and physical transparency. If this description of magnetism is essentially correct, then there is reason to believe that the concept of four-atom JT "negative U" states will shed light on other poorly understood properties of transition metals. Superconductivity seems an obvious candidate, but there are many other possibilities as well.

REFERENCES

1. G. C. Abell, Phys. Rev. B20:4773 (1979).
2. G. C. Abell, Phys. Rev. B22:2014 (1980).
3. G. C. Abell, Phys. Rev. Lett. 44:1264 (1980).
4. P. W. Anderson, Phys. Rev. 124:41 (1961).
5. See W. A. Goddard III, J. Amer. Chem. Soc. 94, 793 (1972), and references therein.
6. P. W. Anderson, in: "Ill-Condensed Matter," R. Balian, R. Maynard, and G. Toulouse, eds, North Holland, Amsterdam (1979).
7. E. N. Economou, private communication.

MÖSSBAUER STUDIES OF TERNARY HYDRIDES*

D. Niarchos, P. J. Viccaro, G. K. Shenoy and B. D. Dunlap

Argonne National Laboratory
Argonne, IL   60439

ABSTRACT

The formation of ternary hydrides is often accompanied by significant changes in the structural, electronic and magnetic properties of the binary host intermetallic. Since Mössbauer spectroscopy is in essence a technique which probes the local structural, electronic and magnetic properties of the employed isotope, we used this technique to study the changes which occur in the ternary hydrides. Data from ^{57}Fe, ^{161}Dy, ^{166}Er will be discussed and compared with results from other techniques.

MÖSSBAUER EFFECT

The Mössbauer effect (ME) is a coupled nuclear and solid state phenomenon in which a nucleus bound in a solid emits or absorbs a gamma-ray without local recoil and, consequently, without energy loss. This means that the uncertainty in the energy of the gamma-ray is the same as that in the energy of the two levels involved as determined by the uncertainty principle. As a result, the gamma-ray resulting from the recoil-free de-excitation of a given nuclei can resonantly excite the same in an absorber. If recoil were present the energy of the emitted gamma-ray would sufficiently alter that resonant absorption could not take place.

---

*Supported by the U.S. Department of Energy.

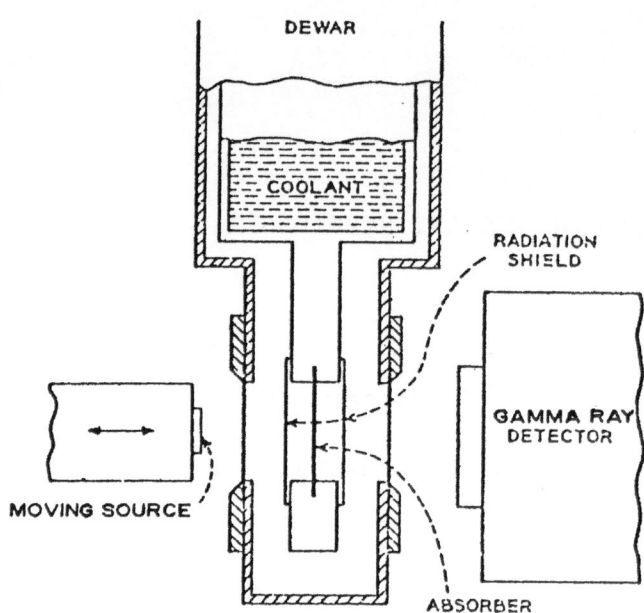

Fig. 1. Schematic diagram for transmission Mössbauer spectroscopy.

The phenomenon of resonant absorption (or emission) is the basis for ME spectroscopy.[1] The usual experimental arrangement is one in which a source of gamma-rays, resulting from the de-excitation of the excited nuclear state of a particular isotope (^{57}Fe, ^{119}Sn, ^{161}Dy, etc.), is passed through an absorber containing the same isotope. The energy of the gamma-rays is Doppler shifted to scan a certain energy interval and the number of gamma-rays transmitted at a given energy is counted (Fig. 1).

In this way, the hyperfine structure of the isotope in the absorber can be investigated. What makes ME spectroscopy so applicable to study materials is the fact that the resonance spectrum is dependent on the electronic, crystallographic and magnetic environment surrounding the specific nuclei. It is also a microscopic probe of these physical parameters sensitive to the local surroundings.

For a given isotope, the two nuclear levels involved in Mössbauer spectroscopy have, in general, several intrinsic properties which enable them to interact with electric and magnetic fields characterizing the environment. The first of these is a nuclear magnetic dipole moment which can interact with external magnetic fields, or with hyperfine magnetic fields produced by the orbital or spin angular momentum of the atomic electrons surrounding the nucleus. In the simplest case, this interaction results in Zeeman splitting of the nuclear levels. Furthermore, a particular level can possess a nuclear electric quadrupole moment which can interact with the electric field gradient, arising from the charge distribution of the surrounding atoms in the lattice and valence electrons of the isotope. This interaction again shifts the nuclear sub-levels. Finally, for many isotopes, the nuclear monopole charge distribution is not equal in the ground and excited states. The Coulomb interaction between these nuclear charge distributions and the electron density at the nucleus results in a change in the energy of the Mössbauer transition. This is proportional to the difference of the magnitudes of the electronic charge density at the nucleus in the source and the absorber lattice and is known as isomer shift (2). Because of its dependence on charge density this shift is a measure of the configuration of the valence electrons and their bonding to the surrounding atoms.

Analysis of ME spectra usually entails determining the magnitudes of the various interactions and extracting information such as local magnetic moments, electronic configurations, crystalline electric fields and electronic densities at the probe site. All of these can be related, in principle, to bulk properties. For the ternary hydrides discussed here, we will show how the Mössbauer technique can be used to obtain information concerning magnetic properties of hydride phases and local structural information, such as possible site locations of hydrogen and their crystallographic configurations. Where isomer shifts are measurable, an estimate of the change in the electronic densities due to the presence of hydrogen will be discussed.

STRUCTURAL CONSIDERATIONS OF TERNARY HYDRIDES

The formation of a particular ternary hydride phase results in a new compound which can have structural properties different from those of the parent intermetallic. It is commonly known that the incorporation of hydrogen in the intermetallic matrix leads to an expansion of the lattice.[3] Often, it also results in a change of structure type.[4] In the first case one considers the hydride structure as one with hydrogen occupying certain interstitial holes in the parent structure. In the second case, a new lattice type made up of metal atoms is considered with the hydrogen again occupying empty interstitial sites. The properties of the ternary hydride in this case are determined both by the new crystal structure and the presence of hydrogen in that structure. In both cases, ME spectroscopy can provide useful information concerning the site occupancies of the hydrogen.

One illustrative example in which ME spectroscopy signals a change in the structure type is found in the FeTi-H system. The parent intermetallic, FeTi, has the body centered cubic (CsCl) structure. In this case, no quadrupole interaction is present at

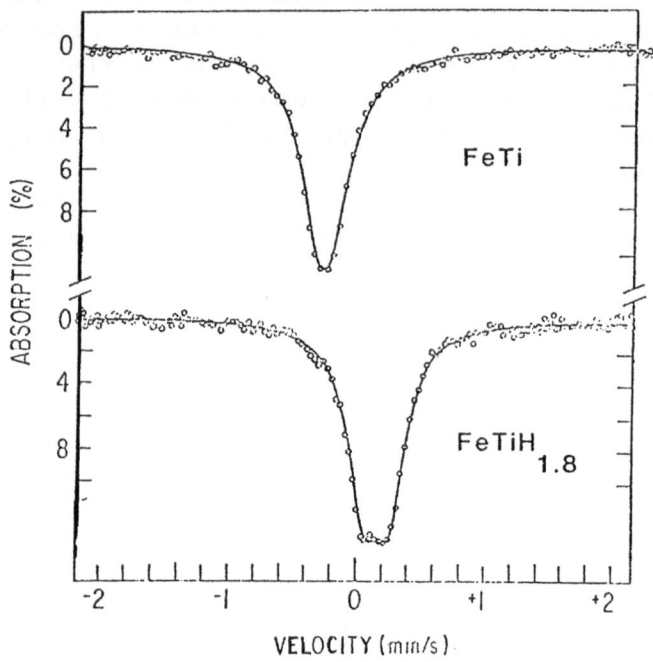

Fig. 2. ^{57}Fe Mössbauer spectra for the FeTiH$_x$ system at 4.2 K.

the Fe site. In addition, no magnetic interaction is present because FeTi is a Pauli paramagnet. As a result the absorption spectrum for ^{57}Fe consists of a single resonance line. The ternary hydride FeTi-H$_{1.81}$ on the other hand has an orthorhombic structure resulting in a lowering of symmetry of Fe atoms, and a consequent quadrupole interaction splits the resonance line (Fig. 2). In the spectra a different shift of the relative centroids is observed which arises from the different isomer shifts present. The relative change in the isomer shift implies a smaller electronic density at the Fe site in the hydride compared to FeTi. This change is related to a change in the electron density[5] associated with a localization of conduction electrons and a rearrangement of electronic charge.   This aspect will be discussed below in conjunction with results from other ternary hydrides.

An example where the hydride phase retains the structure of the parent intermetallic is found in the Laves phase compounds of the type RM$_2$, where R is a rare earth or 4d element and M is a transition metal atom. Most of the Laves phase intermetallics crystallize either in the C15 cubic (Fd3m) or C14 hexagonal

(Pb_3/mmc) structure. The two structures are related and have
similar local environments for the metal atoms.

For the Cl5 series RFe_2, extensive pressure composition isotherms for hydrogen desorption have shown a complex phase diagram
having at least three hydride phases up to the maximum hydrogen concentration of $RFe_2H_{4.5}$.[6] For $RFe_2H_{3.5}$ and lower order hydride
phases, the structure is an expanded version of the cubic Laves
structure. Neutron diffraction results on YFe_2D_2 indicate that
the deuterium occupies a tetrahedral site coordinated by two R
atoms and two Fe atoms.[7] In the parent intermetallic, the R and
Fe atoms each occupy one crystallographic site in the structure.
The ^{57}Fe Mössbauer data show one Fe site in both $DyFe_2$ and $DyFe_2H_{3.5}$.
In the dihydride phase, however, two Fe sites can be discerned even
though the metal atom structure remains the same.[8] The difference
in the two Fe sites is a direct microscopic indication that at least
two distinct hydrogen environments exist around Fe in this hydride.
For the higher order hydride $RFe_2H_{3.5}$ the two Fe sites look identical. Information of this type is presently being used in conjunction with neutron diffraction results in order to determine the
details of the distribution of hydrogen over the available sites
in this material. Preliminary results indicate that the same behavior is found for hydrides of all of the RFe_2 series.[9]

Another case which illustrates the potential of the Mössbauer
effect for distinguishing local hydrogen environments is found with
the Cl4 hexagonal intermetallic $ErMn_2$ and its hydrides. The Er
atom occupies one type of crystallographic site in this structure.
The ^{166}Er Mössbauer spectrum[10] at 4.2 K shows a single five-line
pattern which is characteristic of the presence of nuclear Zeeman
splitting of the 2^+ level (associated with the presence of a magnetic interaction at this temperature) and a quadrupole interaction.
For the hydride $ErMn_2H_{4.0}$, which again has the Cl4 structure, the
^{166}Er Mössbauer spectrum is consistent with two Er sites. For
$ErMn_2H_{4.5}$, only one Er site is observed. The appearance of two Er
sites in the $ErMnH_4$ phase is again microscopic evidence for two
different hydrogen environments. In the Cl4 structure it has been
shown specifically for $ZrMn_2D_3$[11] that the deuterium occupies a
tetrahedral site coordinated by two Zr and two Mn atoms. This is
the equivalent site to that occupied by D (or H) in the hydrides
of the cubic RFe_2 intermetallics discussed above. However, for
the hexagonal Cl4 structure, these tetrahedral sites for hydrogen
are not all equivalent. From the neutron diffraction results for
$ZrMn_2D_3$, it has been shown that different sets of inequivalent
tetrahedral sites have different occupations. Applying these same
ideas to $ErMn_2H_{4.0}$, it was shown[10] that the presence of two Er sites
could be consistent with the occupation either of two tetrahedral
sites. Although distinct, these two types of sites are energetically equivalent from the point of view of available volume and

sites consideration. This distinction is removed for the higher
order phase $ErMn_2H_{4.5}$ where these sites are completely filled.
While this description of hydrogen occupation in the $ErMn_2$ hydrides
may not be unique, it does provide a model to guide future neutron
diffraction experiments in this compound.

MAGNETIC AND ELECTRONIC PROPERTIES

The incorporation of hydrogen into an intermetallic matrix can
significantly alter the magnetic properties of the host. One of the
most effective ways in which to probe these effects is with Mössbauer
spectroscopy. The technique for many isotopes (e.g., those of the
rare earth series) gives a direct measure of the local electronic
magnetic moment of the metal atom in question.[12] It is also capa-
ble to some degree of sensing changes in the magnetic exchange interaction which couples local moments to form the spontaneous bulk
magnetization.

In the case of hydrides, one might expect that the magnetic
properties of a particular intermetallic host would be modified by
hydrogen as a result of several factors. The first of these is a
direct effect of the large volume expansion accompanying hydride
formation. A second effect is a direct perturbation of local
moments through changes in the crystalline electric field. Finally,
a change can occur in the magnetic coupling associated with the
modification of the electronic density in the conduction band.[5]
While Mössbauer spectroscopy alone cannot sort out these effects,
it can often provide unique microscopic information concerning the
nature of the modification that occurs.

One particularly fruitful system where this has been the case,
is the RFe_2 Laves structure intemetallics. Both [57]Fe and several
rare earth isotopes ([161]Dy, [166]Er and [169]Tm, etc.) can serve as
probes for Mössbauer spectroscopy. In intermetallics of this type,
ordered magnetization occurs through a dominant coupling of the Fe
moments via a ferromagnetic exchange interaction. In cases where
the R element possesses a magnetic moment, an antiferromagnetic
exchange coupling between the Fe and R moment also exists. As a
result, intermetallics of this type are ferrimagnetic. Both the
exchange interactions involve a coupling of the moments through the
conduction electrons.[13]

To demonstrate the effects of hydrogen on the magnetic properties we show in Fig. 3 the [57]Fe Mössbauer spectra at 4.2 K for
$ErFe_2$ and three of the known hydride phases. As is evident, the
overall splitting increases with hydrogen concentration up to
$ErFe_2H_{3.5}$, indicating an increase in the Fe moment. This is con-
current with a slight decrease in the Er moment, as measured by
[166]Er Mössbauer spectroscopy, and a reduction in the Er-Fe exchange
interaction.[14] The increase in the Fe moment is most likely as-

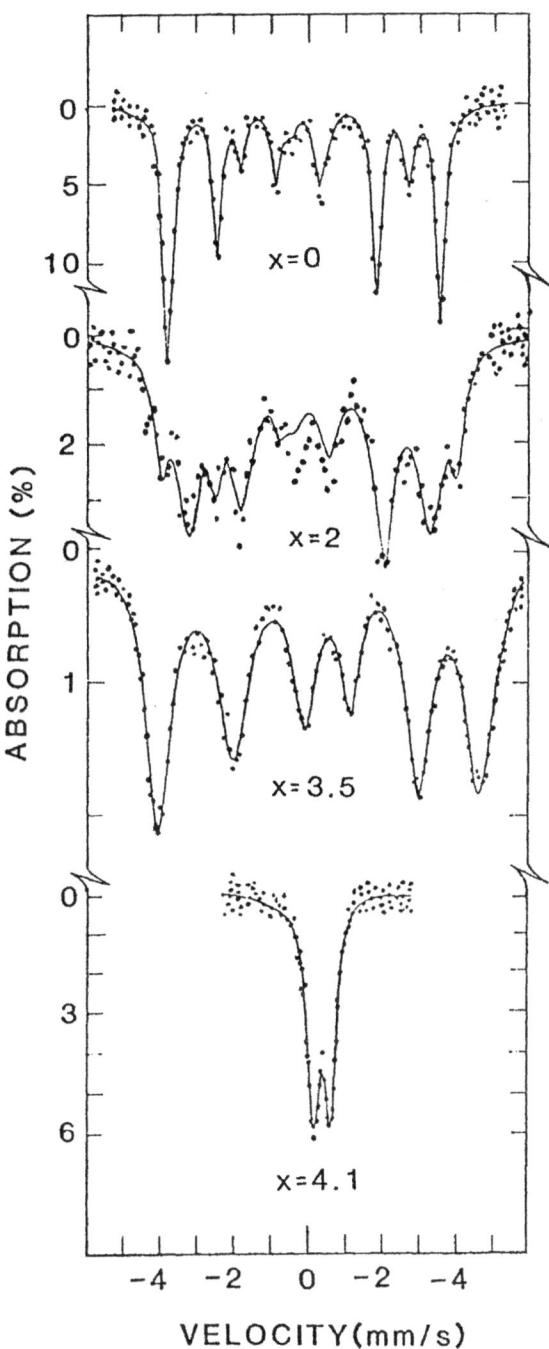

Fig. 3. ^{57}Fe Mössbauer spectra for the $ErFe_2H_x$ system at 4.2 K. The hydrogen concentrations are given (x values).

sociated with a volume effect through a localization of the Fe 3d moment. This result would be expected based on pressure experiments on $ErFe_2$.[15] The decrease in the Er-Fe exchange occurs in spite of the increased Fe moment and indicates that hydrogen has caused a decrease in the conduction electron density. The decrease in the Er moment, on the other hand, appears to be associated with a re-

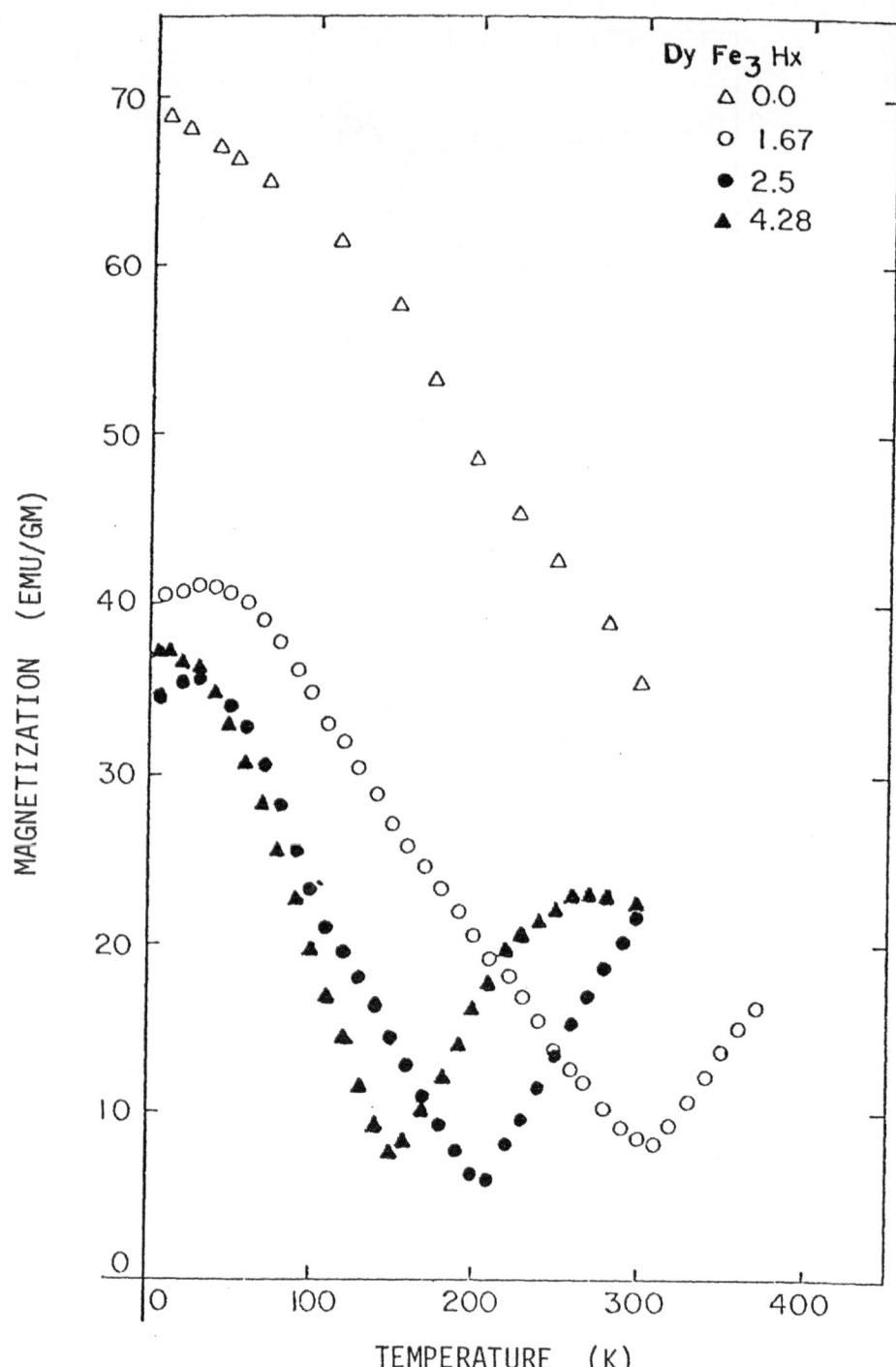

Fig. 4. Variation of the bulk magnetization vs temperature for the $DyFe_3H_x$ system. The minima in the magnetization correspond to the compensation temperatures, T_{comp}, (see text).

duction of the strength of the magnetic exchange relative to the strength of the crystal field interaction. These results appear to be similar for other RFe_2 hydrides studied.[9]

Figure 3 also shows a dramatic decrease in the hyperfine field for the highest order hydride $ErFe_2H_4$ associated with a drastic reduction in the magnetic transition temperature from 608 K (for $ErFe_2$) to less than 2 K. The reduction of the Fe moment is also very large, from 1.5 μ_B for $ErFe_2$ to less than 0.1 μ_B for $ErFe_2H_4$ as deduced from the Mössbauer studies. This implies the absence of an Er-Fe exchange coupling in this phase, leaving only the weak dipole or Er-Er exchange coupling to cause the magnetic order. To a lesser extent, the same happens for the other members of the RFe_2 series at high hydrogen concentrations.[9] Since these effects do not scale with the volume change, we feel that this indicates a severe modification in the conduction electron densities due to the presence of hydrogen. Indeed, from the isomer shift measurements on ^{161}Dy in the $DyFe_2-H_x$ system and also ^{57}Fe measurements across the series, it appears likely that the higher order hydride is more insulator-like than metallic. This conjecture, however, will have to be proven through conductivity measurements.

Effects similar to those observed for the lower order hydride phases of RFe_2 are also seen in some of the RFe_3 phases.[16,17] Specifically for $DyFe_3$ and $ErFe_3$, the hydride phases show weakened R-Fe exchanges. Both these parent intermetallics are ferrimagnetic, with antiferromagnetic coupling between the rare earth and Fe moments. The rare earth moment dominates the magnetization below the compensation temperature (T_{comp}), i.e. the temperature at which the rare earth and iron sub-lattice magnetizations are equal. The ternary hydride phases for $DyFe_3H_x$ for x = 1.67, 2.5 and 4.28 show a monotonic decrease in T_{comp} with increasing x (Fig. 4). In a molecular field approximation[18], T_{comp} is proportional to the Dy-Fe exchange parameter and to a factor depending on the local Dy magnetic moment. The ^{161}Dy and ^{57}Fe results show that the Dy moment decreases from 10 μ_B for $DyFe_3$ to 9.0 μ_B in $DyFe_3H_{4.28}$. In the same concentration range, the Fe moment again appears to increase slightly. These local moment measurements indicate that T_{comp} decreases as a result of the weakening of the Dy-Fe exchange rather than the changes of the local moments. This result also shows that the conduction electron density which couples Dy and Fe moments is perturbed by the presence of hydrogen. This perturbation appears to correlate with the volume expansion (see Fig. 5) but whether this implies an intrinsic volume effect is not clear.

Hydrides of $ErFe_3$ and YFe_3 show results similar to those obtained for $DyFe_3H_x$. At the same time, however, isomer shift measurements for ^{161}Dy and ^{57}Fe in $DyFe_3H_x$ indicate a reduction of conduction electrons, or a charge transfer from the conduction band to the hydrogen. In this sense hydrogen appears to be hydridic, i.e., having a negative charge around the hydrogen atom.

A last example of the effects of hydrogen on magnetic proper-

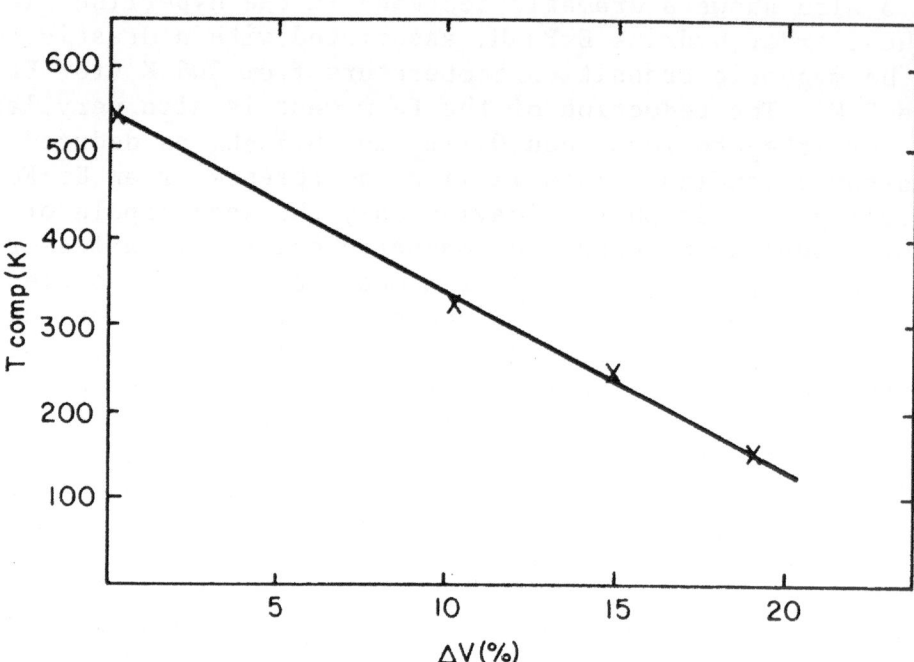

Fig. 5.   Dependence of the spin compensation temperature, T_{comp}, vs volume expansion for the $DyFe_3H_x$ system.

ties comes from the ternary hydrides of the hexagonal C14 Laves structure compound $ScFe_2$.[19,20] Pressure composition isotherms show a phase at $x \simeq 2$ and a supercritical region extending from this value to $x \simeq 3$. An increase in the bulk moment with increasing hydrogen concentrations from 1.15 μ_B for $ScFe_2$ to 2.3 μ_B for $ScFe_2H_{3.2}$ occurs. The [57]Fe Mössbauer data show this by an increase in the [57]Fe hyperfine field. This is a dramatic change which cannot be as easily explained as in the RFe_2 and RFe_3 hydrides, where slight enhancement effects can be correlated with a volume expansion and a localization of the Fe 3d moment. In $ScFe_2$, this increase may be associated with charge transfer to the conduction band as a result of the presence of hydrogen.[19]

ISOMER SHIFT MEASUREMENT

As discussed above, the isomer shift gives a measure of the electronic density at the nucleus of the Mössbauer probe. Excluding relativistic effects, only s-like electrons contribute directly, but shielding of outer s contributions by d or f shells is also important. As a result, the isomer shift contains information concerning non-s-like valence or conduction electrons.

Several rare earth isotopes ([161]Dy, [151]Eu, [155]Gd) and [57]Fe show measurable shifts. Of interest in the ternary hydrides is a comparison of the relative shifts between that of a given inter-

metallic and the hydride phases which that intermetallic forms. In
effect, one monitors the direction of the change in electron density
as a function of hydrogen concentration.

For both the RFe_2 and RFe_3 compounds, the ^{57}Fe isomer shift
becomes more positive with increasing hydrogen concentration. A
positive change in the isomer shift for ^{57}Fe corresponds to a reduction of the total s-like density at the nucleus. This can be
interpreted as either a reduction in 4s conduction band density
or as an increase in the localization of the 3d electrons. Absolute estimates of the actual changes in the electronic density
cannot be made.

The sign of the change is consistent with that expected from
the volume expansion of the lattice through renormalization of the
charge density and localization of the 3d band. However, based on
previous studies of Fe in various metals,[21] the magnitude appears
·to be larger than that expected from volume expansion effects alone.
If this is so, the changes in isomer shifts reflect also a rearrangement of charge densities occurring with the incorporation
of hydrogen in the lattice, perhaps through shielding effects of
the proton charge.

Isomer shift measurements for Dy in the $DyMn_2H_x$ system indicate similar trends in the electron density at the Dy nucleus. The
same ambiguity exists here, as for ^{57}Fe, in determining magnitudes
and relative signs of changes in the Dy 6s and 5d conduction bands.
However, the trend is clear and indicates that a reduction most
probably occurs in selected parts of the conduction band of the
parent as a result of the presence of hydrogen.

More definitive statements concerning the nature of these
perturbations would be possible if systematic data concerning the
effect of pressure on the isomer shift of the parent intermetallics
were known.

REFERENCES

1. G. K. Wertheim, Mössbauer Effect: Principles and Applications,
    Academic Press (1964).
2. G. K. Shenoy, F. E. Wagner, Mössbauer Isomer Shifts, North-
    Holland Publishing Co. (1978).
3. G. K. Shenoy, B. D. Dunlap, P. J. Viccaro and D. Niarchos, Am.
    Chem. Soc. Symp., Houston, Texas, March 24-26, 1980 (to be
    published in Advances in Chemistry).
4. P. Thomson, F. Reidinger, J. J. Reilly, M. L. Corliss and M. J.
    Hastings, J. Phys. F: Metal Phys. 10, L57 (1980).
5. L. J. Swartzendruber, L. H. Bennet and R. E. Watson, J. Phys.
    F 6, L331 (1976).

6. H. A. Kierstead, P. J. Viccaro, G. K. Shenoy and B. D. Dunlap, J. Less-Common Metals 66, 219 (1979).

7. J. J. Rhyne, G. E. Fish, S. G. Sankar and W. G. Wallace, J. Physique 40, C5-209 (1979).

8. P. J. Viccaro, J. M. Friedt, D. Niarchos, B. D. Dunlap, G. K. Shenoy, A. T. Aldred and D. G. Westlake, J. Appl. Phys. 50, 2051 (1979).

9. P. J. Viccaro, D. Niarchos, G. K. Shenoy and B. D. Dunlap, MRS Symposium, Boston, Massachusetts, 16-20 November 1980 (to be published by North-Holland Publishing Co.)

10. P. J. Viccaro, G. K. Shenoy, D. Niarchos and B. D. Dunlap, J. Less-Common Metals 73, 265 (1980).

11. J. J. Didisheim, K. Yvon, D. Shaltiel and P. Fisher, Solid State Commun. 31, 47 (1979).

12. S. Ofer, I. Nowik and S. G. Cohen in Chemical Applications of Mössbauer Spectroscopy (V. I. Goldanskii and R. H. Herber eds.), p. 428, Academic Press, New York (1968).

13. H. R. Kirchmayr and C. A. Poldy in Handbook on the Physics and Chemistry of Rare Earths, Ch. 14 (North-Holland Publ. Co., Amsterdam, 1979).

14. P. J. Viccaro, G. K. Shenoy, B. D. Dunlap, D. G. Westlake and J. F. Miller, J. de Phys. 40, C2-198 (1979).

15. M. Brouha and K. H. J. Buschow, J. Appl. Phys. 44, 1813 (1973).

16. D. Niarchos, P. J. Viccaro, B. D. Dunlap and G. K. Shenoy, J. Appl. Phys. 50, 7690 (1979).

17. D. Niarchos, P. J. Viccaro, B. D. Dunlap, G. K. Shenoy and A. T. Aldred, J. Less-Common Metals 73, 283 (1980).

18. K. H. J. Buschow, phys. stat. sol. (a) 7, 199 (1971).

19. P. H. Smit and K. H. J. Buschow, Phys. Rev. B21, 3839 (1980).

20. D. Niarchos, P. J. Viccaro, G. K. Shenoy and B. D. Dunlap, Inter. Conf. on Hyperfine Interactions, Berlin, Germany, July 21-25, 1980 (to be published in Hyperfine Interactions, 1981).

21. G. M. Kalvius, U. F. Klein and G. Wortmann, J. de Phys. 35, C6-139 (1974).

MAGNETIC PROPERTIES OF COMPOSITIONALLY-MODULATED THIN FOILS AND THEIR HYDRIDES

N. K. Flevaris, D. Baral and J. E. Hilliard

Materials Research Center and Department of
Materials Science and Engineering
Northwestern University
Evanston, Illinois 60201

ABSTRACT

Preliminary measurements have been made on the magnetic properties of compositionally-modulated ferromagnetic thin foils of Pd-Ni and Pd-Co, using ferromagnetic resonance and a torsion balance magnetometer. Qualitative studies have also been made of the effect of the introduction of hydrogen on the structure and magnetization of the foils. The addition of hydrogen changed the amplitude of modulation as well as the state of stress of the foils. The magnetization density and the magnetic anisotropy coefficient were also affected. In foils with and without the hydrogen we observed a peak below 50° K in the FMR linewidth versus temperature curves. When the modulation amplitude was reduced by annealing, a double resonance spectrum was observed. Measurements were also made of the ac susceptibility of a hydrogenated Pd-Ni foil in very low applied magnetic fields.

INTRODUCTION

Compositionally modulated structures were artificially produced for the first time several years ago in the form of thin foils. Several binary compositionally modulated alloys (CMA) have been studied in their structural, diffusion and elastic properties[1]. The ferromagnetic resonance (FMR) measurements on a Cu-Ni CMA recently reported[2], has stimulated a number of both theoretical and experimental studies in this area. In this contribution we report preliminary results on Pd-Ni and Pd-Co modulated foils together with qualitative investigation of the properties of their hydrides.

301

The films were prepared by evaporation of the components from separate crucibles using dual electron beam guns onto mica substrates. A layer of 100nm of Cu was first deposited in order to obtain a good texture along the [111] direction. Prior to the deposition of the modulated alloy a layer of 30nm of Pd was deposited onto the Cu layer in order to achieve a good texture. Then the two components were simultaneously evaporated and the modulation was introduced by alternatively shadowing the substrate from each crucible by means of a horizontally moving shutter. The rate of deposition was controlled to ~ 0.5nm/sec. at the substrate by quartz crystal monitors. The substrate temperature was $500°$ C for the initial Cu layer and $225°$ C for deposition of the modulated foils. The background vacuum was 5×10^{-7} and reduced to 2×10^{-6} torr during evaporation. The approximate parameters for the foils were as follows: total thickness 750nm; modulation wavelength, λ, 1.2nm; and average composition 44 at. % Pd. The texture, amplitude and wavelength of the composition were determined from X-ray diffraction measurements. The intensity of the Bragg peaks for Cu and Pd-Ni(Co) were comparable while their full width at half maximum was less than $1°$, indicating a good texture.

Samples of Pd-Ni foils containing three different amplitudes were obtained by annealing for progressively longer times and higher temperatures ($240°$ -$250°$ C).

Hydrogen was introduced into all of these samples using an electrolytic method at room temperature. X-ray measurements showed that the Pd-lattice constant was expanded by 3.3% to 4.02 A while the effect of the hydrogenization on the amplitude was substantial. We did not determine the exact amount of hydrogen absorbed into the foils and therefore our results are only qualitative.

For the Ni it has been reported that the hydrogen has virtually no effect on the lattice constant and that electrolytic hydrogenization has effects similar to those of local lattice strains. Strain effects of this sort might be the cause of the "reduction" of the modulation amplitude, as determined by X-rays, under hydrogenation.

For Pd-alloys, where the electronic and geometric characteristics of Pd change, it is a question as to whether the hydrogen atoms tend to occupy interstitial sites adjacent to as many Pd atoms as possible. For the Pd-Ni alloys when long-range order is present, the solubility of hydrogen has been found to increase[3]. This, in the case of CMA, might be the driving force for high hydrogen concentrations.

MAGNETISM OF Pd-Ni

Pd-Ni alloys are typical of "giant-moment"-type behavior. The Pd-host, with very highly exchange-enhanced susceptibility, will have a Curie-Weiss type behavior in the presence of local moments of Co (or Fe), and a Pauli-like one when isolated Ni atoms are present[4]. The onset of ferromagnetism occurs at about 2% Ni, where stable giant moments can be produced by three Ni nearest-neighbor atoms[5]. For alloys containing ~ 1.8% Ni although the giant moments are produced, there are nearest-neighbor Ni pairs and isolated Ni atoms and their contribution to the magnetic susceptibility is relatively temperature-independent, an effect that disappears for Ni concentrations higher than 1.9%.

In the samples we used the average Ni concentration was ~ 56 at. % which obviously is free of any of the above critical conditions. Even at an interface, the number of Ni nearest neighbor atoms is large enough for nickel to be ferromagnetic.

In order to obtain a coherent structure of the CMA, coherency strains are introduced (in a modulated form), and, since the magnetic anisotropy energy depends on the degree of strain of the crystal, magnetostriction is very likely to occur. Therefore magnetic anisotropy and magnetostrictive effects[6] are likely to be of importance in a Pd-Ni CMA.

Fischer et al.[7] determined that the orbital magnetic moment of Pd-Ni alloys increases with Pd concentration to a maximum at ~70% Pd. It is also known that both the magnetic anisotropy and magneto-elastic constants have a maximum at the same composition, while the former changes sign at ~ 50% Pd. On the other hand, the spin-orbit parameter of Pd is much larger than that of Ni. Therefore following Van Vleck[8] we may expect information from a study of the magnetic anisotropy of Pd-Ni CMA on the pseudo-dipolar interaction. This is primarily a result of an interplay between orbital valency and the spin-orbit interaction. In our experiments so far we have only examined the effect of a uniaxial anisotropy.

EXPERIMENTAL PROCEDURES

Torsion Balance

A modified torque-magnetometer built by J. Condon[9] was used for the measurements. A torsion balance magnetometer measures the torque about an axis in the plane of the foil and normal to the applied magnetic field when the plane of the foil is twisted an angle θ_o about this axis from the field direction. Calibration was done with an Al disk of known moment of inertia. The sample was placed on a quartz rod hanging between the poles of a rotatable magnet. Essentially the magnetometer is a nulling galvanometer with a smallest detectable torque of 0.005 dyne-cm. At the time these measurements were made the highest field available was 7.5 kOe. Neugebauer's[10] pioneering work was the guide for our interpretation of the torque measurements.

The film was placed at an angle $\theta_0 = 45°$ with the field direction. Its energy is given by

$$E = -M_s H\cos\theta + \tfrac{1}{2} NM_s^2 \sin^2(\theta_0 - \theta) + K \sin^2(\theta_0 - \theta),$$

where θ is the angle between the magnetization vector and the field direction, N (the demagnetization factor) is 4π and K the anisotropy constant which denotes a possibly easy or hard direction of magnetization along the film thickness (K is a measure of the anisotropy for turning the magnetization out of the plane of the foil).

At equilibrium

$$\sin\theta = \frac{-M_s H \pm \left[M_s^2 H^2 + 8\left(2\pi M_s^2 + K\right)^2\right]^{\frac{1}{2}}}{4\left(2\pi M_s^2 + K\right)}$$

the torque on the torsion wire for a film of volume V is $L = M_s VH \sin\theta$.

For very low fields H^2 may be neglected and

$$\frac{L}{H} = -\left[\frac{M_s^2 V}{4\left(2\pi M_s^2 + K\right)}\right] H + \frac{\sqrt{2}}{2} M_s V,$$

and therefore the extrapolation of the (L/H) vs. H curve at 0 field will give $(L/H)_0 = 0.707\, M_s V$.

Typical such curves, as well as those for L vs. θ_0 are shown in Figs. 1 and 2. The curve for Ni shows also how K was calculated. Because of the limitation in field strength we were unable to compare these values with those obtained from the saturation torque.

The (L/H) vs. H curve for a pure Ni foil shown in Fig. 1(a) is very similar to those of Neugebauer's[10], whereas Figs. 1(b and c) for Pd-Ni foils display a departure from the behavior of a simple ferromagnetic. Neugebauer did not give any physical insight for the shape of the (L/H) vs. H curves for very low fields. A ferromagnetic material in this range of fields is by no means saturated and domain wall movement (with possible domain growth) could account for the shape of the curves*. In order to get additional information on the behavior at very low fields we report the experiments described in the next section.

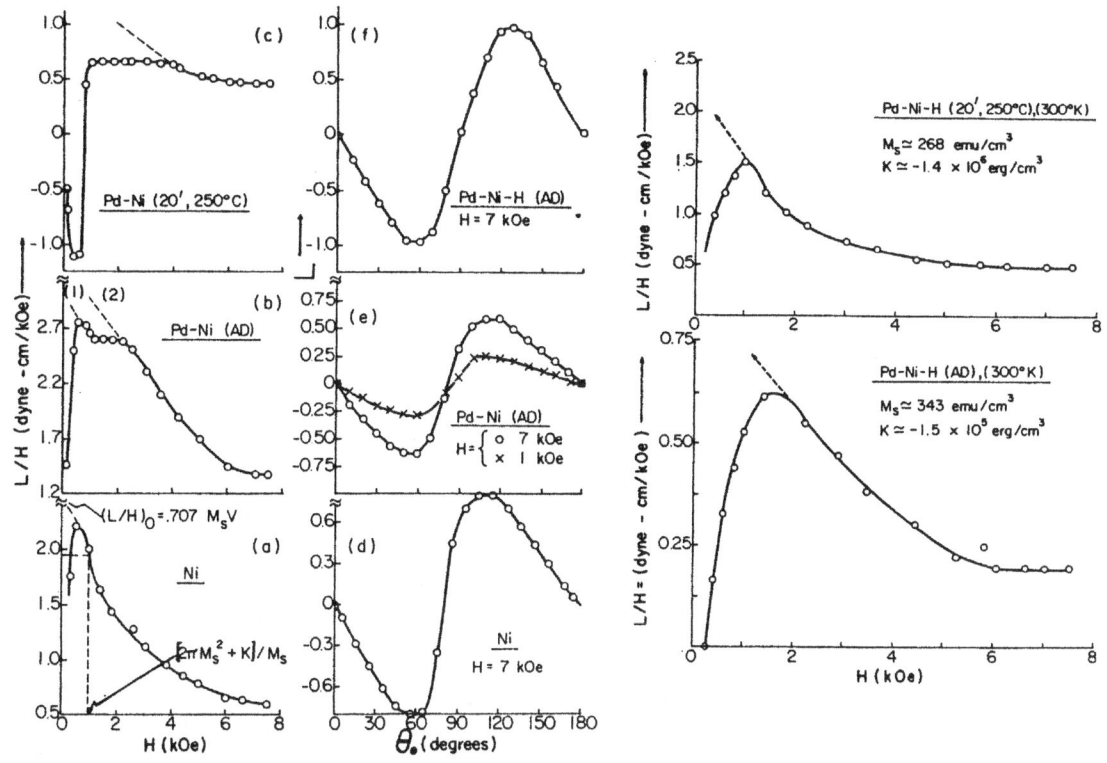

Fig. 1.
Typical (L/H) vs. H curves for: (a) Ni; (b) As-deposited (AD) Pd-Ni CMA annealed for 20 minutes at 250° C; (d), (e) and (f) are θ_0 curves for Ni, Pd-Ni and Pd-Ni-H respectively.

Fig. 2.
The (L/H) vs. H curves for the hydrides of the as deposited and annealed CMA's of Pd-Ni.

ac Susceptibility

In general Pd-M-H (M being a 3-d transition metal) alloys have the behavior of a spin glass (or that of a Kondo system). The hydrides of dilute Pd-Ni alloys are good candidates for a spin-glass-type behavior. Ni, in a Pd environment, will form local spin-fluctuations at very low concentrations as previously mentioned while with noble metals it may form virtual bound states (VBS) and a spin glass transition in Ni-Cu alloys was recently reported[11]. The effects of hydrogenation on the thermopowers of Pd-Ni and Pd-Co alloys was also studied recently[12]

---

*Very recently, we have made computer calculations involving the field dependence of the anisotropy constant K that account for the assymetry of the low-field curves in Figs. 1(d,e and f) as well as for the shape of the (L/H) vs. H curves.

Because of the unique properties of CMA's and the experimental observations shown in Fig. 1, an examination* was made of the ac susceptibility of a Pd–Ni–H CMA at an applied magnetic field of about 5 Oe. The results are shown in Fig. 3.

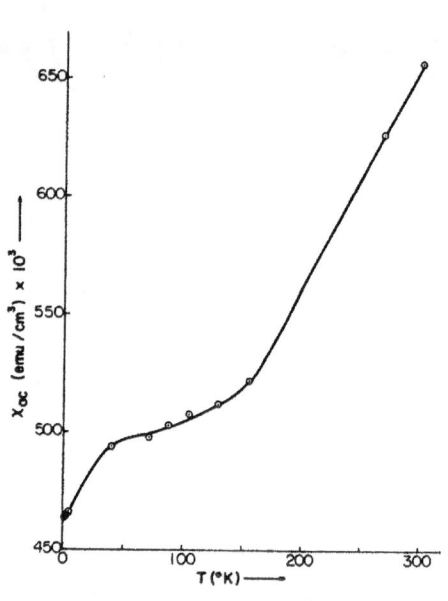

Fig. 3.

The ac susceptibility of a Pd–Ni–H CMA. It will be seen that χ_{ac} increases with T (which is typical of a CMA), but that there is a departure from linear behavior in the region of 100° K.

Ferro-Magnetic Resonance

The FMR technique was used (in the X-band) to determine the magnetization density of our samples. The plane of the foil was held at different angles (θ_0) with respect to the direction of the static magnetic field. By varying θ_0 between 0 and 180° we studied the change in the resonance field H_r. Interpreting the shifts according to MacDonald's theory[13] we found that in H-free samples stresses were present due to the mica substrates and the strains induced by the composition modulation. An even larger degree of stress was observed in the hydrided samples.

For purposes of comparison we performed measurements on Ni foils deposited by the same technique on mica and exhibiting a texture along [111]. We show in Fig. 4 the values of M_s, obtained for $\theta_0 = 0$ and 90°, when Kittel's expressions:

$$\frac{\omega}{\gamma} = H \ (H + 4 \ \pi M_s) \quad \text{for } \theta_0 = 0^\circ, \text{ and}$$

$$\frac{\omega}{\gamma} = H - 4 \ \pi M_s \qquad \text{for } \theta_0 = 90^\circ,$$

were employed. It is obvious that there is a departure from the values[14] of M_s single-crystal nickel. This is due to anisotropies generated by the texture (unidirectional) and the fact that the sample was a thin film (uniaxial).

*We are indebted to W. S. Chou for performing these measurements.

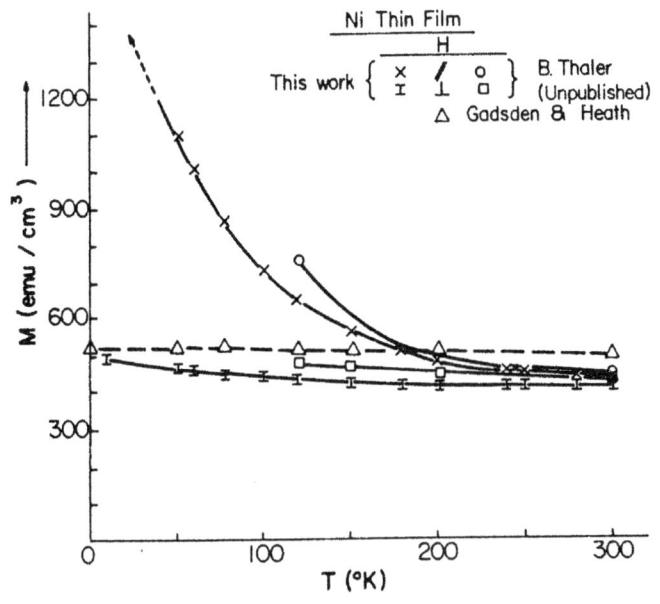

Fig. 4.

The variation of the magnetization density of a Ni thin film as calculated from the Kittel's expressions without anisotropy considerations. Results from Ref. (14) are shown for comparison.

Corresponding plots for Pd-Ni, Pd-Co and their hydrides in the as-deposited foils are shown in Fig. 5. It is possible that the behavior at low T may be accounted for by an anomalous variation of the anisotropy with temperature. This could also account for the behavior of χ_{ac} shown in Fig. 3. We, therefore, examined the change of the peak-peak linewidth (ΔH) with temperature. It will be seen from Fig. 6(a) that there is a peak below 50° K, a behavior similar to that observed[15] for yttrium-iron-garnet ferrimagnetic resonance linewidth. Subsequent investigations[16] failed to establish the source of this effect.

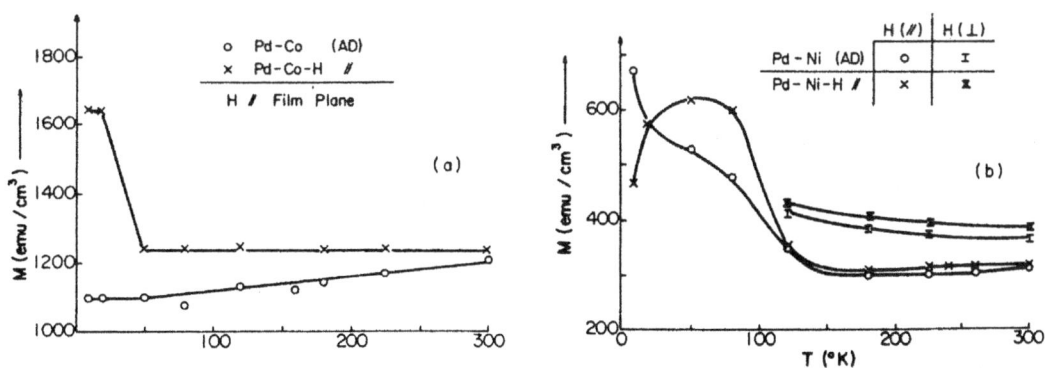

Fig. 5. Curves corresponding to those in Fig. 4 for the as-deposited (AD) Pd-Co and Pd-Ni CMA's and their hydrides.

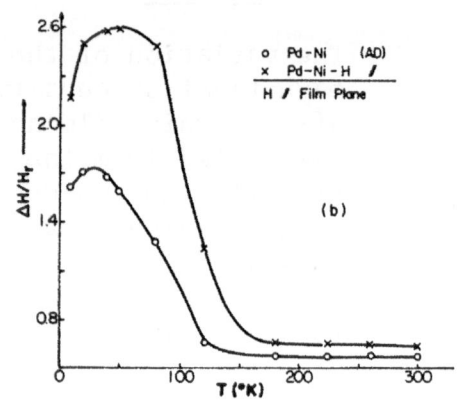
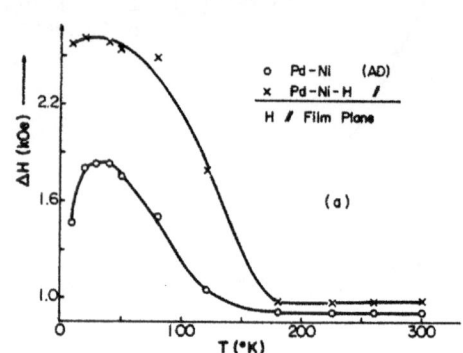

Fig. 6. Temperature dependence of: (a) the linewidth (ΔH) and (b)
($\Delta H/Hr$) for the as-deposited Pd-Ni CMA and its hydride.

We would like to propose a two-fold explanation. First, the
ferromagnetic relaxation frequency has been used[17] to explain ΔH
vs. T curves for Ni and Co that exhibit a roughly similar behavior,
but there is no theory that we are aware of that accounts for the
flat portion of the curves. However, we believe that this may be
attributed to the temperature dependence of the magnetic anisotropy.
The quantity $\Delta H/H_r$ has been[18] correlated with the anisotropy through
the Landau-Lifshitz equation for the ferromagnetic relaxation. From
Fig. 6(b) it will be seen that the effect is more pronounced in the
($\Delta H/H_r$) plot than that for ΔH, which supports this conjecture
which, if true, would require that η in the well known expression:

$$\frac{K(T)}{K(0)} = \left[\frac{M_s(T)}{M_s(0)}\right]^{\eta}$$

be temperature dependent. It is to be noted that η in NiV alloys
changes rapidly with temperature[14].

In our calculations for the magnetization density we have set
g=2.21 and 2.22 for Pd-Ni and Pd-Co respectively on the assumption
that the presence of Pd has no effect on the g values. The anisotropy constants, measured from the torsion-balance experiments,
were inserted into the expression

$$\left(\frac{\omega}{\gamma}\right)^2 = \left[H + (N_x - N_z)M_s + \frac{2K}{M_s}\right]\left[H + (N_y - N_z)M_s + \frac{2K}{M_s}\right],$$

where N_x, N_y and N_z are the demagnetization factors, H the resonance static field applied in the z direction, and K the anisotropy constant.

Finally, in connection to the FMR work, we would like to comment on the differences between the (L/H) vs. H curves for the Ni and the Pd-Ni CMAs. The curves for Pd-Ni exhibited a flat portion whereas those for Ni did not. Furthermore, the extent of the flat portion increased with decreasing amplitude. It was at low amplitudes that the FMR spectrum at $\theta_0 = 90°$ showed a double resonance*, and again this effect increased with decreasing amplitude (Fig. 7).

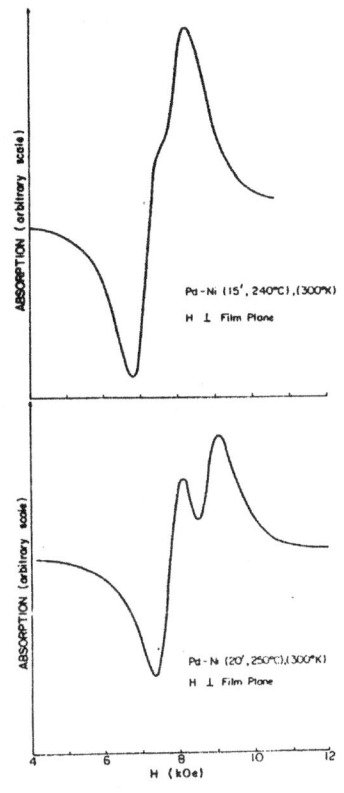

Fig. 7.

The beginning of a double resonance and the evolution with further annealing for the Pd-Ni CMA's.

A similar shape resonance line was observed for the annealed Pd-Co foils. Such a phenomenon has been observed before in thin permalloy films[19-21] and attributed to the coexistence of different magnetization densities[19] or "stratification"[21] of the alloy. It would be reasonable to consider two different magnetization densities and calculate them from the (L/H) vs. H curves from the torsion-balance experiments, and attribute them to differently exchange-enhanced susceptibilities for Pd when the sharpness of the Pd-Ni interface is reduced. But then the question is why the double resonance still exists for the hydrides whereas the flatness of the (L/H) vs. H curves disappears.

---

*The first time we observed double ferromagnetic resonance was for a Co-Ni foil, prepared by H. Itozaki, which in X-rays showed no evidence of modulation whatsoever.

SUMMARY

For the H-free samples, the results indicate that their anisotropy as a function of temperature may be the cause of the effects
shown in Figs. (3-6). It also appears that the composition modulation is the cause of the double resonance that was observed.

We have found a difference between Pd-Ni and Pd-Ni-H in the
orientation of the magnetization vector, magnetic anisotropy and
magnetization density. Taking into consideration the results obtained by Neugebauer[22] for enhancement of the anisotropy constant of
Ni thin films for both hydrogenation and creation of an interface
(with Cu), and his comments on the role of the interface strains on
the anisotropy, we believe that the present results are in agreement with his and, in addition, they have some interesting features
that need further examination. We are in the process of clarifying
the temperature dependence of K by a study of torsion balance curves
at different temperatures.

ACKNOWLEDGMENTS

We would like to thank Mr. J. Anderson for his help during the
FMR experiments and Prof. J. B. Ketterson whose invaluable help made
the torsion balance experiments possible.

This work was supported under the NSF-MRL program through the
Materials Research Center of Northwestern University (grant DMR76-
80847).

REFERENCES

1.   J. E. Hilliard, A.I.P. Conf. Proc. 53, 507 (1979).
     T. Tsakalakos and J. E. Hilliard, (to be published).
     G. E. Henéin and J. E. Hilliard, (to be published).
2.   B. Thaler, J. B. Ketterson and J. E. Hilliard, Phys. Rev.
     Letters 41, 336 (1978).
3.   V. A. Gol'tson et al., Phys. Chem. Mech. Materials 5, 597 (1969).
4.   D. Sain and J. S. Kouvel, Phys. Rev. B 17, 2257 (1978).
     Also a private communication with J. S. Kouvel about recent
     unpublished data.
5.   Considering the problem of interfaces in CMA, this makes Pd-Ni
     quite a bit more interesting than the Cu-Ni system where eight
     nearest neighbors of Ni atoms are necessary for the alloy to be
     ferromagnetic.
6.   T. Tokunaga and H. Fujiwara, J. Phys. Soc. Japan 45, 1232 (1978).
7.   G. Fischer et al., J. Appl. Phys. 39, 545 (1968).
8.   J. H. Van Vleck, Phys. Rev. 52, 1178 (1937).
9.   J. Condon, Ph.D. Thesis, Northwestern University (unpublished),
     1963.
10. C. A. Neugebauer, Structure and Properties of Thin Films,
     p. 358, J. Wiley and Sons, Inc. New York (1959).

11. D. W. Carnegie and H. Clauss, J. Appl. Phys. 50, 1738 (1979).
12. C. L. Foiles, Solid State Commun. 33, 125 (1980).
13. J. R. MacDonald, Phys. Rev. 106, 890 (1957).
14. C. J. Gadsden and M. Heath, J. Phys. F. 7, 1273 (1977).
15. J. F. Dillon, Phys. Rev. 105, 759 (1957).
16. See e.g. J. F. Dillon and J. W. Nielson, Phys. Rev. Letters 3, 30 (1959), E. G. Spencer, R. C. Le Craw and A. M. Glogston, Phys. Rev. Letters 3, 32 (1959).
17. C. J. Gadsden and M. Heath, J. Phys. F. 8, 521 (1978).
18. G. V. Skrotskii and L. V. Kurbatov, Sov. Phys. JETP 35, 148 (1959).
19. P. E. Tannenwald and M. H. Seavey, Phys. Rev. 105, 337 (1957).
20. A. Van Itterbeek et al., J. Phys. Radium 21, 81 (1960).
21. D. Chen and A. H. Morrish, J. Appl. Phys. Suppl. 33, 1146 (1962).
22. C. A. Neugebauer, Z. Angew. Physik 14, 182 (1962).

PHASE TRANSITIONS IN HYDROGENATED NICKEL AND NICKEL ALLOYS

INVESTIGATED BY MAGNETIC METHODS *

Hermann Joh. Bauer

Sektion Physik
University of Munich
Federal Republic of Germany

ABSTRACT

The drastic changes in the spontaneous magnetization of hydride
forming ferromagnetic metals as nickel and nickel alloys combined
with the interstitial absorption of hydrogen can be used for the
analysis of phase transitions in such systems. Further information
may result from the shape of the temperature dependence of spontaneous magnetization including the position of the Curietemperature,
and also from correlated phenomena as the 'Ferromagnetic Resistance
Anomaly'.- A metallographic decoration technique based on a settlement
of magnetic iron particles on the surface of a decomposing hydride
sample allows even a 'visualization' of details during the decomposing process.- The question of stability of the system regarded with
respect to measurements under thermodynamic equilibrium conditions
determines the method of their generation: the electrochemical way
or the application of high pressure gaseous hydrogen.- Results of
the mentioned methods are presented for hydrogenated alloys of nickel
e.g. with Cu, Fe, Mn. Different forms of magnetic behavior are to be
recognized like shifts of the Curie region, an influence of the hydrogen on cluster magnetism or precipitation-like effects.

INTRODUCTION

The possibility of an interstitial hydrogenation of the transition metal nickel by electrochemical means was discovered in 1958 by

---

*Dedicated to Prof. J. Brandmüller on the occasion of his 60th birth-
day.

313

Baranowski and Smialowski[1] using an electrochemical method. Some years later Baranowski et al.[2-4] succeeded in hydrogenating of nickel and of other transition metals in high pressure gaseous hydrogen.

The following presented investigations on electrochemical produced systems have been performed in the University of Munich, the high pressure hydrogen experiments are the outcome of cooperation with B. Baranowski of the Polish Academy of Sciences in Warsaw.

This report deals with mostly published magnetic studies on hydrogenated pure nickel and nickel based alloys, i.e. metal-hydrogen systems whose starting metal-matrix is ferromagnetic, and whose ferromagnetism may be influenced in a distinct manner in proportion to the hydrogen absorbed.[5]

Among the physical properties of nickel which change in consequence to the hydrogenation the saturation magnetization is of particular interest. One of the advantages of this quantity is that especially in case of complete cancelling of the magnetic moment (assumed by d-band filling at atomic ratio $H/Ni \sim 0.7$) the change of the saturation magnetization represents directly the spatial share of the hydrogenated metal matrix. For this reason e.g. measurements of corresponding magnetization changes are an appropriate way for analysis of phase transitions resp. kinetics in such metal-hydrogen systems.[6]

The hydrogen induced changes of the shape of the temperature dependence of the saturation magnetization (resp. the 'spontaneous magnetization') including the Curie region contains further information provided the latter being in the stability region of the metal-hydrogen system. Of special interest is the influence of the hydrogen on correlated – for measurements easy accessible – physical properties as the electrical resistance which mirrors the temperature dependence of the spontaneous magnetization in the 'Ferromagnetic Resistance Anomaly'.[7] Besides this, hydrogenation can further generate new anomalies as e.g. Kondo-like effects in hydrogenated diluted Ni(Fe) alloys.[8]

An important goal must be the exploration of P-T-C-phase diagrams of metal-hydrogen systems which combines H_2-pressure, temperature and the atomic ratio of hydrogen/metal under thermodynamic equilibrium conditions. These conditions can be defined quantitatively in the presented alloys scarcely by electrochemical way but only by application of high pressure hydrogen atmosphere.[4] Basing upon the above mentioned pioneer work of the Polish group concerning the high pressure hydrogenation of nickel and alloys of it, a Soviet high pressure group under E. G. Poniatowskij[9] has successfully extended this research to higher hydrogen pressures and temperatures including the determination of Curietemperatures by a differential transformer method. Nevertheless we are far from a satisfactory knowledge of phase-surfaces, which especially are needed in the field of applica-

tion. At present state we are still testing experimental procedures
e.g. with respect to the stability problem of such metal-hydrogen
systems.

For illustrations look at Fig. 1: a picture of a qualitative
three-dimensional phase-surface of a fictive metal-hydrogen system
with a ferromagnetic starting-matrix. The scheme shows the different
regions of hydrogen concentrations under thermodynamic equilibrium
conditions as the low concentrated α-phase which corresponds (according to the position of the Curietemperature) to the ferromagnetic
state, which is with respect to primary magnetic properties as saturation magnetization and Curietemperature practically not influenced
by the hydrogen. Here secondary magnetic phenomena like hydrogen-
induced aftereffects[10] can take place. The right part of Fig. 1 represents a high concentrated non-ferromagnetic β-hydride-phase. Between these regions of pure α- and β-phase below the 'critical point'
of phase separation coexist α- and β-phase and correspondingly the

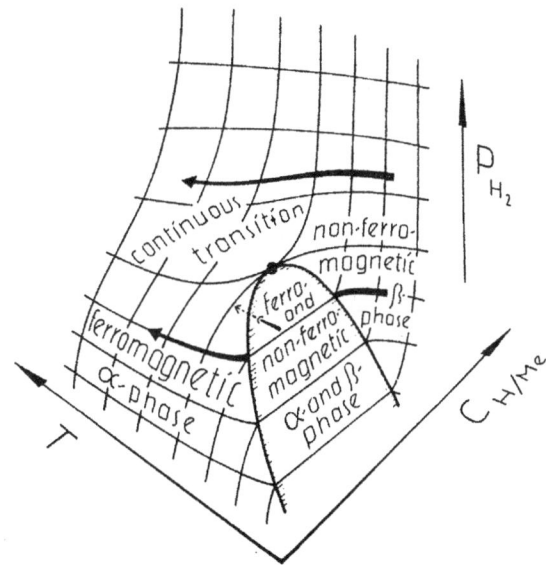

Fig. 1.   Assumed phase surface (qualitative) of a fictive ferromag-
          netic metal-hydrogen system characterized by a ferromag-
          netic α- and a non-ferromagnetic β-phase (corresponding low
          resp. high hydrogen concentrations) and by the critical
          point of phase separation. See text for details.

ferromagnetic and the non-ferromagnetic state. In this area are staged the useful possibilities of magnetic investigations of kinetics of hydride forming and decomposition processes.[6,11] Above the critical point - in the one-phase region (solid solution) - seems to occur a continuous magnetic transition in dependence of the concentration of the hydrogen.[9,12] Especially this region couldn't be experimentally investigated until now because of the difficulties of high pressure - high temperature techniques applied to gaseous hydrogen.

Naturally each alloy-system has its own phase-surface, which varies moreover as a consequence of different processes in the metal-matrix which occur e.g. in the cases of hydrogenation and dehydrogen-ation. The discovery of such phase-surfaces is at present subject of research on nickel based alloys as well as on other substances.

Endeavouring to include the magnetic behavior in investigations of a phase-surface as in Fig. 1 it may be useful simultaneously to have in mind the temperature dependence of the spontaneous magnetiza-tion or the correlated electrical resistance resp. the resulting Curietemperature. The upper arrow in Fig. 1 along an isobaric line of the phase surface shall signify a continuous change of the magnetic state with decreasing hydrogen content in dependence from the corre-sponding electron structure starting here from non-ferromagnetic behav-ior. Following the lowest arrow in thermodynamic equilibrium of the α-region ferromagnetism starts at the coexistance curve of α- and β-phases and decreases while temperature increases up to a Curie-temperature corresponding to the low hydrogen concentration than reached at the existing pressure.

Principally to each hydrogen concentration corresponds (at con-stant pressure) a specific temperature dependence of the spontaneous magnetization. The inclusion of the Curietemperature as a help for analysis of phase transitions resp. phase-surfaces means measurements in a temperature interval requiring a certain time. In case of metal-hydrogen systems this involves the competition between the time for achieving of thermal equilibrium and of equilibrium state of hydrogen in the metal-matrix. A determination of a Curietemperature - necessar-ily accompagnied by changes of temperature - must be performed quick-ly compared to the hydrogen diffusion time in the material in order to avoid changes of hydrogen concentration during the measuring proc-ess: The middle arrow in Fig. 1 shall demonstrate the requirement of a 'snapshot-like' following of the temperature dependence of a physi-cal property as the magnetization under constant pressure and the condition of constant remaining hydrogen concentration. The experi-mental realization needs special measures for apparatuses and sample arrangements (see below).

EXPERIMENTS ON ELECTROCHEMICALLY HYDROGENATED NICKEL AND NICKEL
ALLOYS

In the case of nickel alloys whose Curietemperature T_C lies below the instability temperature of the corresponding alloy-hydrogen
system at normal pressure the saturation magnetization M_S can be used
along the whole M_S-T-curves for analysis of electrochemical hydrogenation. The kinetics of absorption and desorption processes of hydrogen at the regarded systems can be followed quantitatively 'in situ'
in more or less conventional magnetic apparatuses,[13] whereby the temperature region is naturally limited between about 10°C and 40°C.
For recording the complete temperature dependence of M_S of these alloy-hydrogen systems in different states of hydrogen contents the
application of cryostats is necessary.

The hydrogen absorption generally deliver successive lower M_S-T-
curves remaining similar to the starting curve with no change of T_C
corresponding to the reduction of the magnetic effective cross section by a frontal   progression of the non-magnetic hydride in the
sample (that means there is no change of the α-phase in the two-phase
region).[6,14] In contrast to this the hydrogen desorption resp. the
hydride decomposition shows other consequences. Fig. 2 shows the
example of a step by step (respectively at room temperature) decomposing of hydrogenated nickel copper alloys: The arising M_S-T-curves
of a NiCu35at%-H alloy[15] have the characteristic shape of a precipitation-like appearance of α-phase islands in the hydride matrix.[16]
Their growth finally causes the α-phase to assert itself all over
the sample. (Another explanation may be that during the desorption
α-phase islands could preferably be rebuilt in regions with increased
Cu-contents due to sample inhomogenity and correspondingly reduced
hydrogen solubility. In this way Curietemperatures of these regions
may be registered at first.[17]) In detail the decomposition process
and with it the shape of the M_S-T-curves generally depends on temperature during the hydrogen desorption. The influence of hydrogen on
the ferromagnetic coupling leads - as mentioned above - to corresponding changes of resistance anomalies connected with it. Such
resistance measurements simultaneously performed with the magnetic
measurements on the identical sample[15] are shown also in Fig. 3.
For the presentation the temperature coefficient of the electrical
resistance was chosen because it shows more distinct the accordance
of the hydrogen induced consequences on electrical conductivity and
magnetic behavior.

As a further example in Fig. 2 may be chosen the system
NiCu50at%-H whereby the phase transition from the hydride to the
state of complete dehydrogenation can be also regarded as an analytic method with respect to special questions of magnetic structures:
The behavior of electrical resistance of this system - Fig. 2, on the
right - which is connected with a transition from an atomic ratio
H/alloy of n = 0.16 to n = 0, does express the influence of hydrogen

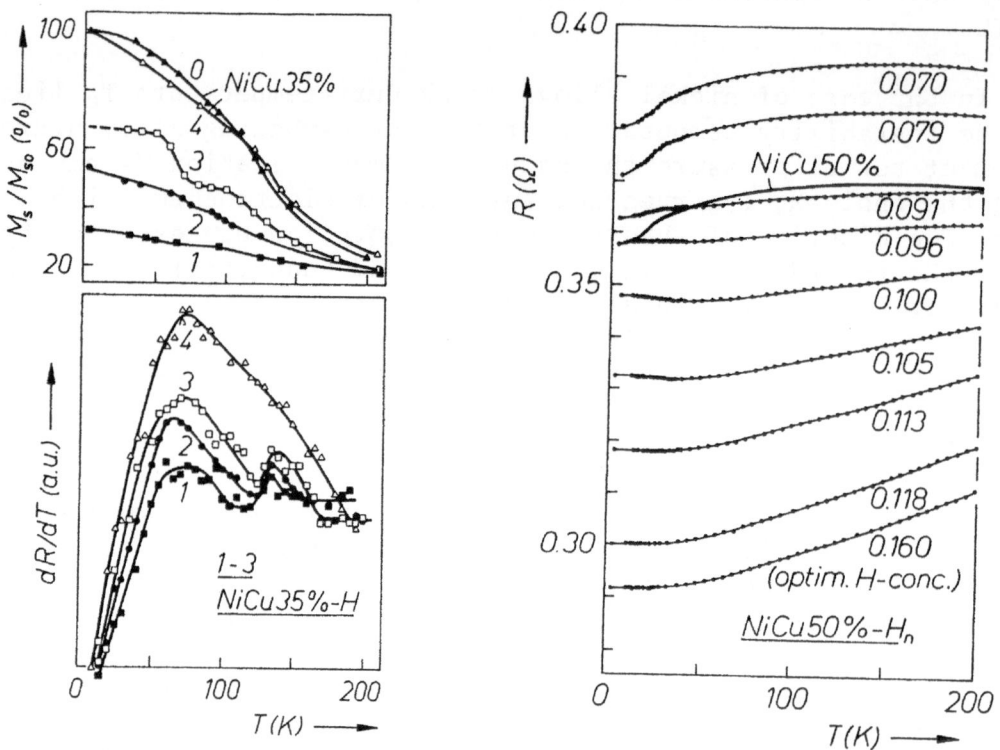

Fig. 2. Change of temperature dependence of saturation magnetization M_s and electrical resistance R resp. temperature coefficient dR/dT at room temperature stepwisely decomposing nickel copper hydrides.[15,17] Left hand: M_{so} saturation magnetization of the hydrogen-free alloy at 4K. Parameter: 0 hydrogen-free start state, 1-4 sequence of desorption states (starting from a hydrogenation of 80% of the cross section of the 12μm thick sample-foil), 4 complete desorption. Right hand: n atomic ratio H/alloy, thickness of the sample foil: 14μm.

on nickel clusters existing in NiCu50. Such 'giant moments' clusters lead with growing temperature (by a decrease of spin disorder scattering) to a decrease in R which in a recognizible manner starts from about 150K. At low temperature behavior of these alloys beginning from about 56at% Cu-content is attributed to a Kondo-like scattering from cluster moments. The hydrogenated alloy NiCu50-H_n shows a monotonic increase of R with growing temperature for n = 0.16 and 0.118 and then a flattening in course of further diminuition of the H-content by desorption. This is caused by the considerable reduction and then the reappearance of cluster magnetism and the spin disorder resistance connected with it. Kondo behavior - e.g. R-minima at low temperatures basing on a logT-term - becomes 'visible' at the NiCu50-H_n alloy for n = 0.113, 0.105 and 0.118, caused by Kondo scattering from Ni-clusters with its arising magnetism. By further lowering the H-concentration occurs a restoration of the intercluster interaction accompanied by the reappearance of Curietemperatures.*

According to the high pressure research of Baranowski et al.[18] the critical point of phase separation is situated at the system NiCu35at%-H above and the NiCu50at%-H below room temperature. This means that in both cases at room temperature stepwisely performed transitions from β- to α-phase take place below resp. above the critical point. Nevertheless a further localization concerning the phase surface is not possible because of the circumstances that the (isothermic) decomposition processes don't occur under thermodynamic equilibrium conditions.

The investigations with respect to the above mentioned different character of hydride-phase forming and decomposition receive a support by metallographic methods applied on the Ni-H-system. Thus cross section pictures of nickel foils, Fig. 3, which have been exposed to hydrogenations for different terms (followed by dehydrogenation)[19] allowed by way of visible crystallite cracks the direct confirmation of the frontal diffusive character of the advancing hydride phase with a diffusion process following the quadratic law[14] concluded from other methods[20,13] (In case of NiCu alloys with higher Cu-contents and correspondingly reduced H-capacity the metallographic effects seem less distinct.[19]) Concerning the decomposition process of hydrides a new application of a magnetic decoration technique[21] using the co-existence of arising magnetic and non-magnetic regions (in the miscibility gap of the phase surface, Fig. 1) allowed directly a detection of (ferromagnetic) α-phase precipitates in a decomposing (non-ferromagnetic) nickel hydride- resp. Ni-H$_β$-matrix. In modification of the well-known Bitter method according to Hutchinson, small ferromagnetic particles produced by evaporating iron in an atmosphere of argon were attracted by the magnetic stray-fields of the sample and settled to its surface. The settlements can be recognized with an optical microscope as blackenings; see Fig. 4. The Figure shows optical micrographs in two only slightly different states of decomposition of the hydride. On the original uniform surface irregularily distributed lighter islands have appeared, Fig. 4a. In order to prove directly the precipitation-like character of the decomposition process, which can be deduced from previous magnetic investigations,[24,25] the magnetic state of the sample was revealed by the above mentioned technique; in Fig. 4b the formation of dark spots of Bitter powder can be recognized being situated only within the light (by further hydrogen desorption a little enlarged) islands. This signifies ferro-

---

*We could find also R-minima (increasing of R during decreasing temperature) below about 80K in case of hydrogenated NiCr alloys and identify its magnetic character by observing changes of the magnetoresistance.[22] In the same way we found a Kondo-like behavior caused by single magnetic Fe moments in diluted Ni(Fe) alloys whose ferromagnetism had been neutralized by hydride formation.[8]

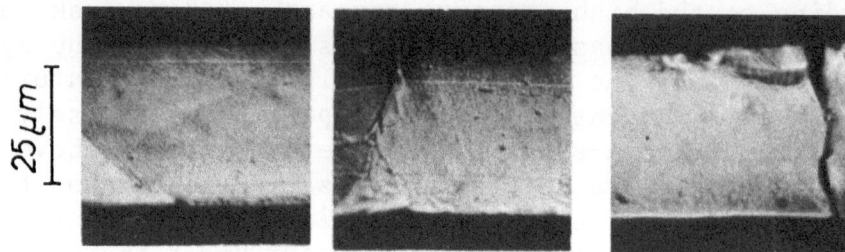

Fig. 3. Optical micrographs of cross sections of Ni-foils after
         electrolytic hydrogenation during 4^h, 16^h, 48^h (at 33.4°C)
         followed by desorption of the hydrogen, showing correspon-
         ding penetration depths of the hydride phase by means of
         frontal crystallite cracks caused by the jump of lattice
         parameters (about 6%[23]) between β- and α-phase.[19]

magnetism, i.e. α-phase in dispersed form, embedded in the homogeneous
decorated non-ferromagnetic β- resp. hydride-phase.- Further details
will not be discussed. This example may be sufficient to show that
these methods of 'visualization' can be used for analysis of local
particularities of phase transitions at metal-hydrogen systems with
ferromagnetic components, especially in the two-phase region of the
phase-surface, Fig. 1.

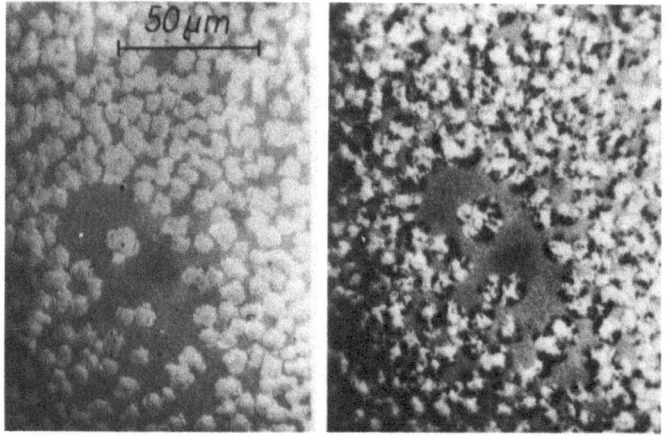

Fig. 4. Detection of α-phase precipitates in decomposing NiH_β by
         Bitter powder pattern. The optical micrographs show the
         (100)-surface in only slightly different α- + β-states of
         desorption of β-NiH: (a) before, and (b) after decoration
         with iron particles.[21]

EXPERIMENTS ON NICKEL AND NICKEL ALLOYS HYDROGENATED IN HIGH
PRESSURE GASEOUS HYDROGEN

A survey of metal-hydrogen systems in high pressure range and of
the corresponding high pressure technique is given in another place
of this volume. The following section reports experiments which have
been performed for some time in cooperation with B. Baranowski in
Warsaw and concerns magnetic investigations.[26,27]

As mentioned above the instability of these nickel-based alloy-
hydrogen systems already near room temperature requires, for further
exploration with respect to cover, to some extent, phase surfaces, the
application of high pressure gaseous hydrogen. This step is - in contrast to electrochemical methods - the only possibility for an extension of the investigations to higher (as well as to lower) temperatures. In this way thermodynamic equilibrium conditions resp. controlled thermodynamic activity can be maintained during an experiment,[4]
and hydride formation as well as decomposition can be followed without additional manipulation. Informative discontinuities in physical
properties attend hydride forming and decomposition. A proved physical property for such studies is the electrical resistance. On the
other hand it is sensitive towards the manifold structural changes
within the material during hydrogenation so that it is desirable to
have a further physical criterion. Especially suitable is the saturation magnetization because of the above discussed advantages.

Besides the necessary small working volume in high pressure apparatuses and its inaccessibility in case of high temperature, arise
new technological problems concerning current leads, sealings or heat
capacity and heat conductivity which as well as the kinetic conditions in the metal-hydrogen system determine the measuring time. As
mentioned, determinations of Curietemperatures require a more or less
extended temperature region and therefore a sufficient stability of
the system studied. This means that for a fast - as much as possible
inertialless - measuring process one has to avoid thermal involvement
of the mass of the high pressure apparatus. Hence miniaturized devices were developed, the last realized type[11,28] see in Fig. 5. An
arrangement of sample, thermocouple and heater (= 50W) is plugged in
a tubular pyrophyllite holder (HL) of only 4.5mm^3 volume. The whole
system is embedded in a working volume of several 1000mm^3 of high
pressure hydrogen gas and is moved by the lever L within the magnetizing field (about 3kOe) of the stationary permanent magnet PM and
the measuring induction coil M delivering the signal for registration
by an electronic integrator. The magnetizing field is correspondingly
controlled by the simultaneous motion of the induction coil C connected with the lever L. The desired lever motion for the measuring
process is actuated by electrical heating of the wire W which by
thermal dilatation releases the compression spring SP.[29]

For demonstration of the ability of the described 'miniature magnetometer' to respond very quickly to high pressure changes in the hydrogen atmosphere may serve, in Fig. 6, a Ni-Fe alloy.11 Even relatively small pressure changes in hydrogen are indicated by a step change in the magnetization cycles. At the regarded alloy with 16at% Fe content the hydride formation pressure amounts to 14-15kbar and one can recognize on the right of Fig. 6 the rapidity at which the hydride phase forms under pressures above the hydride formation pressure and below of it the comparatively slow increase in the magnetization indicating a slow decomposition of the hydride. Obviously 'in situ' magnetization measurements are a definite way for determination of formation as well as for decomposition pressures. Fig. 7 shows a putting together of Ni alloys which completely loses its magnetization in high pressure atmosphere, corresponding increasing formation pressure with increasing Fe content and inversely in case of Mn. The correspondance of electrical resistance and magnetization is shown in the diagrams of NiMn-H systems in Fig. 8 which represents values of equilibrium with the high pressure hydrogen environment.12 Further on, we have an example for direct relation to the phase surface scheme in Fig. 1: Below about 20at% Mn content typical pressure hysteresis exists representing the difference between formation and decomposition pressures caused by a different mechanism (see the Section before),

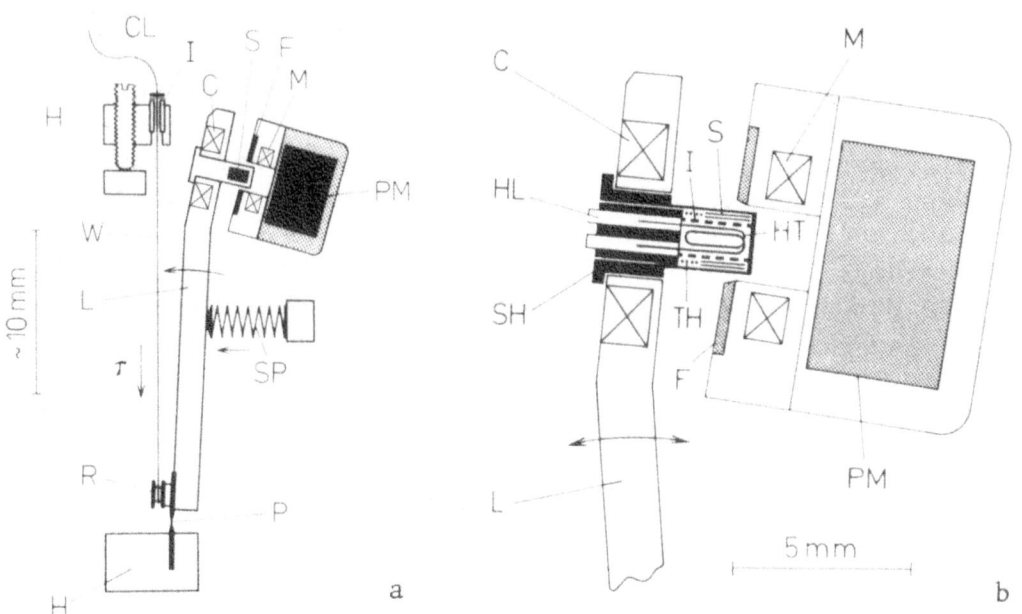

Fig. 5.   Scheme of the magnetic device: 11,28 a) complete, b) enlarged part. S - sample, PM - permanent magnet (SmCo5: Brown, Boveri AG), F - magnetic flux piece, M - measuring induction coil, C-magnetic field control induction coil, L - lever, P - elastic pivot, SP - compression spring, H - holder, W - heating wire (looped around the insulating roller R), τ - thermal dilatation of W, I - isolating tube, CL - current leads, SH - tubular sample holder, HT - sample heater, HL - heater leads, TH - thermocouple.

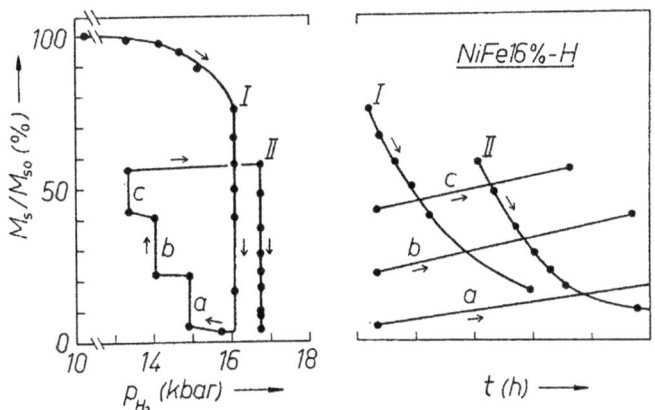

Fig. 6.   Relative saturation magnetization M_s/M_{so} of a Ni-Fe alloy
          versus hydrogen pressure at about 22°C and the correspon-
          ding time sequence of the magnetization at different con-
          stant pressures (I, a, b, c, II). M_{so} saturation magneti-
          zation at normal pressure (before hydrogen absorption).[11]

whereby the phase transitions ($\alpha \to \beta$, $\beta \to \alpha$) occur below the 'critical
point'. The alloys with higher Mn-contents show no pressure hystere-
sis, i.e. the phase transitions seem to occur in a reversible manner
above the critical point. Furthermore it is seen here that in spite
of a complete hydrogenation of the whole sample forming a solid solution confirmed by X-ray analysis[30] the magnetization does not vanish
completely, as a consequence of an increasing prominence of the magnetization in the host lattice.

    In the following Figures some experiments on Ni-H, NiCu-H, NiFe-
H systems[31] are shown which go over to higher temperatures. They are

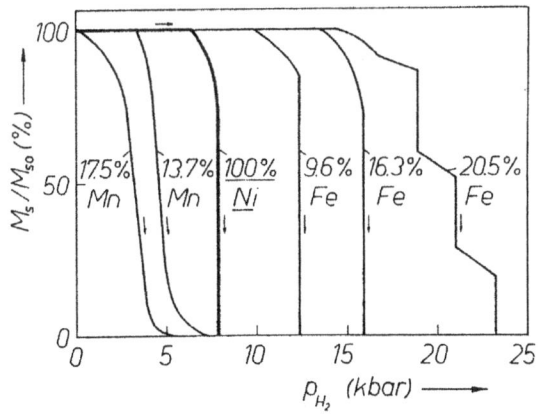

Fig. 7.   Relative saturation magnetization of Ni, Ni-Fe, and Ni-Mn
          alloys versus hydrogen pressure at about 22°C. Further
          items see Fig. 6.

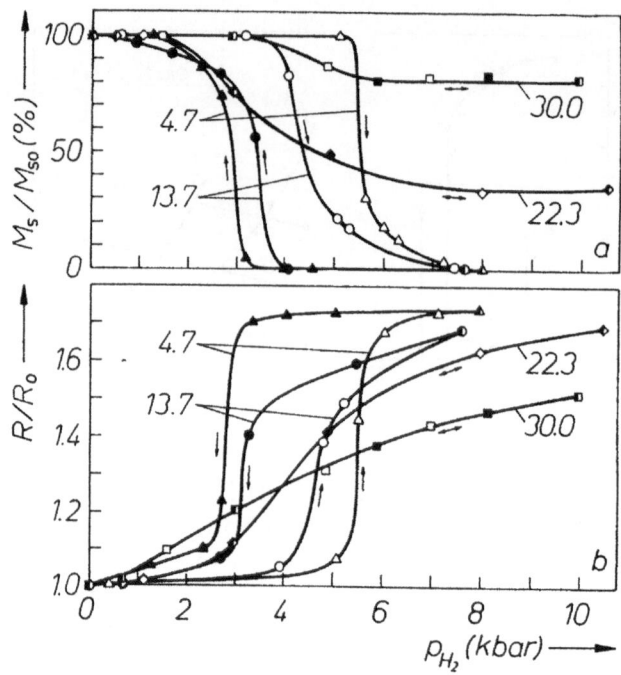

Fig. 8.   Changes of a) relative saturation magnetization M_s/M_{so} and
b) relative resistance R/R_o with increasing (open symbols)
and decreasing (full symbols) hydrogen pressure for NiMn
alloys at about 22°C. M_{so}, R_o saturation magnetization and
resistance at normal pressure; parameter is the Mn content
in at% (no pressure hysteresis above 20at% Mn!).[12]

performed with the miniature magnetometer, Fig. 5, using the 'inner
sample heater' and with a corresponding device for resistance measurements. Although in each case, see Fig. 9, after passing the corresponding formation pressure of hydride[18,32] the magnetism is anni-
hilated H desorption (by pressure reduction or heating) leads to different shapes of the M_s-T-curves. For Ni, decreasing H_2-pressure below
the decomposition pressure yields curves with constant Curietemperature CT without deformation: at the given temperatures and pressures
the phase transitions β→α occur. The splitting of the curves in dependence on the measuring direction is caused by heating out of H at
higher temperatures if the decomposition pressure increases. The extent of the splitting depends on the relation of measuring time and
diffusion rate and on the stage of desorption. In case of NiFe16at%
the 20.0kbar curve lifts up (from $M_s = 0$) between 100 and 200°C: at
this pressure and this temperature the system passes over from the
β-phase to the two-phase region (α,β). Decreasing of H_2-pressure
leads for NiCu (Fig. 9) to a shift of CT accompanied by a M_s-T shape
tendency similar to that observed in the electrochemically produced
system NiCu30at%-H[15-17] at lower temperatures indicating a precipitation-like decomposition process. The behavior of the temperature
dependence of the electrical resistance mirrors in all studied cases

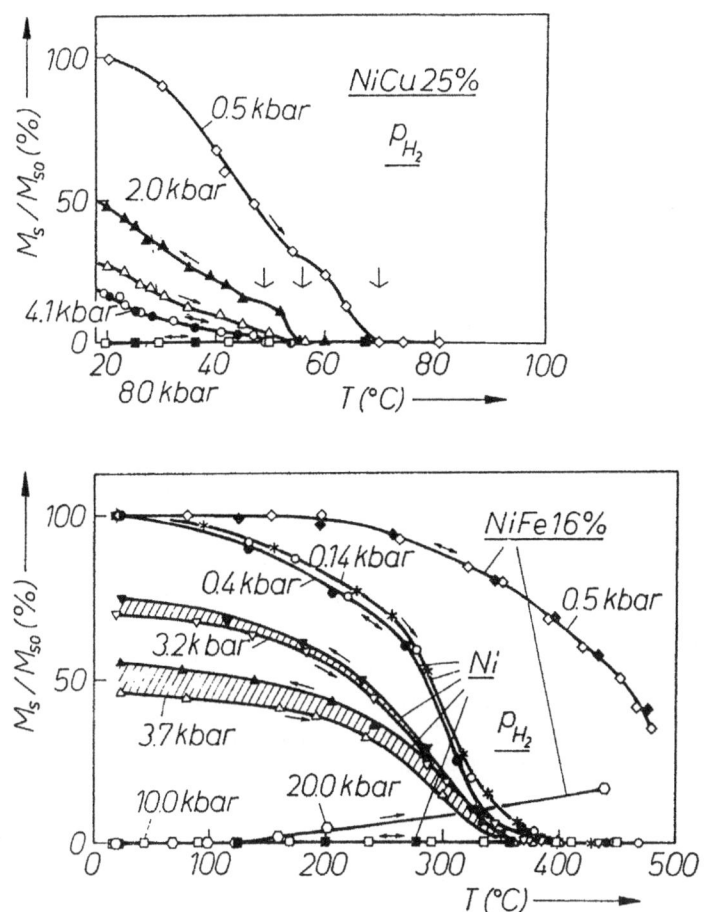

Fig. 9.   Influence of high pressure gaseous hydrogen on the temper-
          ature dependence of the relative saturation magnetization
          M_s/M_{so} of Ni, a Ni-Fe, and a Ni-Cu alloy. M_{so} saturation
          magnetization at normal pressure. Sequences of the station-
          ary pressures: above 0.5, 0.8, 4.1, 2.0kbar; below 0.4,
          10.0, 3.7, 3.2, 0.14 resp. 0.5, 20.0kbar. Arrows (above):
          approx. position of the respective Curietemperature. Meas-
          uring rate: about 50°C min^{-1}.

the magnetic relation, as e.g. at the NiFe-H system, Fig. 10, the
disappearance and reappearance of the 'Ferromagnetic Resistance Anomaly'. The observed shift of CT for about 100°C in the desorption
stage of 10.0kbar H_2 pressure may correspond to the findings of the
Soviet group[9,33] with regard to the position in the phase surface.

The examples may show also that in the high pressure range of
such systems from the sensitive correlation between metallographic
state and magnetic behavior, especially with respect to its temperature dependence, one can expect a basic help for the analysis of
phase transitions and the finding of phase surfaces.

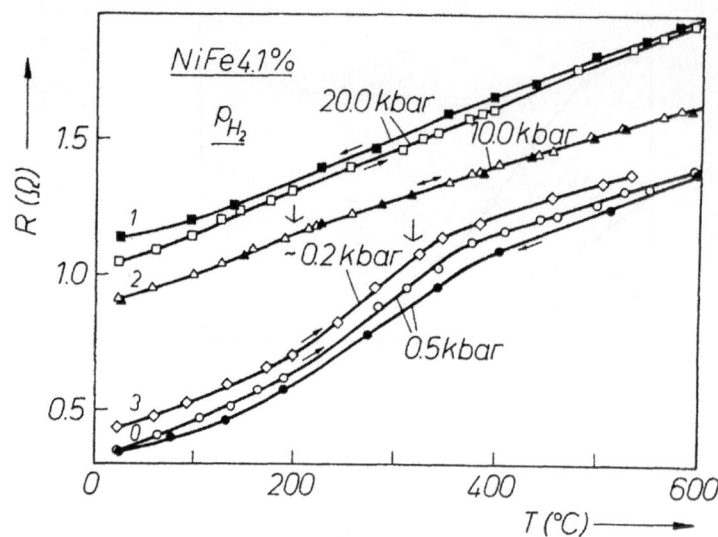

Fig. 10. Influence of high pressure gaseous hydrogen on the tem-
          perature dependence of electrical resistance resp. the
          'Ferromagnetic Resistance Anomaly' of a nickel iron alloy.
          0-3 sequence of the stationary pressures. Arrows: approx.
          position of the Curietemperature at curve 2 and 3. Measur-
          ing rate: about $16^{\circ}C\ min^{-1}$.

ACKNOWLEDGEMENT

     I would like to express my thank to Prof. J. Brandmüller for
his long standing support for our research. The permanently helpful
discussions of M. Becker is gratefully acknowledged. I am also indebted to the mechanical workshop under A. Pfisterer and A. Ebenböck,
and especially to W. Baltes for comprehensive preparatory technical
work. The investigations were supported by the Deutsche Forschungs-
gemeinschaft, the Polish Academy of Sciences, and the Deutscher
Akademischer Austauschdienst.

REFERENCES

1. B. Baranowski and M Smialowski, Bull. Acad. Polon. Sci., Ser.
    sci. chim., geol. geogr 7, 663 (1959).
2. B. Baranowski and K. Bochenska, Roczniki Chem. (Ann. Soc. chim.
    Polonorum) 38, 1419 (1964).
3. B. Baranowski and R. Wisniewski, Bull. Acad. polon. Sci., Ser.
    Sci. chim. 14, 273 (1966).
4. B. Baranowski, Ber. Bunsenges. Physik. Chem. 76, 714 (1972).
5. H. J. Bauer and E. Schmidbauer, Naturwissenschaften 48, 425
    (1961); Z. Phys. 164, 367 (1961).
6. H. J. Bauer, M. Becker, H. Pretsch and M. Zwick, phys. stat. sol.
    (a) 47, 445 (1978).

7. H. J. Bauer, Z. Naturforsch. 22a, 468 (1967).
8. K. A. Kohler and H. J. Bauer, phys. stat. sol. (a) 14, K27 (1972).
9. E. G. Poniatowskij, V. E. Antonov and I. T. Belash, Dokl. Akad. Nauk SSSR 229, 391a (1976).
10. H. Kronmüller, in Hydrogen in Metals I, G. Alefeld and J. Völkl, eds., Springer Verlag Berlin (1978), p. 289.
11. H. J. Bauer and B. Baranowski, in High-Pressure Science and Technology, Vol. 1, K. D. Timmerhaus and H. S. Barber, eds., Plenum Press, New York (1976), p. 248.
12. H. J. Schenk, H. J. Bauer and B. Baranowski, phys. stat. sol. (a) 52, 195 (1979).
13. S. von Aufschnaiter and H. J. Bauer, Z. angew. Phys. 17, 209 (1964).
14. H. J. Bauer and D. Jonitz, Z. angew. Phys. 28, 40 (1969).
15. G. Bacherer, unpublished.
16. H. J. Bauer, Z. angew. Phys. 26, 87 (1969).
17. H.-U. Daniel and H. J. Bauer, J. Magn. Magn. Mat. 6, 302 (1977).
18. B. Baranowski and M. Tkacz, Polish J. Chem. 54, 819 (1980).
19. K. Ludwig, unpublished.
20. H. J. Bauer and U. Thomas, Z. Naturforsch. 21a, 2106 (1966).
21. D. Jonitz and H. J. Bauer, Z. Naturforsch. 33a, 1599 (1978).
22. H. J. Bauer and K. A. Kohler, Phys. Lett. 41A, 291 (1972).
23. A. Janko, Naturwissenschaften 47, 225 (1960); Bull. Acad. Polon. Sci., Ser. sci. chim. 8, 131 (1960).
24. H. J. Bauer and U. Ruczka, Z. angew. Phys. 21, 18 (1966).
25. H. J. Bauer, E. Pfrenger and K. Stierstadt, Z. Naturforsch. 22a, 549 (1967).
26. H. J. Bauer and B. Baranowski, in Proc. Europ. High Pressure Research Group (14th Annual Meeting), W.G.S. Scaife, ed., Typografia Hiberniae, Dublin (1976), p. 3.
27. H. J. Bauer and B. Baranowski, Phys. stat. sol. (a) 40, K35 (1977).
28. H. J. Bauer, J. Magn. Magn. Mat. 15-18, 1267 (1980).
29. H. J. Bauer, J. Phys. E 10, 332 (1977).
30. M. Krukowski and B. Baranowski, J. less-common. Metals 49, 385 (1976).
31. H. J. Bauer, H. J. Schenk and B. Baranowski, Trans. Jpn. Inst. Met., Suppl. 21, 377 (1980).
32. B. Baranowski and S. Filipek, Roczn. Chemii 47, 2165 (1973).
33. E. G. Poniatowskij, W. E. Antonov and I. T. Belash, Dokl. Akad. Nauk SSR 230, 649 (1976).

THE KINETICS OF HYDROGEN ABSORPTION-DESORPTION BY METALS*

M. A. Pick

Brookhaven National Laboratory
Upton, NY 11973

INTRODUCTION

The rate at which hydrogen can be absorbed and desorbed by a particular metal or alloy is an important factor for hydrogen storage. It can be decisive in the question of whether it is a useful material. Many properties influence the kinetics including; the diffusion coefficient of hydrogen; the sample size or surface-to-volume ratio; the occurrence of phase transitions; the heat of solution, and related to that, the rate at which this heat can be extracted or provided during the absorption and desorption cycles respectively. Some of these factors have been dealt with in more or less detail in the past. One factor which has received relatively little attention is the intrinsic rate of hydrogen transfer through the solid-gas interface; the surface. In the present paper I shall review some experimental studies and a theoretical model for the potential energies involved with the hydriding-dehydriding process. I shall first discuss the theoretical model in detail, then, in reviewing the experimental results, show how it applies to the model substances Nb and Ta and how two major predictions of the model are fulfilled. One of the predictions deals with the surface modifications which can enhance the hydrogen uptake rate and a second one with the temperature, coverage and pressure dependence of the sticking coefficient. Nb and Ta were chosen as model substances because most of their bulk metal-hydrogen properties such as the phase diagram, the diffusion coefficient, the solubility, the resistivity as a function of hydrogen concentration etc., have been examined in detail and are well known. Another reason is that these materials can be cleaned easily in UHV.

---

*Work performed under the auspices of the U.S. Dept. of Energy.

329

The paper will also show how the relatively simple experiment of measuring the rate of hydrogen absorption and desorption can be analyzed to yield most of the important parameters related to the solution kinetics including, the energy of chemisorption, the activation energy for chemisorption, the vibrational entropy at the surface, the preexponential of the sticking coefficient and the surface coverage as a function of temperature, as well as the heat of solution and Sievert's constant.

THEORETICAL

The surface barrier model previously described is based on the potential energy diagram shown in Fig. 1(a).[1] The four fluxes to and from the surface are indicated in the figure. This barrier model is characterized by the fact that there is a deep trap for the hydrogen at the surface. This state can be identified as chemisorbed hydrogen. There may be other chemisorbed species if these species do not interact with the absorption process. The model is further characterized by the fact that there is no activation

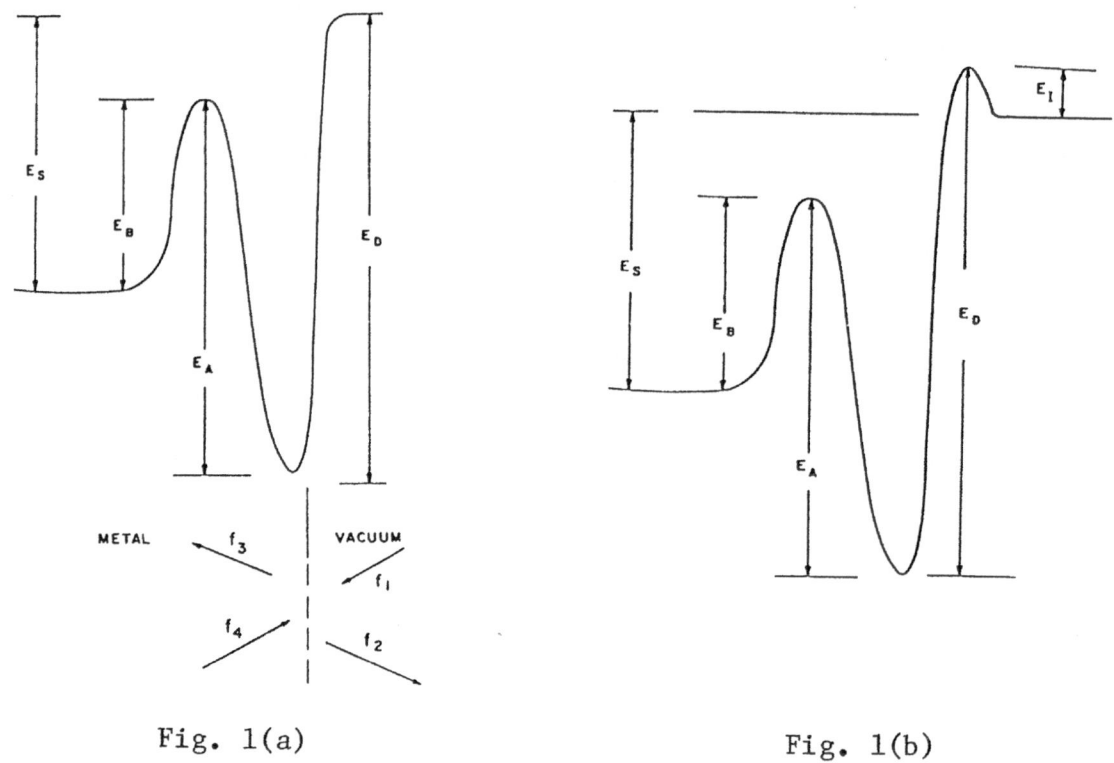

Fig. 1(a)   Fig. 1(b)

Fig. 1(a)   Energies defining the surface barrier model for hydrogen
            absorption/desorption. The indicated fluxes of hydrogen
            to and from the surface are defined in the text.

Fig. 1(b)   The surface barrier model modified to include the activa-
            tion barrier for chemisorption.

barrier for chemisorption and the incoming flux is taken to be:

$$f_1 = \frac{2\Gamma}{N_S} s_o (1-\theta)^2 \tag{1}$$

where Γ is the flux of H_2 molecules impinging on the surface per cm^2 and second. It is given by the kinetic theory expression $\Gamma = 3.5 \times 10^{22} P/(MT)^{1/2}$, where P is in Torr, M in atomic mass units and T the absolute temperature; N_S is the number of metal atoms per cm^2 of surface; θ is the atomic fraction of H atoms on the surface; s_o is, in this case, the temperature independent sticking coefficient of hydrogen on the bare surface. This form for f_1 implies that every single hydrogen molecule which impinges on the surface will dissociate and chemisorb provided it finds two empty sites at the surface; therefore the term $(1-\theta)^2$.

The flux out from the surface into the gas phase is:

$$f_2 = -K\theta^2 \tag{2}$$

K is the rate constant given by:

$$K = K_o \exp (-2E_D/RT) \tag{3}$$

The θ^2 and the $2E_D$ indicate that two hydrogen atoms must associate and leave the surface as an H_2 molecule.

The flux from the surface into the bulk is:

$$f_3 = -\nu\theta \tag{4}$$

ν, the rate constant is given by:

$$\nu = \nu_o \exp (-E_A/RT) \tag{5}$$

The fact that this flux is proportional to the surface coverage θ indicates that the species entering the bulk are individual hydrogen atoms and not molecules.

The flux from the bulk to the surface, finally, is:

$$f_4 = \beta(1-\theta)x \tag{6}$$

and β, the rate constant, is given by

$$\beta = \beta_o \exp (-E_B/RT) \tag{7}$$

x is the atomic fraction of hydrogen atoms in the bulk. The factor $(1-\theta)$ implies that there must be one empty site at the surface for the diffusing hydrogen atom to reach the surface.

This theoretical model satisfies Sievert's Law and is strictly only valid in the region of low hydrogen concentrations or negligible hydrogen-hydrogen interactions. The appropriate experiments should therefore be limited to low concentrations or high temperatures.

In recent experiments on clean Nb it was found that the assumption implicit in f_1, namely, that there is no activation barrier for chemisorption is not strictly obeyed.[2] The barrier model should therefore be modified as shown in Fig. 1(b). The experimental results indicate that s_o obeys an exponential temperature dependence and must be replaced by:

$$S_i = S_o e^{-2E_I/RT} \tag{8}$$

S_i is now the temperature dependent initial sticking coefficient on the bare surface. It has been found, both for Nb[2] and Ta,[3] that S_o is very close to unity and E_I is on the order of 1 kcal/mole H. However slight, this temperature dependence can be dominant, especially for lower temperatures.

Balancing the four fluxes leads to equations (9) and (10):

$$\frac{d\theta}{dt} = (2\Gamma S/N_s)(1-\theta)^2 - K\theta^2 - \nu\theta + \beta(1-\theta)x \tag{9}$$

$$\frac{dx}{dt} = [\nu\theta - \beta(1-\theta)x]N_\ell^{-1}. \tag{10}$$

N_ℓ is the number of layers of metal atoms in the foil. These equations have been generalized to any sample geometry in ref. 4. With the assumption that, past a fast initial transient, x and θ are in quasiequilibrium and that $d\theta/dt$ is small compared to $N_\ell \, dx/dt$, we can solve the equations and obtain for absorption:

$$\frac{1}{2}(1-b)^2 \ln(1+y) - \frac{1}{2}(1+b)^2 \ln(1-y) - b^2 y = at \tag{11}$$

and for desorption:

$$\frac{1}{y} - \frac{1}{y_o} + 2b \ln\left(\frac{y_o}{y}\right) + b^2(y_o - y) = at \tag{12}$$

where;

$$a = 2\Gamma S_i/N_s N_\ell \, x_{max} \tag{13}$$

and

$$b = \left(\frac{\beta}{\nu}\right) x_{max} = [\theta_{max}/(1-\theta_{max})] \tag{14}$$

x_{max} and θ_{max} are the equilibrium values of x and θ at the particular hydrogen pressure and sample temperature. x_{max} is known from thermodynamic measurements and given by Pryde and Titcomb[5] to be:

$$x_{max} = \frac{P^{1/2}}{(3.954 \pm 0.338) \times 10^4} \exp\left(\frac{(4.345 \pm 0.161) \times 10^3}{T}\right) \qquad (15)$$

where the pressure P is measured in Torr. $y = \frac{x}{x_{max}}$ is the normalized concentration and ranges from 0 to 1. y_o is the value of y at the point of dehydriding the sample.

The temperature dependence of the initial sticking coefficient can be due to several different factors. The most straightforward of these has been discussed by Lennard-Jones.[6] He calculated the potential energy of an atom as a function of distance from the surface of a metal, (1) in the form of a molecule and (2) in the form of a dissociated atom. These potential energy curves are shown in Fig. 2. Between the well of curve 1 and that of curve 2 there is an energy barrier designated E_i which the molecules reaching the well of curve 1, the physisorbed state, must overcome to dissociate and fall into the well of curve 2, the chemisorbed state. In order to obtain the temperature dependence as given by equation (8) it is required that the hydrogen reach thermal equilibrium with the surface

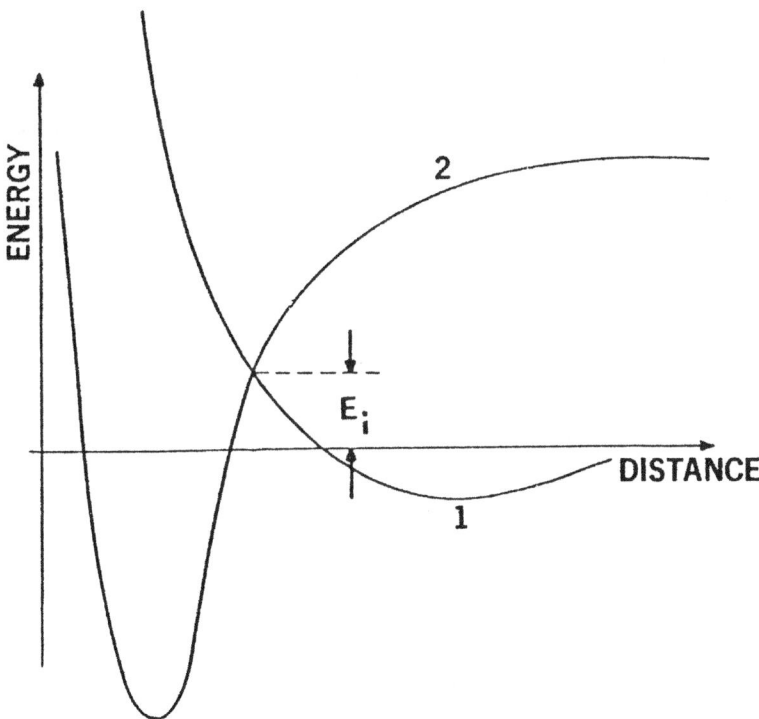

Fig. 2.   The energy of the system versus distance of the absorbate from the surface. (1) Adsorption as a molecule, (2) adsorption as two atoms.

while in state (1) of Fig. 2. It has generally been assumed that for many transition metals this activation energy for chemisorption is very close to zero.[7] Our experiments, however, have shown that these results have to be reexamined carefully. Another factor which can introduce an apparent activation barrier for dissociation is the presence of "active" sites on the surface to which the molecules must diffuse before they can dissociate thereby limiting dissociation by the activation barrier for diffusion. The presence of defects or surface traps can introduce a similar effect on the apparent energy of activation for the sticking coefficient.

The theory, as described, does not take into account the diffusion of hydrogen within the bulk. It can therefore be applied only to those materials and sample dimensions where diffusion does not become a rate determining step. The diffusion coefficient of hydrogen in many of the metal hydrogen systems of interest, e.g. Nb, Ta, V, Pd, is so high that this condition is easily met for the sample dimensions involved.[8] For example, in the case of Nb the diffusion coefficient at room temperature is $D \cong 1 \times 10^{-5}$ cm^2/sec;[8] a hydrogen atom can therefore traverse a 25 μm thick foil in ~1 sec.

EXPERIMENTAL

The experimental procedures used to measure the hydrogen uptake curves have been described in detail elsewhere.[1,9] An important aspect of the experiments is the cleanliness, especially of the surface of the sample. Nb and Ta are ideal substances because they can be cleaned quite readily by heating in oxygen and then in vacuum.[10] The frequent heating and cooling of the Nb and Ta samples cause a recrystallization of the samples. An attached LEED system was used to show that the grains were predominantly oriented with their (110) planes parallel to the surface. Another important experimental requirement is the purity of the hydrogen gas introduced for absorption. Unless careful precautions are taken it is very easy to contaminate originally very pure H$_2$ gas; i.e. the gas manifold must be carefully degassed and of stainless steel. The purity of the gas was monitored by a quadropole mass spectrometer and the contaminant partial pressure was found to be below or comparable to the ambient pressure in the system. The pure hydrogen was obtained either by the decomposition of a metal hydride or by passing hydrogen through a Pd-Ag diffusion cell.

The absolute hydrogen pressure was taken to be 2.2 times the gauge pressure. This value is crucial because it determines the absolute sticking coefficient. This value was chosen after a careful investigation of the literature values[11] plus the fact that the sticking coefficient cannot be greater than 1 as would have been the case for some of the smaller values reported.

The hydrogen absorption and desorption by the sample was monitored by measuring the resistance as a function of time. The increase in the resistivity as a function of hydrogen concentration
is known for both Nb and Ta:

$$\Delta\rho = 0.64 \pm 0.06 \; \mu\Omega cm \; (at \; \% \; H)^{-1} \; for \; Nb[12]$$

$$\Delta\rho = 0.79 \pm 0.06 \; \mu\Omega cm \; (at \; \% \; H)^{-1} \; for \; Ta[13]$$

This increase can be taken to be temperature and concentration independent in the temperature and concentration ranges investigated.[14]
The resistivity of the sample is also used to determine the temperature of the sample assuming a linear increase in the resistivity as
a function of temperature, or by making use of published studies of
the resistivity versus temperature.[15,16]

Metallic overlayers of Pd were deposited on the surfaces of
the samples by evaporation from a heated tungsten coil. The coverage was determined from the Auger line intensities assuming that
those of the substrate would decrease exponentially and those of the
overlayer would increase exponentially.[9] The mean free path of the
electrons is almost independent of energy and metal in this energy
range and can be taken to be approximately 5 Å.[17] This value was
used to calibrate the thickness of the overlayers. The adherence
of the intensities to the exponential model can be taken as an indication that the overlayer is uniform rather than concentrated in
islands. The LEED system was used to follow the growth of the deposited overlayers.

ANALYSIS OF DATA

Fig. 3 shows an example of an absorption-desorption curve. The
change in slope at the time indicated is due to the hydrogen being
pumped out and desorbing from the sample. This is a normalized
curve from which the background has been subtracted. The concentration x was calculated from the resistivity increase and divided
by the value of x_{max} at the pressure and temperature to give y.
The pressure and the voltage across the sample were measured at a
rate of 50 Hz. The line drawn through the points corresponds to a
fit of the theoretical expression (11). The a and b values derived
from the fit to the absorption data in this particular example are:
$a = 2.355 \times 10^{-3}$, b = 48.20 and the temperature, deduced from the
resistivity is 369.4 K. This example shows how the a and b parameters can be deduced by fitting to the absorption part of the curve.
Similarly the a and b parameters can be derived from the desorption
part of the curve by fitting to expression (12). Another method to
extract at least one of the parameters from the curve is to measure
the change in slope of the normalized absorption-desorption curve
at the point where the hydrogen is pumped out of the system. This

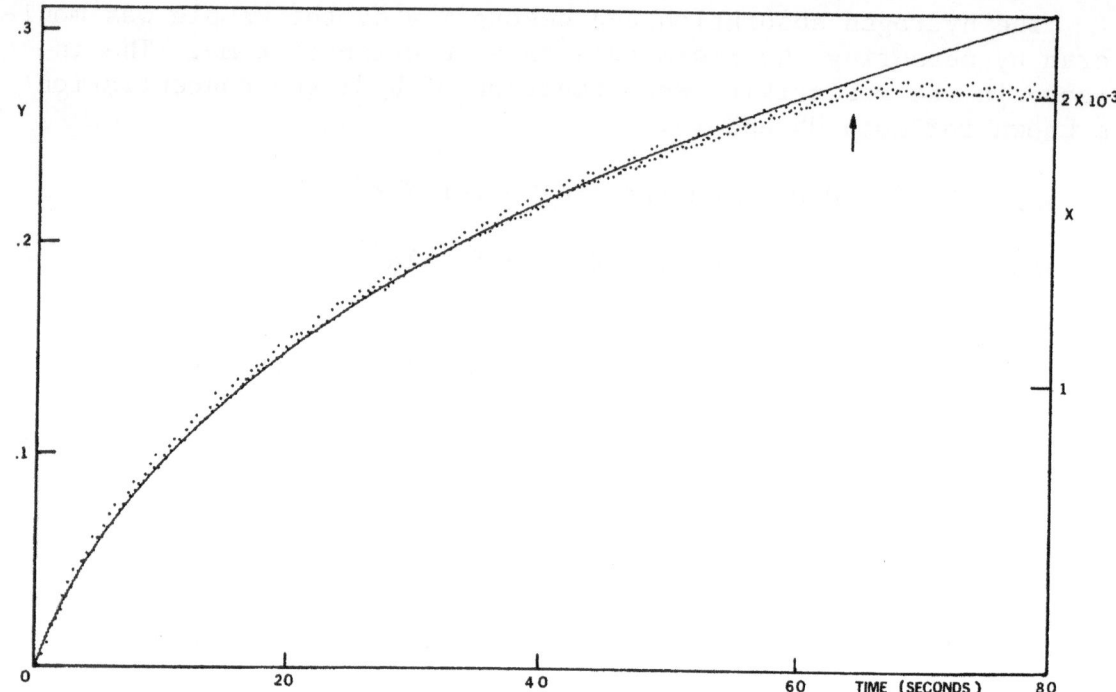

Fig. 3. An absorption-desorption curve. x is the concentration in
         atomic ratio [H/Nb]. y is the normalized concentration
         x/x_{max}. The arrow indicates the point at which the hydro-
         gen was pumped out of the system. T = 369.4 K; P (gauge) =
         = 4.5 x 10^{-5} Torr. The line through the data is the theo-
         retical curve.

slope change is:[2]

$$\Delta \frac{dy}{dt} = \frac{a}{(1 + by_o)^2} \tag{16}$$

The initial slope of the normalized hydriding curve can also be used
to extract the parameter a, because

$$\frac{dy}{dt} = a \quad (\text{for } t \to 0) \tag{17}$$

The initial slope of the original data, the resistance increase as
a function of time, is a direct measure of the initial sticking
coefficient S_i.

By measuring and analyzing such absorption-desorption curves
at several temperatures one can deduce most of the parameters of the
surface barrier. One can measure x_{max}, a, b, and S_i. Plotting the
natural logarithms of these parameters as a function of 1/T will
yield the energies and preexponentials associated with the surface

barrier model. The temperature dependent x_{max} values yield the
solution energy E_S and the preexponential or Sievert's Constant.
The b parameter yields the chemisorption energy E_D and the preexponential K_o. The S_i values as well as the a parameter yield the
activation energy for chemisorption E_I, and S_o, the high temperature limit for the sticking coefficient on the bare surface.

RESULTS AND DISCUSSION

The Clean Surface

Figs. 4, 5, and 6 show the temperature dependence of the
initial sticking coefficient S_i as well as that of the parameters
a and b for clean Nb. From Fig. 4 we obtain:

$$S_o = 0.915 \text{ and } E_I = 1.28 \text{ kcal/mole H}$$

From Fig. 6 we obtain:

$$E_D - E_I = 13.52 \text{ kcal/mole H}$$

$$\text{or} \quad E_D = 14.8 \text{ kcal/mole H}$$

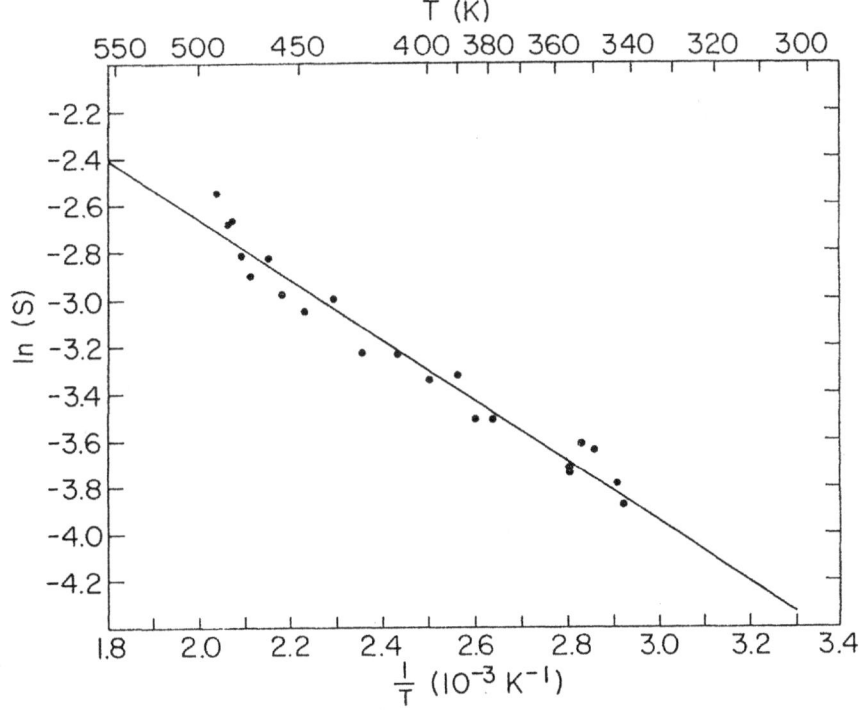

Fig. 4. The temperature dependence of the initial sticking coef-
        ficient.

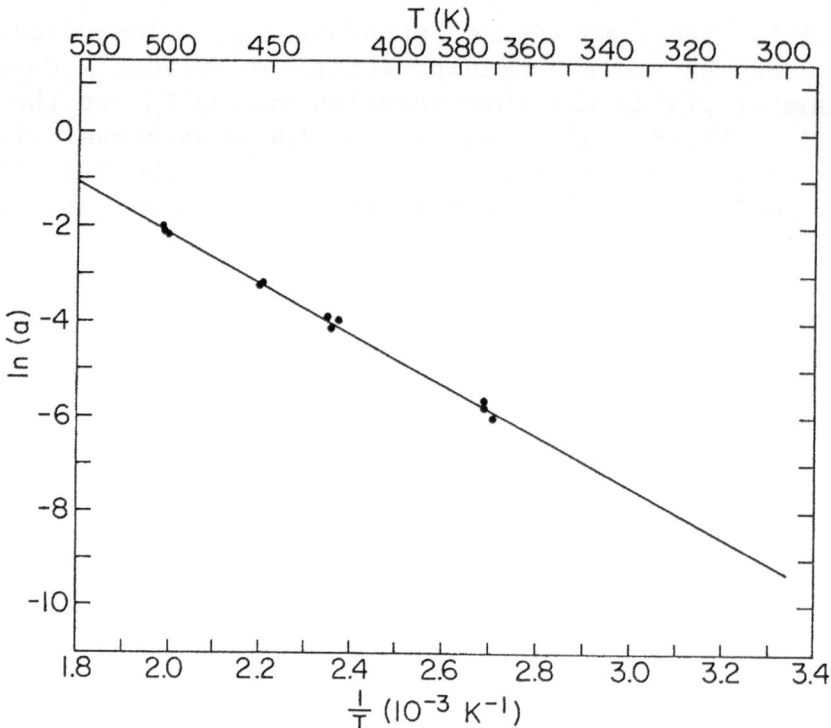

Fig. 5.   The temperature dependence of the parameter a.

From Fig. 5 we obtain:

$$E_S - 2E_I = -10.65 \text{ kcal/mole H} \quad .$$

from which, using the thermodynamic value for E_S given by Pryde and Titcomb[5] $E_S = -8.633$ kcal/mole H, we can deduce a second, less direct, value for E_I;

$$E_I = 1.01 \text{ kcal/mole H} \quad .$$

A similar activation barrier for chemisorption has been found for Ta. However, in this case the barrier appears to be smaller leading to higher initial sticking coefficients than Nb at the same temperature.[3]

In a recent publication[4] we showed that there is a good qualitative agreement between the temperature dependence of the sticking coefficient calculated on the basis of our model and the previous experimental results of Ko and Schmidt.[18] This result is illustrated in Fig. 7. In this figure the sticking coefficient on Nb is plotted as a function of temperature at a constant coverage of 15 monolayers of hydrogen. The pressure, as well as the other experimental parameters are taken from ref. 17. The sticking coefficient increases with the temperature, reaches a maximum, and then falls to zero when the 15 monolayers represent the equilibrium concentra-

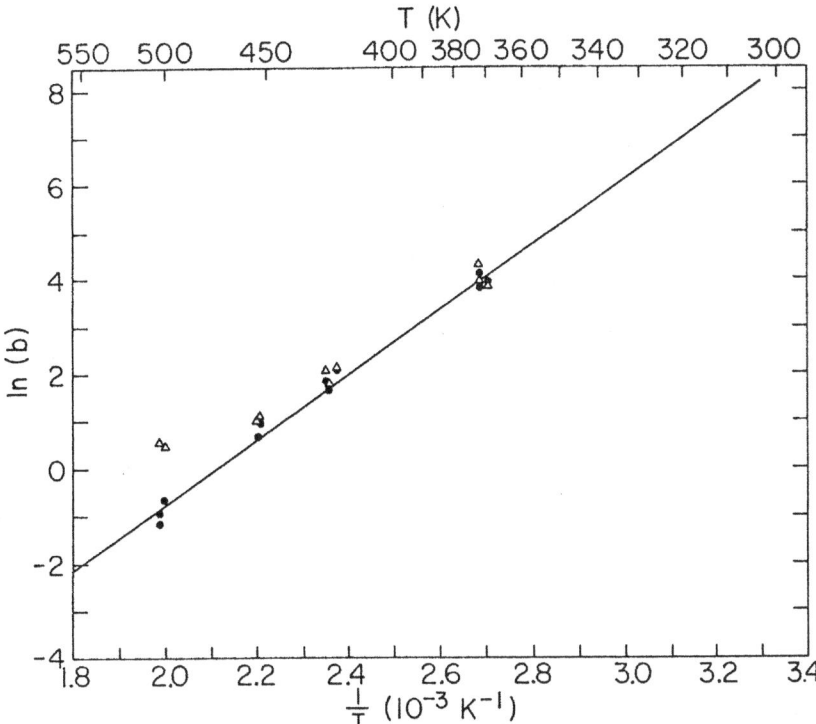

Fig. 6.   The temperature dependence of the parameter b.   The full
         dots are the values derived from absorption curves; the
         triangles represent the values derived from the desorption
         curves.

tion x_{max} at the particular experimental conditions.   This qualitative temperature dependence will remain unchanged by the inclusion
of the temperature dependence of the initial sticking coefficient
itself, as described above, which was not included in ref. 4.

The Modified Surface

     Knowing that the kinetics of hydriding are primarily governed
by the presence of the deep trap for hydrogen at the surface leads
one immediately to the possibility of enhancing the kinetics by
modifying the surface.   Depositing a material on the surface which
has a smaller trap i.e. a smaller heat of chemisorption and which
also allows hydrogen to diffuse through it readily, should increase
the uptake rate.   The result of depositing a thin layer of Pd on
both Nb and Ta is illustrated in Figs. 8 and 9.[9]   In the case of
Nb a Pd layer less than one monolayer thick appears to have very
little effect on the initial sticking coefficient and may even decrease it somewhat.   This changes drastically at one monolayer.   At
this point the sticking coefficient increases rapidly and actually
reaches unity for Pd layers of the order of three monolayers thick.
A similar effect occurs in the case of Pd on Ta, Fig. 9.   Here the

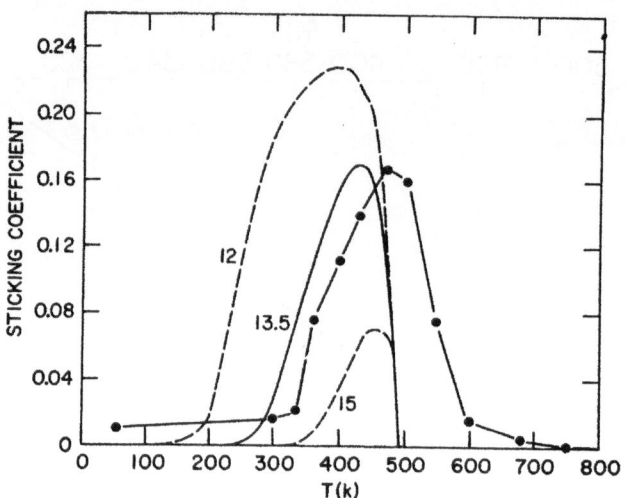

Fig. 7. The temperature dependence of the sticking coefficient of
hydrogen on Nb at a coverage of 15 monolayers and a pressure of 1.3×10^{-7} Torr. The full circles represent data
taken by Ko and Schmidt.[18] The full and dotted lines represent theoretical curves for various energies of chemisorption, E_D, in kcal/mole H.[4]

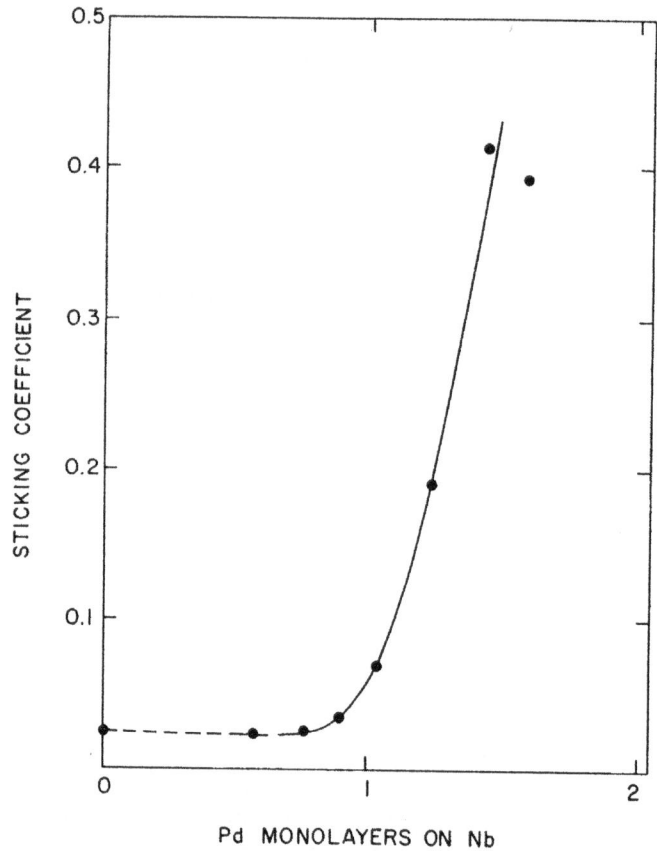

Fig. 8. The sticking coefficient of hydrogen on niobium as a function of the thickness of a palladium overlayer (temperature,
approximately 470 K; 2.2×10^{-5} Torr H_2).

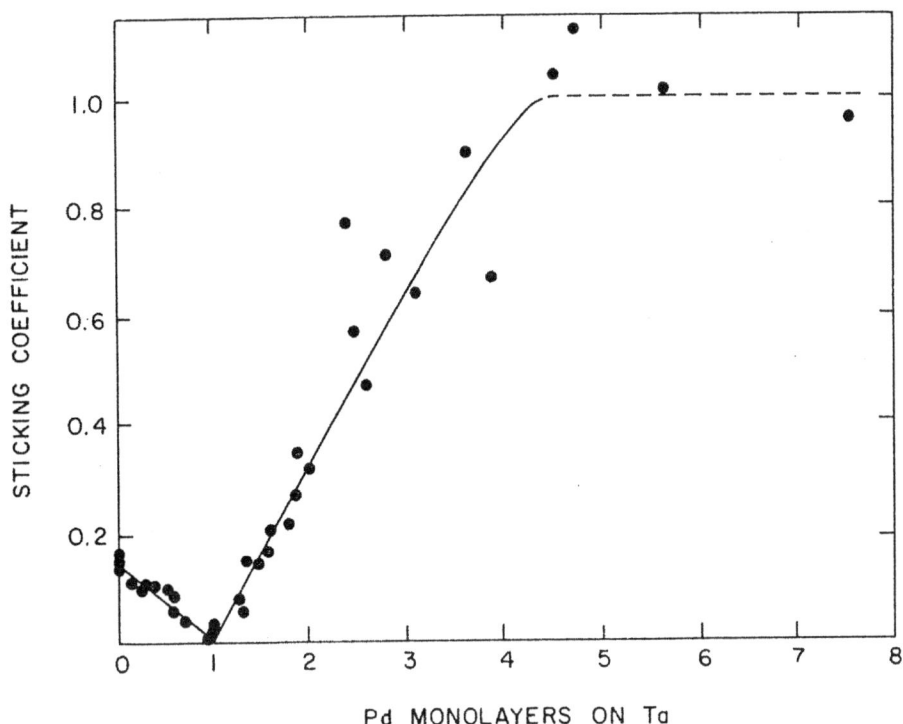

Fig. 9. The sticking coefficient of hydrogen on tantalum as a function of the thickness of a palladium overlayer (temperature,
approximately 500 K; 2.2 x 10^{-5} Torr H$_2$).

decrease in the sticking coefficient for layers less than one monolayer thick is clearly measurable. For thicker layers of Pd the
sticking coefficient becomes unity within experimental error. Every
single hydrogen molecule reaching the surface will dissociate and
enter the bulk! In both cases, Nb and Ta, this high sticking coefficient is sustained until the concentration reaches x_{max}. The
theoretical expressions for absorbtion/desorption (Eqs. (11) and
12)) still fit the data and the results indicate that the heat of
chemisorption has acquired the value of that of (111) Pd, leading
to low surface concentrations θ and therefore numerous open sites
for dissociation and chemisorption.

The change in the sticking coefficient at one monolayer of Pd
has been found to be correlated to a phase transition of the overlayer. It was discovered by low energy electron diffraction (LEED)
investigations that the palladium changes from a submonolayer film
commensurate with the (110) surface of Nb or Ta to an incommensurate
(111) palladium layer for layers more than one monolayer thick.[19]
This phase change has been investigated in detail by angular resolved photoemission experiments to try to understand the electronic
changes which accompany the phase change and which are the underly-
ing reason for the changes in the sticking coefficient.[20,21]

There are two possible reasons for the decrease in the initial sticking coefficients for the submonolayer Pd layers. On the one hand the decrease could be due to an increase in the energy of chemisorption leading to a higher surface concentration θ thereby decreasing the initial uptake rate by $(1-\theta^2)$. On the other hand the decrease in the sticking coefficient could be due to the presence of a larger activation barrier for chemisorption resulting in a decrease of the probability of dissociation. The results of the photoemission experiments indicate that it is the existence of a substantial activation barrier for chemisorption which decreases the uptake of hydrogen in the case of the submonolayer Pd layers. The electronic structure of the commensurate (110) Pd layer shows a surprising similarity to that of the noble metals such as Cu, with a low density of states near the Fermi energy. The phase change to the incommensurate Pd (111) is accompanied by a change of the electronic structure to that of a normal transition metal with a high density of states at the Fermi energy. Pd in the (110) structure of Nb, therefore, acts very much like Cu which also possesses a large activation barrier for the chemisorption of hydrogen.[22] The fact that Pd (111) enhances the uptake to such an extent that the sticking coefficient reaches unity indicates that the chemisorption of hydrogen on this surface must be strictly nonactivated.

Surface layers of other metals such as Pt and Ni,[9] as well as recently also Cu[3] have been studied. Although increases in the sticking coefficient and, therefore also the uptake rates have been observed they remain less than those obtained with Pd overlayers. In the case of Pt the solution of hydrogen is endothermic and the solubility is very small thereby limiting the permeation rate through the surface layer. Nickel on Nb enhances the uptake rate by a factor of approximately five which is less than the effect of Pd. This increase is followed by a gradual decrease for nickel layers thicker than about two monolayers. In this case it is apparently also the permeation rate through the overlayer which becomes rate limiting.

CONCLUSIONS

The reviewed work had led to significant progress in the understanding of the process of hydrogen transfer from the gas phase through the surface into the bulk of materials which dissolve hydrogen exothermically. The absorption/desorption kinetics are governed by the existence of strongly bound chemisorbed hydrogen atoms at the surface, the energy of which is lower than that in the bulk, as well as by the existence of an activation energy for this chemisorption. It has been shown that it is possible to modify the surface, changing the energy of chemisorption and the activation energy for chemisorption, by depositing a layer of several monolayers of another material on the surface. This type of composite system can be used to enhance the uptake rate to such an extent that every hydrogen molecule hitting the surface will dissociate and enter the bulk. It could also be used to achieve other beneficial properties including immunity to gases which normally poison the surface or to make the surface impermeable to hydrogen transfer. The result has important implications for the problem of hydrogen storage because it suggests means for improving one of the important factors: kinetics.

REFERENCES

1. M. A. Pick, J. W. Davenport, M. Strongin, and G. F. Dienes, Phys. Rev. Lett. 43:386 (1979).
2. M. A. Pick, (in publication).
3. M. A. Pick, (unpublished).
4. M. A. Pick and M. G. Greene, Surf. Sci. 93:L129 (1980).
5. J. A. Pryde and C. G. Titcomb, J. Phys. C. 5:1293 (1972).
6. J. F. Lennard-Jones, Trans. Faraday Soc. 28:333 (1932).
7. R. Speiser, in:"Metal Hydrides," W. Müller, J. P. Backledge, and G. G. Libowitz, eds., Academic Press, New York (1968).
8. J. Völkl and G. Alefeld, in: "Diffusion in Solids: Recent Developments," A. S. Nowick and J. F. Burton, eds., Academic Press, New York (1975).
9. M. A. Pick, M. G. Greene, and M. Strongin, J. of Less-Common Met. 73:89 (1980).
10. H. Wenzl and F-M. Welter, in: "Current Topics in Materials Science," Volume 1, E. Kalids, ed., North Holland Publishing Company, New York (1978).
11. D. R. Denison, in: "Vacuum Physics and Technology," G. L. Weissler and R. W. Carlson, eds., Academic Press, New York (1979).
12. G. Pfeiffer and H. Wipf, J. Phys. F. 6:167 (1976).
13. K. Rosan and H. Wipf, Phys. Stat. Sol. (a) 38:611 (1976).
14. K. Watanabe and Y. Fukai, J. Phys. F. 10:1795 (1980).
15. J. M. Abraham and B. Deviot, J. of Less-Common Met. 29:311 (1972).
16. V. E. Peletskii, Teplofiz. Vys. Temp. 14:295 (1976).
17. C. C. Chang, in: "Characterization of Solid Surfaces," P. F. Kane and G. B. Larrabee, eds., Plenum Press, New York (1974).
18. S. M. Ko and L. D. Schmidt, Surf. Sci. 42:508 (1974).
19. Myron Strongin, M. El-Batanouny, and M. A. Pick, Phys. Rev. B. 22:3126 (1980).
20. G. P. Williams, M. El-Batanouny, F. Colbert, E. Jensen, and T. N. Rhodin, (in publication).
21. M. El-Batanouny, M. Strongin, G. P. Williams, and F. Colbert, (in publication).
22. M. Balooch, M. F. Cardillo, D. R. Miller, and R. E. Stickney, Surf. Sci. 46:358 (1974).

THERMODYNAMICS AND KINETICS OF HYDROGEN

ABSORPTION IN AMORPHOUS NiZr-ALLOYS

F.H.M. Spit, J.W. Drijver, W.C. Turkenburg and
S. Radelaar †

Technical Physics Dept., State University Utrecht,
P.O. Box 80.000, 3508 TA Utrecht, The Netherlands

†Also at:
Laboratory of Metallurgy, Technical University, Delft
Rotterdamseweg 137, 2628 AL Delft, The Netherlands

ABSTRACT

The hydrogen absorption in amorphous NiZr-alloys will be discussed and compared with the absorption of crystalline alloys of similar composition. The hydrogen to metal ratio appears to be slightly smaller for amorphous alloys, and no pressure plateau is observed.

The results of kinetic studies using both the conventional volumetric technique and a novel combined calorimetric-gasdetection technique will be presented.

Preliminary results of surface studies by means of R.B.S. (Rutherford Back Scattering) show changes in surface composition after repeated hydrogen absorption and desorption cycling.

INTRODUCTION

A large number of alloys can be obtained in the amorphous state by rapid quenching from the melt. In most cases a very high cooling rate, of the order of 10^6 K·s^{-1}, is required. Several types of metallic glasses exist, the two most prominent types are glasses consisting of approximately 80 percent of a transition or noble metal and 20 percent metalloïd (e.g. $Pd_{80}Si_{20}$, $Fe_{80}B_{20}$) and glasses consisting of an early and a late transition metal (e.g. CuZr, NiNb). Alloys of first type can only be made amorphous in a rather narrow range of compositions ($\sim$ 10 at.perc.). On the

345

other hand alloys of the second type can be made amorphous in a
relatively large range of compositions. In favorable systems (e.g.
CuZr) the composition range may extend from 30 to 70 at.perc.
Since the early transition metals have a large affinity for hydrogen one might expect that these amorphous alloys can absorb a
fair amount of hydrogen. This proved to be the case [1,2]. Amorphous alloys are of interest because of several reasons:

Firstly, it is not known a priori whether hydrogen-rich amorphous materials at sufficiently low temperatures will separate
into a hydrogen-poor and a hydrogen-rich amorphous "phase".

Secondly, the mechanical properties of amorphous materials
are usually much better than those of crystalline materials of
similar compositions. Therefore one would expect that the frag-
mentation that occurs during hydrogen sorption in crystalline intermetallic compounds would be less severe in amorphous materials.

Thirdly, for a given combination of metals the range of homogeneous materials that becomes accessible for investigation is
greatly increased. The intermetallic compounds usually have a
rather limited range of composition variation.

In this paper we will describe the results of investigations
on amorphous $Ni_{64}Zr_{36}$ and related crystalline compounds. Results
on other alloys will be briefly described.

2. STABILITY OF THE AMORPHOUS STATE

Amorphous alloys are metastable with respect to the crystalline state. Amorphous alloys usually crystallize close to the
glass-transition temperature, the temperature at which the vis-
cosity of the amorphous material decreases below 10^{12} $N \cdot s \cdot m^{-2}$
(10^{13} Poise). The question arises what is the influence of the
presence of large amounts of hydrogen on the relative stability
of the amorphous alloy with respect to the crystalline state.
Five different processes can be envisaged:
(i)   The amorphous alloy remains amorphous during hydrogen ab-
      sorption. No amorphous hydride is formed, the hydrogen re-
      mains in solution:

$$AB_x(a) + yH_2 \rightarrow AB_x(a) + 2y[H]$$

      In the temperature range investigated amorphous $Ni_{64}Zr_{36}$ is
      an example of this case.
(ii) The amorphous alloy forms an amorphous hydride:

$$AB_x(a) + yH_2 \rightarrow AB_xH_{2y}(a)$$

      No examples of this type of reaction have been found so far.
(iii) The amorphous alloy decomposes into an amorphous hydride
      and an amorphous alloy (disproportionation):

(iii) $AB_x(a) + yH_2 \rightarrow A(a) + B_xH_{2y}(a)$

No examples have yet been found.
(iv)   The amorphous alloy decomposes and either one or two of the new phases crystallize:

$AB_x(a) + yH_2 \rightarrow A(a \text{ or } c) + B_xH_{2y}(a \text{ or } c)$

Numerous examples exist: E.g. CuZr, CuTi.
(v)   The amorphous alloy crystallizes during hydrogen absorption:

$AB_x(a) + yH_2 \rightarrow AB_xH_{2y}(c)$

An example of this type of reaction is provided by amorphous $Ni_{50}Zr_{50}$ [3].

Very little is known about the stability against crystallization of amorphous alloys with large amounts of dissolved hydrogen. We believe that it is safe to assume that the chemical properties of an amorphous alloy differ only slightly from those of a homogeneous crystalline alloy of the same composition. This means that the driving force for the disproportionation reaction will be approximately the same. Since the decomposed state is the most stable state, the occurrence of the reactions (i) to (v) will be governed by the kinetics of the various processes. Up till now we have found no examples of reaction (iii). It is likely that, if the atomic mobility is high enough for decomposition in the amorphous state, crystallization will also take place.

Another uncertainty is the existence of amorphous hydrides (reaction (ii)). As will be shown below up till now we have found no alloys which show a pressure plateau upon hydriding. It has been claimed that this may be due to the fact that contrary to the crystalline state there exists a large variety of interstitial sites in an amorphous alloy. However, similar arguments could be used against phase separation in the liquid state of which numerous examples exist.

3. EXPERIMENTAL DETAILS

The alloys investigated had the following nominal compositions: $Ni_{63.7}Zr_{36.3}$, $Ni_{58.9}Zr_{41.1}$ ($Ni_{10}Zr_7$) and $Ni_{71.3}Zr_{28.7}$ (Ni_5Zr_2). The preparation of the alloys and the method to measure the pressure-composition isotherms have been described in an earlier paper [1].

The heats of formation of the crystalline alloys were determined with a liquid-bath calorimeter [4]. From the heats of solution of Zr, Ni and the alloys mentioned above in liquid Ge, the heats of formation of the alloys were obtained.

The kinetics of the hydrogen absorption reaction was measured in two ways. (i) The pressure change during each step of the

PC-isotherm measurements was recorded. (ii) The heat effect that occurred during hydriding was recorded with a calorimeter (Perkin-Elmer, DSC-2).

The surface composition was analysed by several techniques: Rutherford Back Scattering, Auger spectroscopy and X-ray micro-analysis. Some preliminary results will be presented.

4. PHASE DIAGRAM NEAR THE EUTECTIC COMPOSITION $Ni_{64}Zr_{36}$

There exists considerable confusion on the composition and structure of the intermetallic compounds adjacent to the eutectic composition $Ni_{63.7}Zr_{36.3}$. Table I shows some relevant data from the literature and in fig. 1 the part of the NiZr-phase diagram in concern, according to Elliott [5], is depicted.

Table I. Literature Data of Phases on
both Sides of eutectic $Ni_{64}Zr_{36}$

Reference	Intermetallic compounds
Kirkpatrick and Larsen [6]	$Ni_{10}Zr_7$ and Ni_5Zr_2
Meny and Champigny [7]	Ni_3Zr_2 and Ni_3Zr
Kubaschewski [8]	$Ni_{10}Zr_7$ and Ni_2Zr

In addition Sweeney and Batt [9] report that they did not observe Ni_5Zr_2 in their diffusion experiments, but Pet'kov, Markiv and Gorskiy [10] did not observe Ni_5Zr_2 in as-cast specimens. However, after annealing these specimens at $900^{\circ}C$ for 200 hours the Ni_5Zr_2 phase had disappeared. To check the phase diagram we prepared specimens with nominal compositions $Ni_{10}Zr_7$, $Ni_{63.7}Zr_{36.3}$ and Ni_5Zr_2. Microprobe investigation showed that the as-cast $Ni_{63.7}Zr_{36.3}$ sample consisted of a fine-grained mixture of $Ni_{10}Zr_7$ and Ni_5Zr_2. The $Ni_{10}Zr_7$ sample consisted mainly of $Ni_{10}Zr_7$ but besides contained less than 2 percent of the eutectic mixture. The as-cast Ni_5Zr_2 sample consisted of three phases: a primary phase $Ni_{77}Zr_{23}$ a secondary phase Ni_5Zr_2 and thirdly an eutectoid mixture with overall composition $Ni_{59}Zr_{41}$ (this deviation from the eutectic composition may be due to the rapid cooling of the specimen). After annealing for 4 hours at $1030^{\circ}C$ the sample consisted of Ni_5Zr_2 and for about 5% of $Ni_{10}Zr_7$. Further annealing this sample at $900^{\circ}C$ during 215 hours caused no difference in the composition, contrary to the results of Pet'kov et al. mentioned above. But our results are in agreement with the phase diagram given by Elliott (5; see fig. 1) and for the rest of this paper we consider Elliott's diagram as the correct one.

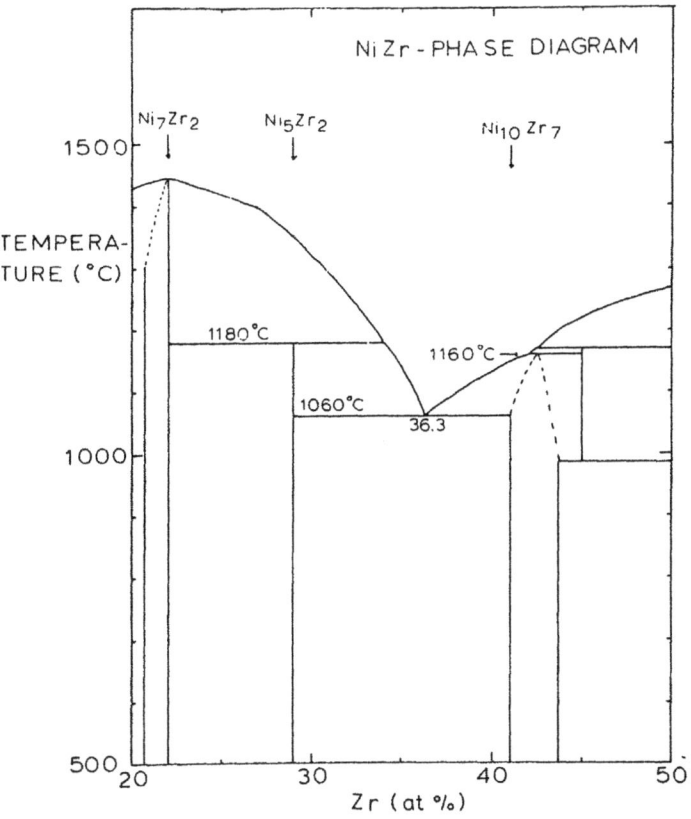

Fig. 1. Part of the NiZr Phase Diagram according to Elliott [5].

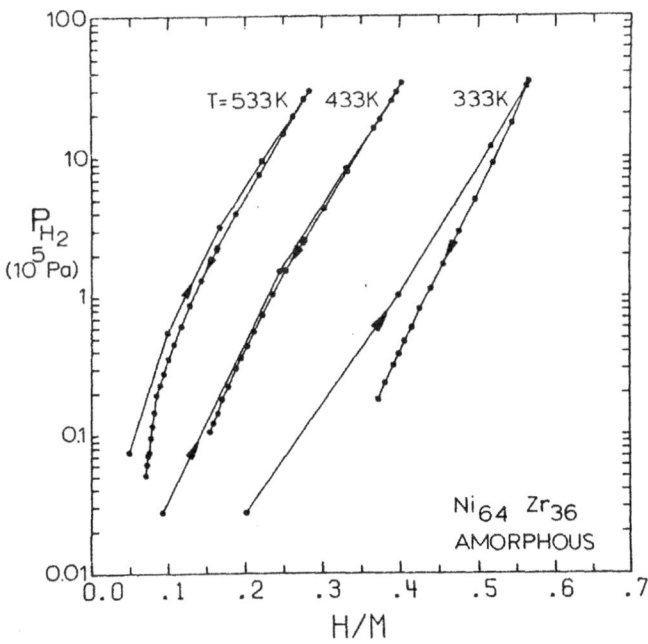

Fig. 2. Isotherms of amorphous $Ni_{64}Zr_{36}$ at 533 K, 433 K and 333 K.

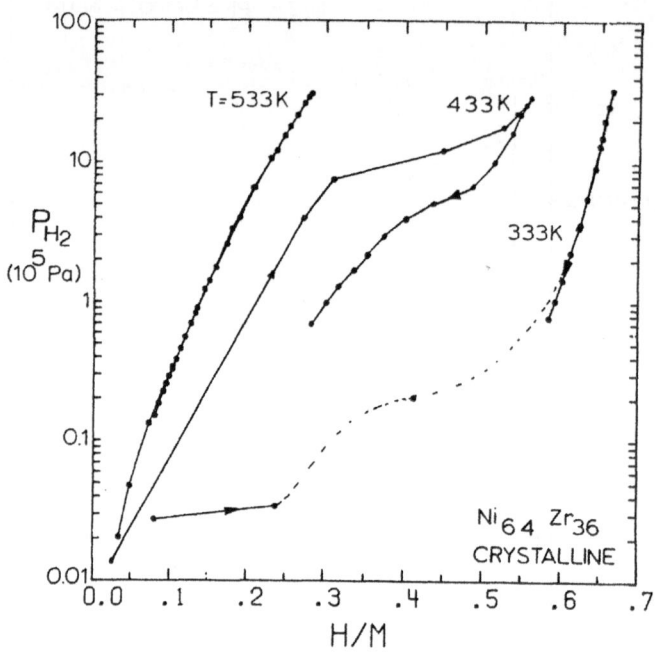

Fig. 3. Isotherms of crystalline $Ni_{64}Zr_{36}$ at 533 K, 433 K and
333 K.

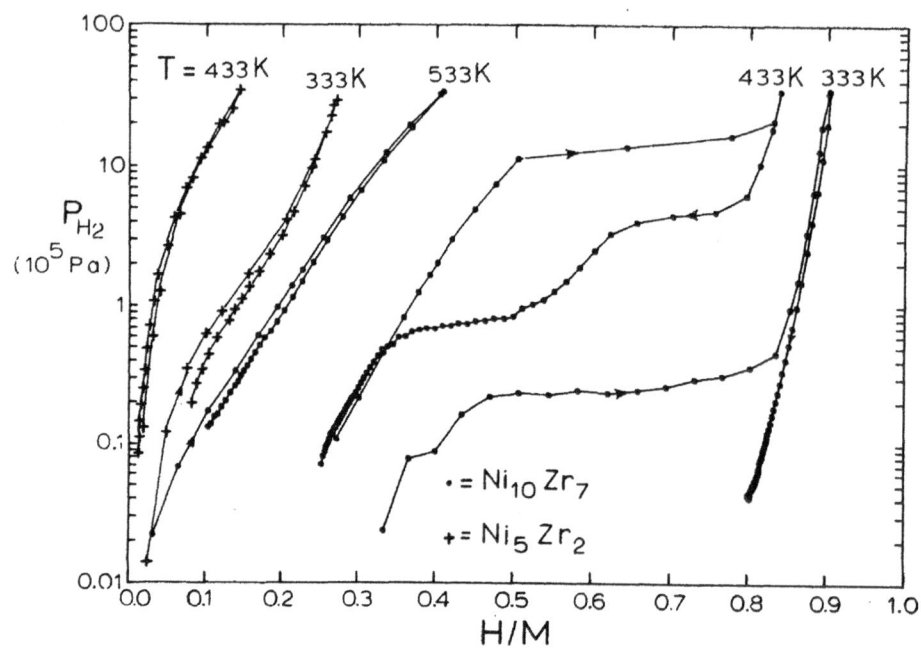

Fig. 4. Isotherms of $Ni_{10}Zr_7$ at 533 K, 433 K and 333 K and of
Ni_5Zr_2 at 433 K and 333 K.

5. PRESSURE-COMPOSITION-ISOTHERMS

Below a certain critical temperature the pressure-composition-isotherms of many hydride forming metals and alloys show a so-called pressure plateau: within a range of hydrogen concentration the pressure remains constant. Such a plateau corresponds with the coexistence of a saturated solution of hydrogen in metal, the α-phase, and a metal hydride, the β-phase. However, none of the isotherms of amorphous $Ni_{64}Zr_{36}$ measured at 333 K, 433 K and 533 K showed a pressure plateau (see figure 2).

This phenomenon can be due to a rather low critical temperature of amorphous $Ni_{64}Zr_{36}$ hydride or there may be no phase separation at all, even at low temperature. The low temperature range is difficult to investigate because of prohibitively long equilibration times. At room temperature for example the equilibration time amounted to several days, moreover no plateau was registered.

There is a clear difference between the hydrogen sorption behaviour of amorphous and crystalline $Ni_{64}Zr_{36}$ (compare figs. 2 and 3). Whereas the 533 K isotherms of both alloys are almost similar, the 433 K isotherms of crystalline $Ni_{64}Zr_{36}$ suggest the presence of a pressure plateau. Because of the limited number of data the existence of a plateau in the 333 K isotherm of crystalline $Ni_{64}Zr_{36}$ is not clearly defined but this isotherm differs considerably from the 333 K curve of amorphous $Ni_{64}Zr_{36}$. Our eutectic crystalline sample $Ni_{64}Zr_{36}$ consisted of $Ni_{10}Zr_7$ and Ni_5Zr_2 as discussed above. The 433 K and 333 K isotherms of crystalline $Ni_{64}Zr_{36}$ can be constructed from the corresponding isotherms of $Ni_{10}Zr_7$ and Ni_5Zr_2 (fig. 4) by adding 61.3% of the H/M ratio of $Ni_{10}Zr_7$ and 38.7% of this ratio of Ni_5Zr_2 at the corresponding pressures (fig. 5). But unfortunately a direct comparisson of the isotherms of amorphous $Ni_{64}Zr_{36}$-hydride with the isotherms of a single phase crystalline $Ni_{64}Zr_{36}$-hydride is not possible.

For crystalline NiZr-alloys some data about the hydrogen sorption characteristics are available from the literature. Tannenbaum et al. [11] give values ranging from 0.9 to 1.1 for the maximum hydrogen to metal ratio of $Ni_{10}Zr_7$-hydride at room temperature. We found a H/M ratio of 0.9 at 333 K. The H/M-ratio's of crystalline $Ni_{64}Zr_{36}$ and Ni_5Zr_2 measured at 333 K are somewhat smaller than the ratio's which Tannenbaum et al. found at room temperature.

6. THERMODYNAMICS

From the P-C curves an estimation of the partial molar enthalpy of solution of hydrogen (heat of solution, ΔH_s^o) and of the partial molar enthalpy of formation of a hydride (heat of formation, ΔH_f^o) can be made. In case of hydride formation the

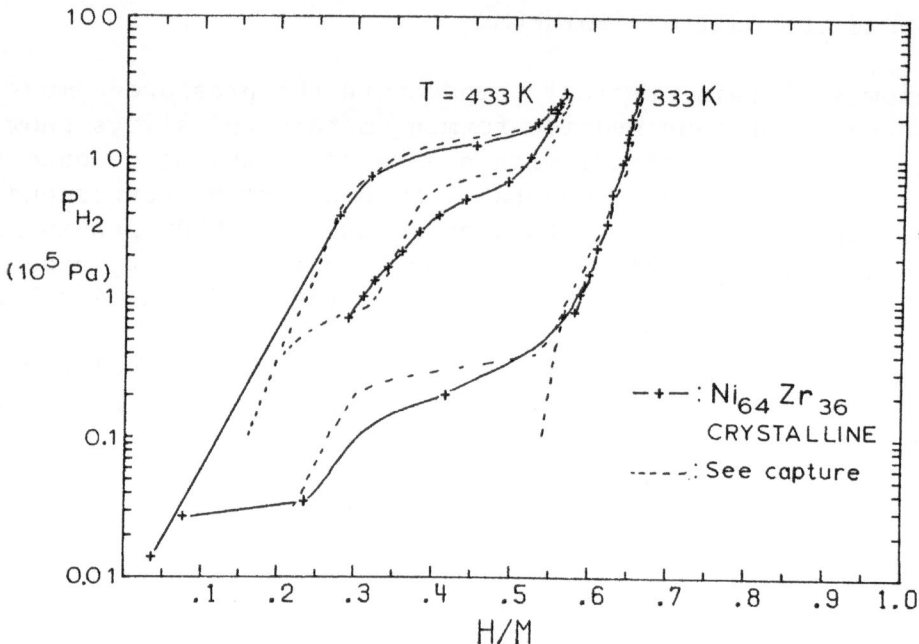

Fig. 5. Isotherms of crystalline $Ni_{64}Zr_{36}$ at 433 K and 333 K and the isotherms (dashed lines) constructed from the curves of Ni_5Zr_2 and $Ni_{10}Zr_7$ as described in the text.

dependence of the plateau pressure $P_{H_2}^*$ on ΔH_f^o and the change in partial molar entropy ΔS_f^o per mole H_2 is described by $\ln P_{H_2}^* = \Delta H_f^o/RT - \Delta S_f^o/R$ (1). In case of an ideal solution of hydrogen the formula $\ln P_{H_2} = \Delta H_f^o/RT - \Delta S_s^o/R + 2 \ln x_H$ (2) holds where P_{H_2} is the equilibrium pressure and x_H the number of dissolved H-atoms per metal atom. Assuming that ΔH_f^o and ΔS_f^o, respectively ΔH_s^o and ΔS_s^o are independent of the temperature in a limited range, ΔH_f^o, respectively ΔH_s^o can be calculated from a plot of $\ln P_{H_2}^*$ or $\ln P_{H_2}$ versus T^{-1}. In the case of $Ni_{10}Zr_7$ and crystalline $Ni_{64}Zr_{36}$ equation (1) and in the case of Ni_5Zr_2 equation (2) is used to calculate the value of ΔH^o.

These values of ΔH^o are compared in Table II with the values calculated from the alloy model developed by Miedema and coworkers [12-14]. In this model it is assumed that upon the formation of a ternary hydride from a binary alloy AB_n, bonds between A and B atoms are broken by hydrogen atoms. It is supposed that A-A bonds do not occur in B-rich alloys. So in this cellular model, the heat of formation of the ternary hydride is taken to be the sum of the heats of formation of the two binary hydrides minus the heat of formation of the binary alloy B_nA [14]; e.g.:
$\Delta H^o(AB_5H_5) = \Delta H^o(AH_2) + \Delta H^o(B_5H_3) - \Delta H^o(AB_5)$ and
$\Delta H^o(AB_3H_4) = \Delta H^o(AH_2) + \Delta H^o(B_3H_2) - \Delta H^o(AB_3)$
In a refinement on this theory [15] Miedema takes into account that a number of A-A bonds will exist in the alloys with a higher

A-content, and he corrects this by subtracting only part of the heat of formation of the original alloys, e.g.:

$$\Delta H^o(AB_2H_3._5) = \Delta H^o(AH_2) + \Delta H^o(B_2H_1._5) - 0.7 \Delta H^o(AB_2)$$
$$\Delta H^o(ABH_2._5) = \Delta H^o(AH_1._5) + \Delta H^o(BH) - 0.5 \Delta H^o(AB).$$

For these calculations we require therefore the heats of formation of ZrH_2, NiH, $Ni_{64}Zr_{36}$, Ni_5Zr_2 and $Ni_{10}Zr_7$. Notice that in the formula AB_n the stable-hydride forming element is A (here:Zr) and B (here:Ni) does not form stable hydrides or only at very high hydrogen gas pressures. We used for the binary hydrides of Zr and Ni values of $- 163$ kJ/mol H_2 [16] and $- 8$ kJ/mol H_2 [17] respectively. As mentioned above the heats of formation of $Ni_{64}Zr_{36}$, Ni_5Zr_2 and $Ni_{10}Zr_7$ were determined by liquid bath calorimetry. For the heat of formation of amorphous $Ni_{64}Zr_{36}$ we used the value of crystalline $Ni_{64}Zr_{36}$ minus the heat of crystallization of amorphous $Ni_{64}Zr_{36}$ [1]. Table II shows the results obtained.

Table II. Heats of Formation of binary Alloys, of ternary Hydrides as calculated from the PC-isotherms ΔH^o(exp), and of ternary Hydrides as calculated from the Miedema-model, ΔH^o (model).

Alloy	ΔH^o(exp) of binary alloy (kJ/mol)	ΔH^o(exp) of ternary hydride (kJ/mol)	ΔH^o(model) of ternary hydride (kJ/mol H_2)
Ni_5Zr_2	$- 45 \pm 4$	$- 39 \pm 4$	$- 20$
$Ni_{64}Zr_{36}$ (crystalline)	$- 40 \pm 5$	$- 41 \pm 4$	$- 48$
$Ni_{64}Zr_{36}$ (amorphous)	$- 44 \pm 4$	$- 44 \pm 6$	$- 48$
$Ni_{10}Zr_7$	$- 48 \pm 4$	$- 47 \pm 5$	$- 52$

In view of the rather crude approximations made in the calculation of both experimental and theoretical ΔH^o-values of the ternary hydrides, the agreement between the two sets of values is good. Only the Ni_5Zr_2-values differ considerably. In this material however only a small amount of hydrogen is absorbed (fig. 4) and therefore the Miedema-theory may not be applicable.

7. KINETICS

     Several processes take place during the hydriding of a material. If no oxide layers or other surface barriers are present hydrogen molecules are adsorbed on the surface of the sample and dissociated into hydrogen atoms. These atoms are then absorbed in the outer layers of the sample and diffuse into the bulk of the material. Each of these processes can be the rate-limiting step for the absorption of hydrogen. In most cases however, the situation is even more complex because of the role which surface oxide layers and metallic precipitates play in the absorption process [18,19]. Thus for a quantitative description of the ab-

sorption process information about the surface conditions is of
paramount importance. Preliminary results of surface analysis
will be presented in the next section.

One of the most prominent features of the existence of surface barriers is the fact that as-quenched material does not absorb
hydrogen unless the samples have been activated. Activation can
be achieved in several ways. One way is to heat the samples in a
hydrogen atmosphere. In the differential scanning calorimeter during heating of amorphous $Ni_{64}Zr_{36}$ at a rate of 40 K/min, the
exothermic heat effect caused by the absorption of hydrogen started only at 600 K. A minor improvement was found by etching a
sample (in a weak HCl solution) in order to remove an oxide layer,
or by plating the surface of a sample with a Pd layer (in a $PdCl_2$
solution). The best preparation method was to grind the as-quenched samples, in which case hydrogen absorption started already
at 400 K in the DSC. If after the absorption of hydrogen the pur-
ge gas of the DSC was changed from hydrogen to argon, desorption
of hydrogen was observed. Below 470 K the desorption of hydrogen
was not complete: at increasing the temperature, more hydrogen
evolved from the sample. Repeating the procedure of (partially)
hydriding and dehydriding, it was noticed that the sorption rate
increased steadily. Only after about 15 cycles was the rate constant. The time dependence of the heat effect during absorption
of hydrogen at 5 temperatures between 380 K and 470 K is shown
in fig. 6. Using the straight parts of the curves of figure 6,

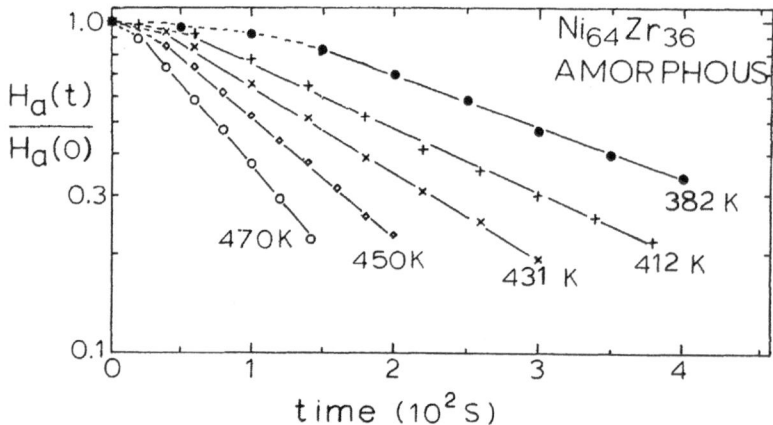

Fig. 6. The heat of absorption of hydrogen by amorphous $Ni_{64}Zr_{36}$
        as function of time and temperature.

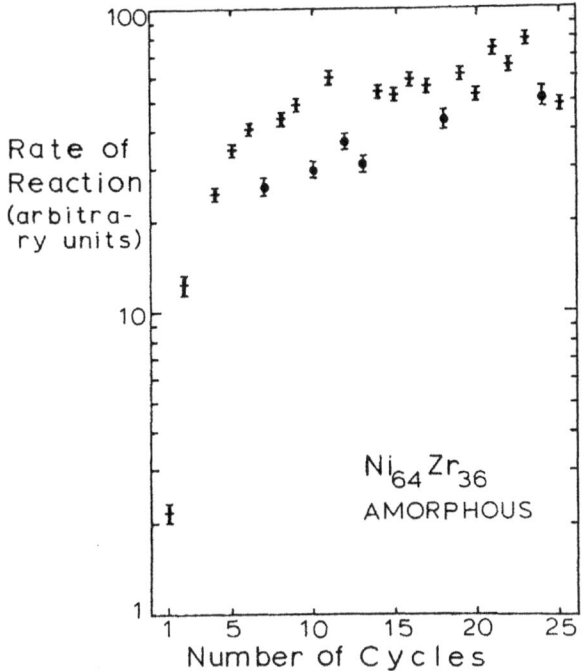

Fig. 7.
The hydrogen absorption
rate as function of the
number of hydrations.
For the meaning of the
symbols see the text.

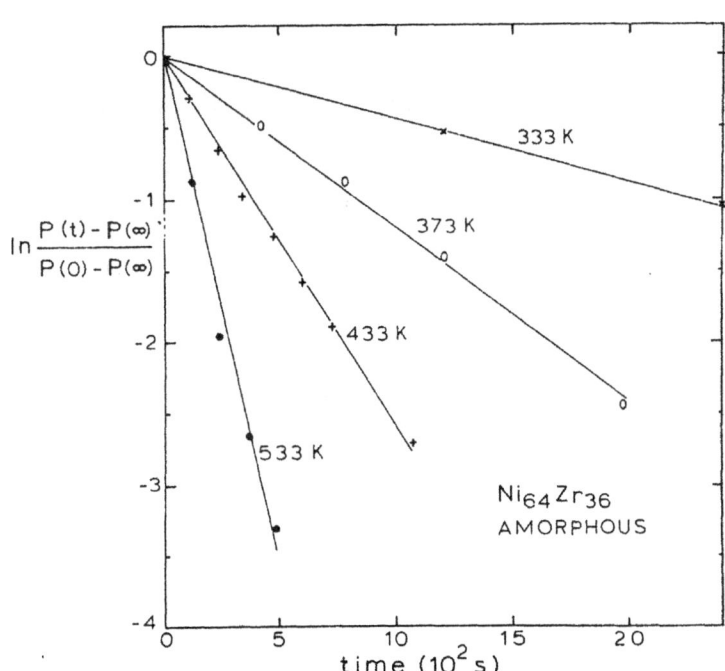

Fig. 8. The pressure drop during the hydriding of amorphous
Ni$_{64}$Zr$_{36}$ as function of time and temperature.

the activation energy of the absorption process was estimated at
25 ± 3 kJ/mol.

The absorption rate of amorphous $Ni_{64}Zr_{36}$ samples was determined also by monitoring the pressure drop. In this
case the activation of the samples was induced by cycling between
a H_2 pressure of about 30 bar and vacuum (10^{-2} Torr) at an elevated temperature (370 K - 530 K). Here also the absorption rate increased with the number of absorption-desorption cycles as can be
seen in figure 7. This figure shows the absorption rate as a function of the number of cycles. A plot of $\ln(p(t) - p(\infty))$ against t
was made where $p(t)$ is the hydrogen pressure at time t and $p(\infty)$
the equilibrium pressure. As a measure for the absorption rate the
mean slope of the first 30% of the pressure drop was used. The
absorption rate increased considerably during the first 3 sorption
cycles and much less during the next cycles. Six of the data
points (marked with a different symbol) were obtained after ope-
ning the reaction vessel or after prolonged pumping. Under these
circumstances oxidation of both Ni and Zr can take place. The
first process can be detrimental to the absorption rate since metallic nickel functions as a catalyst for the dissociation of hydrogen molecules. The increase of the hydrogen sorption rate during the first sorption cycles can not be ascribed to an increase
of surface area since no desintegration of the specimen has been
observed. Therefore the increase of the sorption rate must be due
to the catalytic effect of metallic Ni precipitates at the surface.

The absorption rate was determined at 533 K, 433 K, 373 K
and 333 K. In fig. 8 the logarithm of the relative decrease of
the pressure, $\ln((p(t) - p(\infty)))$, with time is shown. The slope
of these curves was plotted logarithmically against $1/T$. From the
Arrhenius plot the value of the activation-energy was estimated
at 20 ± 5 kJ/mol, which is in good agreement with the value of
25 ± 3 kJ/mol found by calorimetry.

8. CHANGES IN SURFACE COMPOSITION

It is well-known that repeated hydrogen absorption and desorption induces changes in the properties of metallic hydrides.

A first indication of such changes in amorphous $Ni_{64}Zr_{36}$
samples was obtained from measurements of the magnetization of
the samples after a varying number of absorption and desorption
cycles. We found that whereas at room temperature the as-quenched
samples were paramagnetic, samples which had undergone 100 cycles
were superparamagnetic (fig. 9). This observation points to
segregation of part of the Ni in the form of small magnetic clusters. The measured value of the saturation magnetization corre-
sponded to the segregation of 1.7% of the Ni-atoms.

Since it is likely that this segregation takes place predominantly at the sample surface we investigated the surface by
means of Rutherford Back Scattering (R.B.S.) and other surface

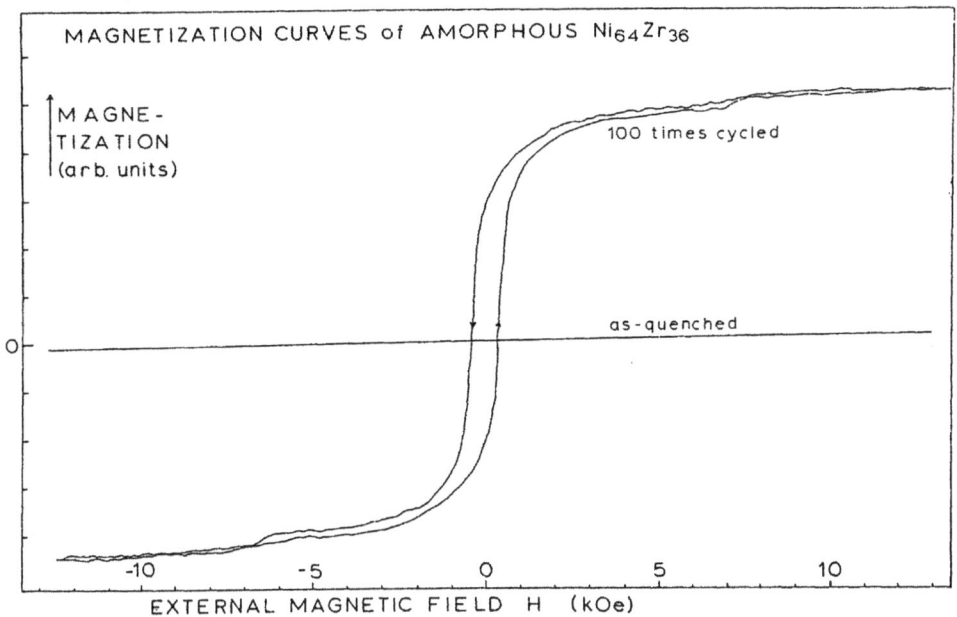

Fig. 9. Magnetization curves of a 100 times cycled and an as-
quenched amorphous $Ni_{64}Zr_{36}$ sample. The hysteresis is of
instrumental origin.

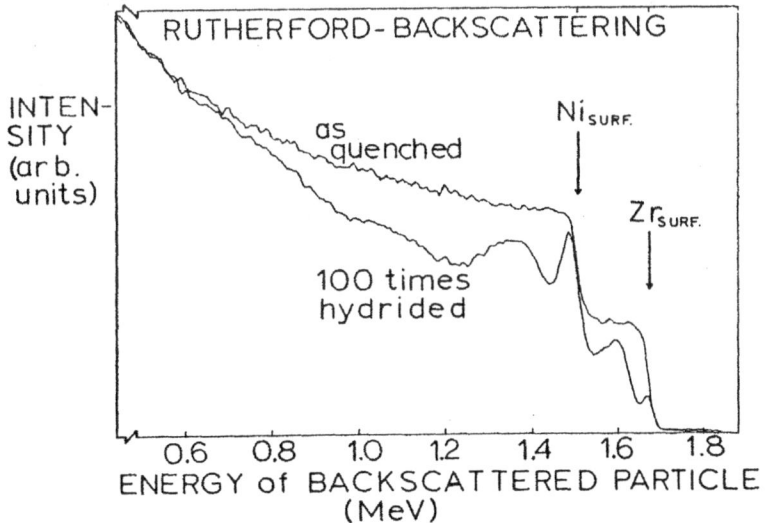

Fig.10. Rutherford-backscattering of 2 MeV He^{+}-ions on an as-quenched and a 100 times hydrided $Ni_{64}Zr_{36}$ sample. The energy-
scale corresponds with a depth-scale dependent on the mass
of the atom, by which the He^{+}-ion is backscattered.

techniques. In R.B.S., specimens are bombarded with energetic particles (in our case 2 MeV He$^+$-ions). The energy of the back-scattered particles is measured. This energy depends both on the mass of the atom with which the particle has collided and on the depth where the collision has taken place. R.B.S. spectra for both an as-quenched and a 100 times cycled specimen are shown in fig. 10. The energies of particles backscattered from surface Ni- and Zr-atoms are indicated. It is evident from this figure that a layer just beneath the surface is slightly enriched in nickel and depleted in Zr. The existence of such a layer was con-firmed by means of Auger spectroscopy. From the electronic stop-ping power, which is about 500 eV/nm^{-1}, one can estimate the thickness of the Ni-enriched layer. We found a thickness of about 0.05 μm. Since the total thickness of the sample is ∿ 30 μm the amount of Ni segregated near the surface is estimated to be less than 0.5%. Although the precision of both the magnetization and the surface composition measurements is not very high the discre-pancy between both results indicates that part of the superpara-magnetic Ni particles were formed somewhere in the bulk of the amorphous material.

The driving force for the segregation of Ni may be the same as for the disproportionation reaction (iv). An alternative ex-planation may be provided by the selective oxydation of Zr due to the presence of oxygen or water vapour in the hydrogen gas [20, 21].

Indeed, x-ray microanalysis and preliminary results of ESCA- and Auger measurements showed the presence of oxygen on the sur-face of the hydrided specimens. Thus the occurrence of supermag-netic Ni-particles at the surface can be explained by the reac-tion:

$$Ni_x Zr_y + yO_2 \rightarrow xNi + yZrO_2$$

The same reaction could occur in the bulk of the amorphous mate-rial if the solubility of oxygen in the material is high. It is known that the solubility of oxygen in pure zirconium is as high as 30 at.perc. Unfortunately no data are available on the solu-bility of oxygen in amorphous NiZr, but even if the oxygen solu-bility is much lower internal oxydation is possible in principle. A more detailed description of the surface layer analysis and the results obtained will be given elsewhere [3].

9. CONCLUSIONS

The isotherms of amorphous $Ni_{64}Zr_{36}$ do not show a pressure plateau in the temperature range investigated in contrast with the crystalline compound $Ni_{10}Zr_7$. The critical temperature for amorphous $Ni_{64}Zr_{36}$ hydride, if it exists at all, lies below room temperature. It is of interest to note that the 333 K isotherm for crystalline Ni_5Zr_2 already shows a clear inflection point whereas no such effect is observed in the 333 K isotherm for the amorphous alloy.

The heats of formation of alloys for which a pressure plateau is observed is in good agreement with the values obtained from the Miedema model. The same holds for the heat of solution obtained for amorphous $Ni_{64}Zr_{36}$. Agreement for the crystalline alloy Ni_5Zr_2 is rather poor. However this alloy absorbs very little hydrogen. Therefore the applicability of Miedema's model to this case is probably not justified.

Changes in surface composition and improvement of absorption kinetics of amorphous $Ni_{64}Zr_{36}$ are most probably due to selective oxydation of zirconium by small amounts of oxygen or water vapour in the hydrogen gas. The remaining Ni-rich clusters probably serve as catalytic sites for the dissociation of hydrogen molecules.

REFERENCES

1. F.H.M. Spit, J.W. Drijver and S. Radelaar, Zeitschrift für Physikalische Chemie Neue Folge Bd 116 (1979) 225
2. A.J. Maeland,"Proc. Int. Symp. on Hydrides for Energy Storage". Int. J. of Hydr. Energy, Pergamon Press, 1978
3. F.H.M. Spit et al. To be published.
4. K. Tang and R. Castanet, J. of Less Common Metals 51 (1977) 125
5. R.P. Elliott, Constitution of binary alloys, 1st Suppl., McGraw Hill (1965)
6. M.E. Kirkpatrick and W.L. Larsen, Transactions of the ASM, vol. 54 (1961) 580
7. L. Meny, M. Champigny. CEA-R-3517, Serv. Centr. de Doc du CEA (1968)
8. O. Kubaschewsky (ed.). Atomic Energy Rev. Special issue no. 6
9. W.E. Sweeney, Jr. and A.P. Batt, J. Nucl. Mat. 13, no. 1 (1964) 87
10. V.V. Pet'kov, V.Ya. Markiv and V.V. Gorskiy, Russ. Metallurgy (1972) 137
11. I.R. Tannenbaum, W.L. Korst, J.S. Mohl and G.G. Libowitz, 144th meeting of the Amer. Chem. Soc., Los Angeles, Calif., April 1963 (NAA-SR-7132)
12. A.R. Miedema, R.Boom, F.R. de Boer, J. Less Common Metals 41 (1975) 283
13. A.R. Miedema, J. Less Common Metals 46 (1976) 67
14. A.R. Miedema, K.H.J. Buschow, H.H. van Mal, J. Less Common Metals 49 (1976) 463
15. A.R. Miedema, private communication
16. D.R. Frederickson, R.L. Nuttal, H.E. Flotow and W.N. Hubbard, J. Phys. Chem. 67 (1963) 1506
17. B. Baranowski, Hydrogen in Metals II. Ed.: G. Alefeld and J. Völkl, Springer Verlag, Berlin (1978)
18. L. Schlapbach, A. Seiler, F. Stucki and H.C. Siegmann, J. Less Common Metals 73 (1980) 145

19. G.D. Sandrock and P.D. Goodwell, J. Less Common Metals
    <u>73</u> (1980) 161
20. G.S. Krinchik, L.V. Nikitin, V.V. Lunvin and P.A. Chernavskii,
    Sov. Phys. Solid State <u>21</u> (1979) 354
21. W.E. Wallace, R.F. Karlicek and H. Imamara, J. Phys. Chem.
    <u>83</u> (1979) 1708

ACKNOWLEDGEMENTS

Ir. D. Schalkoord, Dr. P. Kool and co-workers (Delft) are acknowledged for performing X-ray micro-analysis, Auger spectroscopy and ESCA-measurements.
Thanks are due to Prof. J.C. Mathieu, Dr. R. Castanet and co-workers (Marseille) for putting a high temperature calorimeter at our disposal.
The authors are grateful to their students K. Blok, W. Claassen, E. Hendriks, M. Kramer, H. Wevers and G. Winkels for their contributions to this study.

THERMODYNAMICS OF METAL-HYDROGEN SYSTEMS

Ted B. Flanagan

Department of Chemistry
University of Vermont
Burlington, VT   05405

In this paper some topics on the thermodynamics of metal-hydrogen systems will be discussed. The principal experimental techniques used to determine thermodynamic data in these systems are described and compared. A new technique, "hysteresis scan calorimetry", is described and its application to metal-hydrogen systems is discussed. The thermodynamics of the solvus are developed. The effect of uniform and non-uniform stress on hydride precipitation is given and finally the application of the phase rule to metal-hydrogen systems is considered.

INTRODUCTION

If solid hydrides are to be used for energy storage purposes, the thermodynamics of these systems will be of paramount importance. In this article the thermodynamics of hydrogen absorption (or desorption) will be discussed with reference to hydride formation (or decomposition) and to solution in single phase regions. The latter is important because the dissolved state is the precursor to hydride formation. Topics such as low temperature heat capacities will not be considered here. Generally the thermodynamic development presented here is equally applicable to hydrogen absorption by metals, alloys, or intermetallic compounds. It should be realized, however, that the various solids should not be thermodynamically characterized without recognizing the special features of each type of system. For example, almost all intermetallic hydrides are thermodynamically unstable with respect to disproportionation to the hydride phase of the constituent element which forms a hydride phase most exothermically and to the elemental state of the remaining element(s). At low temperatures this process is slow kinetically and therefore at these temperatures the systems can be considered

361

to be a psuedo-binary system.[1] However at elevated temperatures, which will be specific ones for each system, the disproportionation reaction occurs. Failure to realize this may lead to an incorrect interpretation of thermodynamic data. If simple statistical thermodynamic models are applied to these systems, the differences between the various solids must also be recognized. For example, the ideal partial configurational entropy depends upon the number of interstices available for occupation. For an alloy not all of the interstices may be energetically equivalent due to the different nearest neighbor environments of the interstices and therefore the expression for the partial configurational entropy will not be a simple one as in the case of pure metals.

Experimental Methods and Derived Thermodynamic Quantities. Reaction Calorimetry

The two principal types of techniques are calorimetric ones and those based on pressure-composition-temperature, p-c-T, data. Surprisingly the method of reaction calorimetry, as opposed to measurements of the heats of combustion or solution of solid hydrides, has been a relatively neglected one until recently. Kleppa, Boureau and coworkers[2-8] have recently exploited this powerful method for pure metal and alloy-H systems. Small increments of hydrogen gas are absorbed or desorbed by the sample and the heat change is determined whilst simultaneously the mols of hydrogen absorbed or desorbed are determined from pressure changes. If small increments of hydrogen are employed, the measured changes approach relative partial molar values. A correction must be made for the work of expansion of the gas into (or out of) the calorimeter vessel. There has been some discussion about this correction term[9] but apparently the one employed many years ago by Nace and Aston[10] is the correct one. Relative partial molar enthalpies are defined as

$$\Delta H_H = H_H - \tfrac{1}{2}H^o_{H_2} = (\partial H/\partial n_H)_{T,p,n_M} - \tfrac{1}{2}H^o_{H_2} \text{ ------------------1}$$

where n_H and n_M are mols of hydrogen and metal, respectively and the standard designation refers to unit fugacity. This is sometimes termed the relative partial molal enthalpy but molar is the recommended S.I. terminology. If instead of n_H we employ, r where r = mol H/ mol metal, for the compositional variable then,

$$(\partial H/\partial r)_{T,p,n_M} (dr/dn_H)_{T,p,n_M} = (\partial H/\partial r)_{T,p,n_M} 1/n_M = (\partial H/\partial n_H)_{T,p,n_M}$$

$$,\text{i.e., } (\partial H_m/\partial r)_{T,p,n_M} = H_H \text{ where } H_m = H/ n_M. \text{ ---------2}$$

From the equilibrium hydrogen pressure measured at the same hydrogen content where the enthalpy has been determined calorimetrically, $\Delta\mu_H$ may be obtained; in practice, the average hydrogen pressure is employed which obtains at an average hydrogen content before and after the increment of hydrogen has been added (or removed). Thus

$$\Delta\mu_H = \mu_H - \tfrac{1}{2}\mu^o_{H_2} = RT \ln p^{\frac{1}{2}}_{H_2} \text{(equil)} \quad\text{------------------------}3$$

since $\mu_H = \tfrac{1}{2}\mu^o_{H_2} + RT \ln p^{\frac{1}{2}}_{H_2}$ (equil). From the calorimetric measurements values of ΔH_H are determined and combination of these values at a given H-content with values of $\Delta\mu_H$ gives values of ΔS_H. Thus all three thermodynamic parameters can be determined as a function of r from absorption runs (or desorption) at a given temperature.

In the two solid phase region, absorption of hydrogen (per mol $\tfrac{1}{2}H_2$) corresponds to the integral reaction

$$\tfrac{1}{2}H_2(g, 1 \text{ atm}) + \frac{MH_{a'}}{(b'-a')} \rightarrow \frac{MH_{b'}}{(b'-a')} \quad\text{----------------------}4$$

where a' and b' refer to the H-to-metal atom ratios for the coexisting H-saturated metal and the hydride phase, respectively (Fig. 1). The reverse process corresponds to desorption

$$\frac{MH_b}{(b-a)} \rightarrow \frac{MH_a}{(b-a)} + \tfrac{1}{2}H_2(g, 1 \text{ atm}) \quad\text{------------------------}5$$

where a and b refer to the same quantities as for reaction 4 but due to hysteresis the values differ slightly from a' and b' (Fig. 1). Enthalpies of hydrogen absorption, reaction 4, or desorption, reaction 5, can be determined from the calorimetric measurement of the heat changes for these reactions where relatively large amounts of hydrogen can be added or removed because the relative enthalpy change is constant in the two solid phase coexistence region. Because of hysteresis, the magnitude of the relative enthalpy change for absorption should be slightly smaller than that for desorption. This difference has not been detected so far in calorimetric studies[8,11]

A recent calorimetric technique for solid-H systems has been developed by the author and coworkers[12,13] - "hysteresis scan calorimetry". This technique measures the enthalpy changes along hysteresis scans (Fig. 1) and its usefulness derives from the fact that it allows the values of the relative partial enthalpies of the coexisting single phases to be determined more precisely than by direct measurement in the single phase regions. According to C. Wagner[14] during a hysteresis scan hydrogen is added or removed

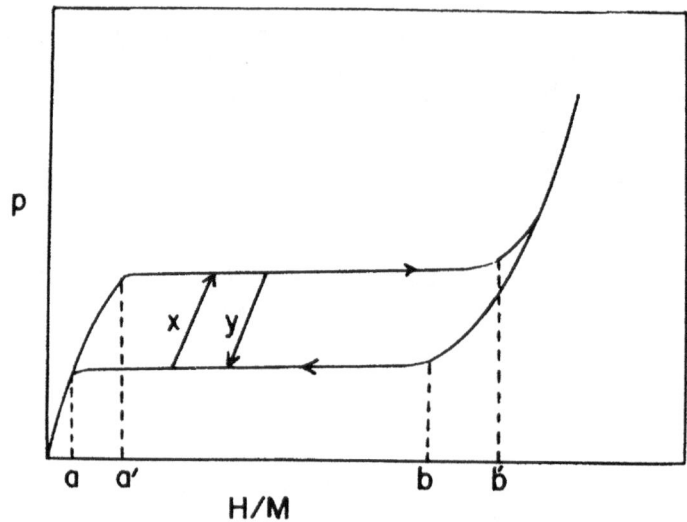

Fig. 1.   Schematic representation of hysteresis loop in a metal-
hydrogen system. a' and a represent the α-phase boundaries during absorption and desorption of hydrogen,
respectively and b' and b represent the hydride phase
boundaries for absorption and desorption of hydrogen,
respectively. X represents an absorption hysteresis scan
commencing from the desorption branch of the hysteresis
loop and y represents a desorption scan commencing from
the absorption branch of the hysteresis loop.

from both phases, e.g., for an absorption scan, a → a' and b → b', Fig. 1, without forming any additional amounts of either phase. Curiously, the validity of this mechanism had not been examined until recently[12]. It is easy to test whether this mechanism is valid because it is a reversible mechanism, i.e., it predicts that hydride phase neither forms nor decomposes. Irreversibility is always associated with hydride phase formation or decomposition in metal-H systems. If the scan is reversible, the Wagner mechanism is the only one which obtains. For well-annealed palladium this mechanism is valid until either plateau pressure branch is closely approached. However, for palladium which has been previously subjected to the hydride phase transition and then degassed without an anneal, this reversible mechanism predominates only until about half-way between the two pressure plateaus and then an irreversible component enters. The importance of this irreversible component increases as the final pressure plateau is approached[12].

For the reversible portions of the scans, the technique of hysteresis scan calorimetry is applicable. It follows from this mechanism that the measured relative partial enthalpy for a scan should be given by

$$\Delta H_H^s = f_\alpha^* \, \Delta H_H^\alpha + f_\beta^* \, H_H^\beta \, \text{--}6$$

where f_α^* and f_β^* are the fractions of hydrogen entering the α and β phases, respectively, at any hydrogen content and ΔH_H^α and ΔH_H^β are the relative partial molar enthalpies for the coexisting α and β phases, respectively. Scan enthalpies are measured across the two phase coexistence region and as the lower phase boundary is approached, $f_\alpha^* \to 1$ and $f_\beta^* \to 0$, and therefore $\Delta H_H^s \to \Delta H_H^\alpha$ and similarly as the hydride phase boundary is approached, $f_\beta^* \to 1$ and $f_\alpha^* \to 0$ and $\Delta H_H^s \to \Delta H_H^\beta$. In this way values of ΔH_H^α and ΔH_H^β have been determined for $LaNi_5$-H[13].

Pressure-Composition-Temperature Techniques

From equation 3, values of $\Delta \mu_H$ can be derived from p-c data for each hydrogen content and from the temperature dependence of these values, ΔS_H can be derived, i.e.,

$$(\partial \Delta \mu_H / \partial T)_{r,p} = - (\partial \Delta S / \partial n_H)_{T,p,n_M} = -\Delta S_H \, \text{------------------}7$$

and from values of ΔS_H and $\Delta \mu_H$, values of ΔH_H can be derived. For the case of pure metal and alloy-H systems the calorimetric and p-c-T technique appear to give identical results[2,4-6].

A useful variant to the p-c-T method is the isochoric technique[15]. In this method a large sample is used with a small sample container volume. The pressure measuring device is within the

sample vessel. The equilibrium pressures are then recorded as a
function of temperature under conditions where a minimum amount of
hydrogen enters or leaves the sample; the hydrogen content of the
sample is therefore essentially constant (corrections can be made
for the small amounts of hydrogen actually transferred). This
method has two advantages over the conventional isotherm method for
certain regions of hydrogen content. In the conventional method
cumulative errors in hydrogen content occur in each isotherm and
these become especially important in the region of large hydrogen
contents where the pressures vary markedly with hydrogen content;
the isochoric method is free from these cumulative errors. An
additional advantage is that a small amount of hydrogen is transferred following the temperature changes and therefore equilibrium
can be rapidly achieved even at relatively low temperatures.

Data obtained in the two solid phase region is a special case
of an isochoric method; in this region the hydrogen contents of
the coexisting phases are maintained constant by the phase equilibria
so that large amounts of hydrogen can be transferred without changing the coexisting phase's hydrogen contents.

Sieverts' Law of Ideal Dilute Solubility

In order to best derive Sieverts' law one result from statistical thermodynamics is needed, i.e.,

$$S_H^c = - R \ln_{(\frac{r}{\beta - r})} \text{--8}$$

where S_H^c is the ideal partial configuration entropy and β is the
number of interstices per metal atom or per formula unit if r is
expressed as H atoms per formula unit. S_H^c is derived by considering the partition function for assignment of H atoms to interstices
in a random manner such that an interstice is either occupied or
empty. Thus using equation 3 we obtain,

$$\Delta\mu_H = \mu_H - \tfrac{1}{2}\mu_{H_2}^o = H_H - T(S_H^o + S_H^c) - \tfrac{1}{2}\mu_{H_2} =$$

$$H_H - TS_H^o + RT \ln \frac{r}{(\beta - r)} - \tfrac{1}{2}\mu_{H_2}^o = RT \ln p_{H_2}^{\frac{1}{2}} \text{----------------9}$$

where S_H^o is the non-configurational entropy. Solving for r/β-r we
obtain

$$r/(\beta-r) = p_{H_2}^{\frac{1}{2}} \exp - (\Delta H_H - T\Delta S_H^o)/RT) \text{----------------------10}$$

Now as r → 0, $\Delta H_H = H_H^o - \tfrac{1}{2}H_{H_2}^o \rightarrow \Delta H_H^o$ and β-r→β, and equation 10
reduces to

$$r = \beta \, p_{H_2}^{\frac{1}{2}} \exp - (\Delta H_H^o - T\Delta S_H^o)/RT = K_s p_{H_2}^{\frac{1}{2}} \quad \text{------------------11}$$

We thus obtain Sieverts' law of ideal dilute solubility (Henry's law for a dissociating solute). Thus while ΔH_H^o can be obtained without knowledge of β, ΔS_H^o requires that β be known.

The standard designation refers to the values at infinite dilution of hydrogen without the ideal partial configurational term for the entropy, e.g., $\Delta H_H^o = H_H^o - \frac{1}{2}H_{H_2}^o$ (1 atm), $\Delta S_H^o = S_H^o - \frac{1}{2}S_{H_2}^o$ (1 atm).

Excess Thermodynamic Functions

Excess thermodynamic functions can best be defined for metal-H interstitial solutions with reference to the ideal chemical potential[16],

$$\mu_H^{id} = \mu_H^o \ (p^o, V^o, r \to 0) + RT \ln r/(\beta - r) \quad \text{------------------12}$$

Thus μ_H^E is given by

$$\mu_H^E = \mu_H - \mu_H^{id} \quad \text{--13}$$

Therefore

$$H_H^E = H_H - H_H^o \quad \text{and} \quad S_H^E = S_H - S_H^o + R \ln r/(\beta - r) \quad \text{------------14}$$

This definition of S_H^E differs from some earlier ones[17] but it results from the above, reasonable definition of the ideal interstitial solution.

Regular Interstitial Solution

A comprehensive review of the regular interstitial solution has been recently given by Oates and the author[18] and so just a brief review will be given here. For this approximation

$$\mu_H = \mu_H^o + RT \ln r/(\beta - r) + H_{HH}r/\beta \quad \text{------------------------15}$$

This approximation leads to a number of simple results. For example, the interaction energy is related to the critical temperature, T_c, by

$$H_{HH} = - 4RT_c \quad \text{--16}$$

and from the equality of the chemical potentials of the metal and hydrogen in the coexistence region we can obtain

$$T/T_c = 2(2\underline{a} - 1)/\ln \, (a/(\beta - a)) \quad \text{--------------------------17}$$
$$\underline{\beta}$$

where $(\frac{a}{\beta})$ is the fraction of filled interstices for the lower phase boundary. The effect of hysteresis will be neglected in this section and a and b will be used to denote the phase boundaries. It can also be shown that $(\frac{a}{\beta}) + (\frac{b}{\beta}) = 1$ where $(\frac{b}{\beta})$ is the fraction of filled interstices at the upper phase boundary. For reaction 4 we obtain

$$\Delta H(\tfrac{1}{2}H_2 + a \rightarrow b) = \Delta H_H^o + \tfrac{1}{2}H_{HH} \quad\text{------------------------------18}$$

$$\Delta S(\tfrac{1}{2}H_2 + a \rightarrow b) = \Delta S_H^o \quad\text{------------------------------------19}$$

Thermodynamic Parameters Derived from Phase Boundaries: the Solvus

A useful equation has been given by Flanagan and Lynch[19],

$$[d(\Delta\mu_H/T)/dT^{-1}]_{a,b} = (\partial(\Delta\mu_H/T)/\partial T^{-1})_{a,b} +$$

$$(\frac{\partial(\Delta\mu_H/T)}{\partial r})_{a,b} (dr/dT^{-1})_{a,b} \quad\text{------------20}$$

where $d(\Delta\mu_H/T)/dT^{-1}$ refers to the change of $\Delta\mu_H/T$ with T^{-1} in the α or β phases adjacent to the phase boundaries. This must have the same value as ΔH for the reaction $\tfrac{1}{2}H_2 + a \rightarrow b$, since $\Delta\mu_H$ for the two solid phase coexistence region must change with T^{-1} just as does $\Delta\mu_H$ for the single phase regions adjacent to the phase boundaries. $((\partial\Delta\mu_H/T)/\partial T^{-1})$ is the value of ΔH_H for an average constant composition in either the α or β phase near the respective phase boundary. The term $(dr/dT^{-1})_{a,b}$ represents the change of the phase boundary composition with temperature. Equation 20 can be rewritten as

$$\Delta H(\tfrac{1}{2}H_2 + a \rightarrow b) = \Delta H_H \text{ (at a, b)} + (\frac{\partial(\Delta\mu_H/T)}{\partial r})(\frac{dr}{dT^{-1}})_{a,b} \quad\text{-----21}$$

Now $\Delta\mu_H/T = R \ln p_{H_2}^{1/2}$ (equil) and $((\partial\Delta\mu_H/T)/\partial r) = R(\partial\ln p_{H_2}^{1/2}/\partial r)_T$. Thus by determining values of $R(\partial\ln p_{H_2}^{1/2}/\partial r)_T$ in single phase regions adjacent to the coexisting phases at a given temperature all terms in equation 21 are known provided that the two enthalpies have been determined, except $(dr/dT^{-1})_{a,b}$. The use of equation 21 can be illustrated by Pd-H at 298 K. The values of $R(\partial\ln p_{H_2}^{1/2}/\partial r)$ are 151 J$(K \text{ mol } H)^{-1}$ and 547 J$(K \text{ mol } H)^{-1}$ for the β and α phases, respectively[12]. $\Delta H(\tfrac{1}{2}H_2 + \bar{a} \rightarrow \bar{b}) = -19.9$ kJ$(\text{mol } H)^{-1}$, $\Delta H_H = -10.5$ kJ$(\text{mol } H)^{-1}$ and -23.5 kJ$(\text{mol } H)^{-1}$, for the α and β single phase regions adjacent to the boundaries, respectively[15]. Thus from equation 21 we obtain $(da/dT^{-1}) = -1.7 \times 10^{-2}$ K and $(db/dT^{-1}) = 2.4 \times 10^{-2}$ K. These values indicate the degree of symmetry of the phase diagram in this temperature range. If the phase diagram is symmetric about the critical composition, then $(da/dT^{-1}) + (db/dT^{-1}) = 0$. It is clear from equation 21 that if there are no discontinuities in the thermodynamic parameters at the phase boundaries, e.g., if $\Delta H(\frac{1}{2}H_2 + a \rightarrow b) = \Delta H_H(at\ a,b)$, then the phase boundaries would be infinitely steep with temperature.

If, in equation 21, we substitute Z for the composition variable r and solve for $d\ln Z/dT^{-1}$, we obtain

$$-d\ln Z/dT^{-1} = \frac{-\Delta H(\frac{1}{2}H_2 + a \rightarrow b) + \Delta H_H(at\ a)}{Z(\partial\Delta\mu_H/T)/\partial Z} = \Delta H_{sol}/R \text{-----}22$$

at the lower phase boundary, where $Z = a/(\beta - a)$ and ΔH_{sol} is the experimental solvus enthalpy. In the very dilute region where ideal behavior obtains, $(\partial(\Delta\mu_H/T)/\partial Z) = R/Z$, and equation 22 reduces to the situation where $\Delta H_{sol} \rightarrow \Delta H^{\circ}_{sol}$ and this corresponds to the transfer of one mol of H from the hydride phase to the hydrogen-saturated metal. This is the usual interpretation of ΔH_{sol} but as Flanagan and Oates have pointed out[20] non-ideality can be important for finite values of a and this changes the interpretation of ΔH_{sol}.

The solvus free energy change is zero since the two solid phases are in thermodynamic equilibrium at the solvus concentration, a. The excess solvus free energy can be derived in the following manner. The solvus reaction corresponds to the transfer of one mol of hydrogen from the hydride phase to the saturated solid solution and the free energy change for this is given by

$$\Delta G_{sol} = -\Delta\mu(\frac{1}{2}H_2 + a \rightarrow b) + \Delta\mu_H(at\ a) \text{----------------------}23$$

In order to derive the excess solvus free energy the ideal relative chemical potentials must be added to the first term and subtracted from the second term of the right-hand-side of equation 23, i.e.,

$$\Delta G^E_{sol} = -\Delta\mu(H_2 + a \rightarrow b) + \Delta\mu^{\circ}_H + \Delta\mu_H(at\ a) - (\Delta\mu^{\circ}_H + RT\ln Z) =$$

$$-RT\ln Z \text{--}24$$

since $-\Delta\mu(\frac{1}{2}H_2 + a \rightarrow b) + \Delta\mu_H(at\ a) = 0$ and the ideal relative chemical potential of the phase change reaction is simply $\Delta\mu^{\circ}_H$, the result on the right-hand-side of equation 24 is obtained. Since the solvus enthalpy is obtained from ΔG^E_{sol}, it could also be designated ΔH^E_{sol} but $\Delta H_{sol} = \Delta H^E_{sol}$. The excess solvus entropy is obtained from

$$\Delta S^E_{sol} = (\Delta H_{sol} + RT\ln Z)/T = \Delta S_{sol} + R\ln Z \text{------------}25$$

where ΔH_{sol} refers to the experimental slope of $-d\ln Z/dT^{-1}$ (equation 22) and ΔS^E_{sol} is therefore given by

$$\Delta S^E_{sol} = \frac{-\Delta S(\frac{1}{2}H_2 + a \to b) + \Delta S_H(at\ a)}{R^{-1}(\partial(\Delta\mu_H/T)/\partial\ln Z)_{a,T}} + R\ln Z \quad \text{----------26}$$

and therefore ΔS^E_{sol} refers to the transfer of one mole of hydrogen from the hydride phase to the solid solution without the configurational contribution to the partial entropy of the solid solution only as $a \to 0$. For finite values of a, ΔS^E_{sol} contains contributions from non-ideality arising from the denominator of equation 26 and as $a \to 0$, $\Delta S^E_{sol} \to \Delta S^o_{sol}$.

The RIS solution approximation can be useful for the interpretation of solvus results because these are often obtained at low temperatures where p-c-T data are not available for precise evaluation of the non-ideality. Now it can be shown from equations 15, 18 and 22 that for the RIS

$$\Delta H_{sol} = \frac{-(\frac{1}{2} - \frac{a}{\beta})\ H_{HH}}{\frac{\beta}{\beta-a} + \frac{aH_{HH}}{\beta RT}} \quad \text{------------------------------27}$$

which for small values of a reduces to

$$\Delta H_{sol} = -\frac{1}{2} H_{HH} \quad \text{--------------------------------28}$$

Combination of this result with equation 25 leads to $\Delta S^o_{sol} = 0$ for a RIS. Under conditions where a is small we obtain for the RIS approximation

$$\Delta S^E_{sol} = \frac{-R\ln Z}{1 + \frac{a\ H_{HH}(\beta-a)}{\beta RT \quad \beta}} + R\ln Z \quad \text{--------------------------29}$$

From the data of Westlake and Ockers[21] we find for Nb-H $\Delta S^E_{sol} = -3.6\ J(K\ mol\ H)^{-1}$ as an average value from $T = 167$ to 308 K. If equation 29 is employed with $H_{HH} = -22.93\ kJ(mol\ H)^{-1,2}$, we obtain for ΔS^E_{sol}: $+0.12\ J(K\ mol\ H)^{-1}$, $a = 0.04$; $-2.0\ J(K\ mol\ H)^{-1}$, $a = 0.01$; and $-3.2\ J(K\ mol\ H)$; $a = 0.001$. Thus the degree of non-ideality affects the calculated values and undoubtedly also affects the experimental value. The negative experimental value of ΔS^E_{sol} indicates that there is more order in the terminal hydrogen solution than in the hydride phase. For Nb-H[21] $\Delta H_{sol} = 11.5\ kJ(mol\ H)^{-1}$ and the approximate value from equation 28 is 11.46 kJ(mol H)$^{-1}$ which is probably a fortuitous result since the RIS model should not work that well for Nb-H although ΔS^E_{sol} is also not that different from the predicted value of zero.

The Effect of Stress on Hydride Precipitation

The effect of stress on hydride precipitation is an important factor in a mechanism which has been proposed for stress-induced hydride fracture[22-28]. Two different stress situations will be discussed here; the second situation is relevant to stress — induced hydride fracture. The two situations are: (1) a uniformly stressed metal-hydrogen system and (2) a metal-hydrogen system with a small volume subjected to a uniform stress in thermodynamic equilibrium with a large reservoir of unstressed material.

For the first case the chemical potential of hydrogen in both phases is lowered by uniform tensile stress by the amount σV_H (Fig. 2). The solvus concentration will not be affected by this uniform stress provided that V_H, the partial molar volume, is independent of r. The partial molar volume has a meaning over the whole range of hydrogen contents only for miscibility gap systems, e.g., Pd-H. For structural transformation systems there are two sets of partial molar volumes corresponding to the two crystallographic symmetries of the metallic matrix. However, if the total volume of the hydride phase is not much different from that predicted by the relation, $V = n_H V_H^o + n_M V_M^o$, where the superscript zeros refer to the solid solution as $r \to 0$, then it is apparent that the volume change for the solvus reaction is minimal. This is in fact a good approximation for many systems[30,31]. We will assume that we are dealing with miscibility gap systems where V_H is independent of r. The sum of reactions 30 and 31 correspond to the solvus reaction as $a \to 0$, i.e.,

$$\tfrac{1}{2}H_2 (g,\ 1\ atm) \to [H]_a; \quad \Delta\mu_H = \Delta\mu_H^o + RT\ \ln Z - \sigma V_H \quad\text{----------30}$$

$$MH_b/b \to M/b + \tfrac{1}{2}H_2 (g,\ 1\ atm); \quad \Delta\mu_H = -\Delta\mu(\tfrac{1}{2}H_2 + a \to b) + \sigma V_H \quad\text{--31}$$

$$MH_b/b \to M/b + [H]_a; \quad \Delta G_{sol} = -\Delta\mu(\tfrac{1}{2}H_2 + a \to b) + \Delta\mu_H + RT\ \ln Z \text{--32}$$

or for the excess solvus free energy change, $\Delta G_{sol}^E = -RT\ \ln Z$, where $[H]_a$ refers to one mol of hydrogen dissolved in the α-phase at $r = a$, and it can be seen that stress effects cancel out in the solvus reaction for the case of V_H independent of r. For structural transformation systems there may be a small effect of stress arising from the difference between the volume of one mol of hydrogen in the solid solution and in the hydride phase.

For the case of non-uniform stress reaction 30 is not affected by stress but reaction 31 is and we therefore obtain for the excess free energy change for reaction 32

$$\Delta G_{sol}^E (\sigma > 0) = \Delta G_{sol}^E (\sigma = 0) + \sigma V_H = -RT\ \ln Z' \quad\text{-----------33}$$

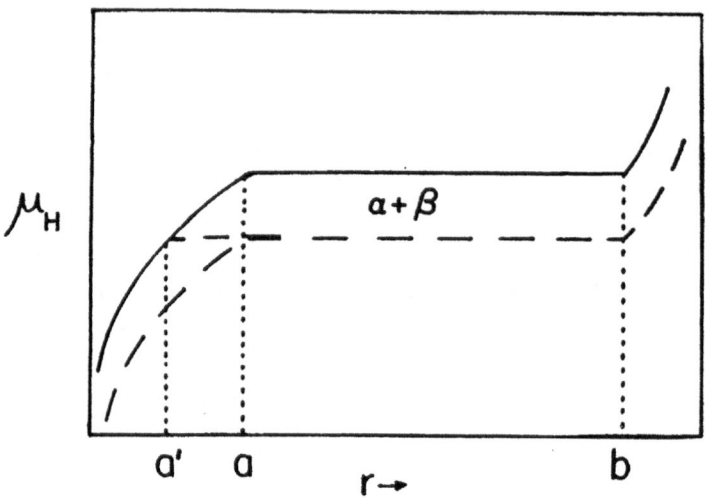

Fig. 2.   Schematic representation of the effect of stress on the
chemical potential of dissolved hydrogen. Solid curve
represents the chemical potential of dissolved hydrogen
during absorption in the absence of stress. The dashed
curve represents the chemical potential of dissolved
hydrogen in the presence of homogeneous tensile stress.
The solvus, a, is unaffected by a homogeneously applied
stress. In the presence of a non-homogeneous stress
the "solvus" of the unstressed reservoir is a' whilst
that in the small stressed region is unchanged at a.

or

$$\ln Z' \approx \ln a' \text{(as } a' \to 0) = \Delta S^E_{sol}/R - \Delta H_{sol}/RT - \sigma V_H/RT \quad \text{----34}$$

Again it is assumed that V_H is independent of r. The solvus concentration a' now refers to the concentration in the reservoir, i.e., it is the minimum value of r in the reservoir at which the hydride phase can precipitate in the small stressed volume. The r value within the small stressed volume at which the hydride precipitates is unchanged at a (Fig. 2). It has been pointed out that this thermodynamic model[29] is appropriate for stress-induced hydride precipitation and fracture. The hydride phase can precipitate in tensile stressed regions in front of crack tips and the crack will propagate through the brittle hydride phase and the process will continually repeat at the tips of newly extended cracks until fracture results.

Hydride precipitation in an open system via reaction 4 will be affected by either uniform or non-uniform stress. Thus a uniform stress affects reaction 4 (Fig. 2)

$$\Delta\mu(\sigma > 0; \tfrac{1}{2}H_2 + a \to b) = RT \ln p^{\frac{1}{2}}_{H_2} (\sigma > 0) =$$

$$\Delta\mu(\sigma = 0; \tfrac{1}{2}H_2 + a \to b) - \sigma V_H \quad \text{----------------------------35}$$

for V_H independent of r. A non-uniform stress allows reaction to occur in the stressed regions at $p^{\sigma>0}_{H_2}$ as given by equation 35 before it precipitates in the stress-free reservoir.

Hysteresis and the Phase Rule

For solid-H systems hysteresis (Fig. 1) appears to be an almost universal phenomenom and therefore it must be "lived with" by workers in this research area. From a practical point of view hysteresis leads to losses of efficiency in hydride storage systems. Hysteresis also introduces uncertainty into the thermodynamic results obtained for the two solid phase coexistence region since reactions 4 and 5 are not reversible. Because hysteresis is ignored, the application of the phase rule to solid-H systems is usually considered to be trivial, i.e., it is considered simply that within the two solid phase coexistence region there are two components and three phases and therefore f = 1, as observed. Alloy-H and intermetallic-H systems are considered to be two component systems too because of the "frozen in" solid configurations. It should be remembered that this is only true at low temperatures.

The application of the phase rule to real metal-H systems will now be considered, i.e., those exhibiting hysteresis. Carl Wagner[14] has stated that the origin of hysteresis in palladium-H is due to the fact $\mu_{Pd}^{\alpha} \neq \mu_{Pd}^{\beta}$. Presumably he intended this remark to apply to all solid-H systems which exhibit hysteresis and, for example, $\mu_{LaNi_5}^{\alpha} \neq \mu_{LaNi_5}^{\beta}$. Now Wagner's remark has often been cited as an explanation for the origin of hysteresis but the authors have then proceeded to offer their own explanations of hysteresis[32] without relating their explanations to Wagner's remark. This suggests that the consequences of Wagner's comment about the lack of metal atom equilibrium have not been fully appreciated.

An attempt will be made here to expand on Wagner's comment. It is clear that within the hysteresis loop (Fig. 1) the phase rule is apparently disobeyed. Following Wagner[14], the reason for this is because $\mu_M^{\alpha} \neq \mu_M^{\beta}$. Within the metastable region the only restraints from the equality of chemical potentials are: $\mu_H^{\alpha} = \mu_H^{\beta} = \mu_H^{g}$. In this metastable region there are two compositional variables, r_{α} and r_{β}, and six pressure and temperature variables. There are two restraints from the above hydrogen chemical potential equalities and four from the temperature and pressure equalities giving f = 2 as observed. When either plateau pressure is reached, another restraint appears since experimentally, f = 1. This restraint must arise from reaction 4 or 5 but this restraint does not derive from the equality of the metal atom chemical potentials because reactions 4 and 5 are irreversible and the usual condition of chemical equilibrium, i.e., $\Delta\mu = 0$, does not apply at either $p_{\alpha \to \beta}$ or $p_{\beta \to \alpha}$. These pressures are not true equilibrium pressures. Nonetheless the irreversible reactions, 4 or 5, must still constitute restraints. We can write for reaction 4, where now $P_{H_2}^{\frac{1}{2}}$ ($\frac{1}{2}H_2 + a' \to b'$) and $P_{H_2}^{\frac{1}{2}}$ ($b \to a + \frac{1}{2}H_2$) refer to the psuedo equilibrium pressures and not to 1 atm as in reactions 4 and 5,

$$\mu_H - \frac{1}{2}\mu_{H_2}^{o} - RT \ln P_{H_2}^{\frac{1}{2}} \; (\tfrac{1}{2}H_2 + a' \to b') + \frac{(\mu_M^{b'} - \mu_M^{a'})}{(b' - a')} =$$

$$\Delta\mu' (\tfrac{1}{2}H_2 + a' \to b') \text{ --36}$$

Now since $\mu_H^{g} = \mu_H^{\alpha} = \mu_H^{\beta}$, equation 35 reduces to

$$\frac{\mu_M^{b'} - \mu_M^{a'}}{(b' - a')} = \Delta\mu' (\tfrac{1}{2}H_2 + a' \to b') \neq 0 \text{ -----------------------37}$$

applying similar arguments to reaction 5 we obtain,

$$\frac{-(\mu_M^{b} - \mu_M^{a})}{(b-a)} = \Delta\mu(b \to a + \tfrac{1}{2}H_2) \neq 0 \text{ ------------------------38}$$

Each of these differences of metal atom chemical potentials has a specific value and therefore comprises a restraint even though $\mu_M^\alpha \neq \mu_M^\beta$. It is clear that if $\mu_M^\alpha = \mu_M^\beta$, then hysteresis will disappear as suggested by Wagner[14]. We have recently identified the inequality of the metal atoms chemical potentials with the requirement to produce dislocations[33]. Dislocations are non-equilibrium defects and if they are produced during a chemical reaction, the reaction is perforce a non-equilibrium one.

Formally we will write

$$\frac{\mu_M^a - \mu_M^b}{(b-a)} = \Delta H_{disl} \quad \text{and} \quad \frac{\mu_M^{b'} - \mu_M^{a'}}{(b'-a')} = \Delta H_{disl} \qquad \text{------------------39}$$

where it must be realized that these differences of chemical potentials are associated with the energies needed to create dislocations and it is not meaningful in this context to discuss the individual metal atom chemical potentials. For the cycle: $a' \to b' \to b \to a$, we have

$$2\Delta H_{disl} = RT \ln \{p^{\frac{1}{2}}(\tfrac{1}{2}H_2 + a' \to b')/p^{\frac{1}{2}}(b \to a + \tfrac{1}{2}H_2)\} \qquad \text{------40}$$

and it has been shown for Pd-H[33] that the energy stored in the dislocation network is compatible with the magnitude of the hysteresis gap.

It is interesting that the interstially dissolved hydrogen behaves quite predictably during hysteresis. For example, from the data of Wicke and Nernst[34] at 303 K, a = 0.009 and a' = 0.014. The ideal configurational change of the chemical potential of dissolved hydrogen between these two values is RT ln (0.014/0.009) = 1.1 kJ (mol H)$^{-1}$ and the width of the experimental hysteresis loop in terms of the chemical potentials is given by equation 40 as 0.96 kJ(mol H)$^{-1}$.

Aside from complications introduced by hysteresis, there are two other applications of the phase rule to metal-hydrogen systems which can be discussed. When the two solid phases are in equilibrium with the gas phase consisting only of gaseous hydrogen, the variables are: p^α, p^β, p^g, T^α, T^β and T^g and the two composition variables: r_α and r_β. The restraints are $p^\alpha = p^\beta = p^g, T^\alpha = T^\beta = T^g$ and $\mu_H^\alpha = \mu_H^\beta = \mu_H^g$ and the chemical reaction with gaseous hydrogen discussed above. This leads to f = 1, i.e., if we choose T, $p^g = p_{H_2}$ (equil) and r_α and r_β have unique values. Now if an inert gas is added, the variables are: $p^\alpha, p^\beta, p^g, T^\alpha, T^\beta, T^g, r_\alpha, r_\beta$, and $X_H^g = p_{H_2}/p^g$ where p^g is the total pressure of the gas phase. Since there are still seven restraints, there are now two degrees of freedom instead of one. If we choose T, we are free to choose p^g

but now r_α, r_β and X^β must have unique values, which means that p_{H_2} is again unique.

A commonly encountered case is that of two solid phases in coexistence in the absence of gaseous equilibrium. This frequently occurs in metal-hydrogen systems, e.g., group VB-H systems, where the surfaces can be readily poisoned by exposure to the atmosphere. In this case the variables are: $p^\alpha, p^\beta, T^\alpha, T^\beta, r_\alpha$ and r_β and the restraints are: $p^\alpha = p^\beta$; $T^\alpha = T^\beta$, $\mu_H^\alpha = \mu_H^\beta$ and one from the chemical reaction. This gives two degrees of freedom. If the pressure is fixed, at say, one atmosphere, the temperature can be varied whilst maintaining the two solid phase equilibrium. r_α and r_β have unique values at each temperature. The reaction is the phase change $\alpha \rightarrow \beta$ or vice versa and this cannot occur under isothermal conditions. Hysteresis is also present in this case and presumably the same considerations apply here with respect to hysteresis as in the case of the three phase equilibrium.

Acknowledgements

The author thanks the NSF for financial support of his research on metal-hydrogen systems.

References

1. F. A. Kuijpers, RCo_5-H and related systems, Phil. Res. Repts. Suppl., No. 2, (1973).

2. O. J. Kleppa, P. Dantzer and M. E. Melnichak, High-temperature thermodynamics of the solid solutions of hydrogen in b.c.c. vanadium, niobium and tantalum, J. Chem. Phys., 61:4048 (1974).

3. P. Dantzer, O. J. Kleppa and M. E. Melnichak, High-temperature thermodynamics of the Ti-H_2 and Ti-D_2 systems, J. Chem. Phys., 64:139 (1976).

4. G. Boureau, O. J. Kleppa and P. Dantzer, High-temperature thermodunamics of palladium-hydrogen. 1. Dilute solutions of H_2 and D_2 in Pd at 555K. J. Chem. Phys., 64:5247 (1976).

5. G. Boureau and O. J. Kleppa, High temperature thermodynamics of palladium-hydrogen. II. Temperature dependence of the partial molar properties of dilute solutions of hydrogen in the range 500 to 700K, J. Chem. Phys., 65:3915 (1976).

6. G. Boureau, O. J. Kleppa and K. C. Hong, A thermodynamic study of dilute solutions of hydrogen and deuterium in $Pd_{0.9}Ag_{0.1}$ at 555K and 700K, J. Chem. Phys., 67:3437 (1977).

7. C. Picard, O. J. Kleppa and G. Boureau, High temperature thermodynamics of the solutions of hydrogen in palladium-silver alloys, J. Chem. Phys., 70:2710 (1979).

8. C. Picard, O. J. Kleppa and G. Boureau, A thermodynamic study of the palladium-hydrogen system at 245-352°C and at pressures up to 34 atm., J. Chem. Phys., 69:5549 (1978).

9. G. Boureau and O. J. Kleppa, Significance of thermal effects associated with solid-gas reactions in the Tian-Calvet calorimeter, J. Chem. Thermodyn., 9:543 (1977).

10. D. M. Nace and J. G. Aston, Palladium Hydride. I. The thermodynamic properties of Pd_2H between 273 and 345K, J. Am. Chem. Soc., 79:3619 (1957).

11. B. S. Bowerman, C. A. Wulff and T. B. Flanagan, Calorimetric enthalpies for solution of hydrogen in the $LaNi_5$-H system, Z. Physile. Chem., 116:197 (1979).

12. B. S. Bowerman, G. E. Biehl, C. A. Wulff and T. B. Flanagan, Calorimetry within hysteresis loops of metal/hydrogen systems: Application to Pd/H, Ber. Bunsenges Physik. Chem., 84:536 (1980).

13. B. S. Bowerman, C. A. Wulff, G. E. Biehl and T. B. Flanagan, Calorimetry within hysteresis loops: Application to $LaNi_5$-H, J. Less Common Met., 73:1 (1980).

14. C. Wagner, Hysteresis phenomena in the system palladium-hydrogen and in rotational transitions, Z. Physik. Chem., 143:386 (1944).

15. J. D. Clewley, T. Curran, T. B. Flanagan and W. A. Oates. Thermodynamic properties of hydrogen and deuterium dissolved in palladium at low concentrations over a wide temperature range, J.C.S. Faraday Trans. I, 69:449 (1973).

16. W. A. Oates and T. B. Flanagan, The ideal interstitial solution and the effect of conditions of constant pressure and constant volume on non-ideality, Scripta. Met., 12:759 (1978).

17. T. B. Flanagan and W. A. Oates, Thermodynamics of metal-hydrogen systems, Ber. Bunsenges Physik Chem., 76:706 (1972).

18. W. A. Oates and T. B. Flanagan, The regular interstitial solution, J. Materials Science, submitted.

19. T. B. Flangan and J. F. Lynch, The thermodynamics of a gas in equilibrium with two non-stoichiometric condensed phases, J. Phys. Chem., 79:444 (1975).

20. T. B. Flangan and W. A. Oates, Interpretation of the solvus line for metal-hydrogen systems, Scripta Met., 12:873 (1978).

21. D. G. Westlake and S. T. Ockers, The isotope effect and the influence of interstitial impurities on the hydrogen solubility limit in niobium and vanadium, Met. Trans. A, 6A:399 (1975).

22. N. E. Paton, B. S. Hickman and D. H. Leslie, Behavior of hydrogen in α-phase Ti-Al alloys, Met. Trans., 2:2791 (1971).

23. D. G. Westlake, A Generalized Model for hydrogen embrittlement, Trans. ASM, 62:1000 (1969).

24. H. K. Birnbaum, M. L. Grossbeck and M. Amano, Hydride precipitation in Nb and some properties of NbH, J. Less-Common Met., 49:357 (1976).

25. S. Gahr, M. L. Grossbeck and H. K. Birnbaum, Hydrogen embrittlement of Nb. I. Macroscopic behavior at low temperatures, Acta Met., 25:125 (1977).

LIST OF CONTRIBUTORS

Abell, G. C.
Mound Facility
Monsanto Research Corporation
Miamisburg, Ohio   45324   USA

Asano, H.
Institute of Materials Science
University of Tsukuba
Ibaraki 305   JAPAN

Baranowski, B.
Inst. of Physical Chemistry
Polish Academy of Sciences
01-224 Warsaw   POLAND

Bauer, J. H.
Sektion Physik
Universität München
D-8 München 22   WEST GERMANY

Bowman, R. C., Jr.
Noyes Laboratory, M/C 127-72
Calif. Institute of Technology
Pasadena, CA 91125 USA

Burger, J. P.
Laboratoire de Physique
      des Solides
Universite Paris Sud
91405 Orsay   FRANCE

Economou, E. N.
Department of Physics
University of Virginia
Charlottesville, VA 22901 USA

Flanagan, T. B.
Department of Chemistry
University of Vermont
Burlington, VT 05405 USA

Flevaris, N. K.
Mat. Sci. and Eng. Department
The Technological Institute
Northwestern University
Evanston, Il   60201 USA

Gupta, M.
CECAM Bât. 506
Université d'Orsay
91405   Orsay FRANCE

Hirabayashi, M.
Res. Inst. for Iron, Steel &
      Other Metals
Tohoku University
Sendai 980   JAPAN

Maeland, A. J.
Materials Res. Division
Allied Chemical Corp
Morristown, NJ   07869 USA

Niarchos, D.
Solid State Division
Argonne Nat. 'l Lab.
Argonne, Il 60439   USA

Papaconstantopoulos, D. A.
Code 6480
Naval Research Lab.
Washington, D. C.   20375   USA

379

Pick, M. A.
Brookhaven National Lab.
Upton, NY 11973´ USA

S. Radelaar
Technical Physics Dept.
State Univ. Utrecht
P. O. Box 30.000
3508 TA Utrecht THE NETHERLANDS

Völkl, J.
Physik-Department der Tech.
    Universität München
D-8046 Garching   WEST GERMANY

Wallace, W. E.
Chemistry Department
University of Pittsburgh
Pittsburgh, PA 15260 USA

Westlake, D. G.
Materials Science Division
Argonne National Laboratory
Argonne, IL 60439   USA